Parametric Time–Frequency Domain Spatial Audio

Parametric Time–Frequency Domain Spatial Audio

Edited by

Ville Pulkki, Symeon Delikaris-Manias, and Archontis Politis
Aalto University
Finland

WILEY

Registered Offices
John Wiley & Sons, Inc., 111 River Street, Hoboken, NJ 07030, USA
John Wiley & Sons Ltd, The Atrium, Southern Gate, Chichester, West Sussex, PO19 8SQ, UK

Editorial Office
The Atrium, Southern Gate, Chichester, West Sussex, PO19 8SQ, UK

For details of our global editorial offices, customer services, and more information about Wiley products visit us at www.wiley.com.

Wiley also publishes its books in a variety of electronic formats and by print-on-demand. Some content that appears in standard print versions of this book may not be available in other formats.

Library of Congress Cataloging-in-Publication Data

Names: Pulkki, Ville, editor. | Delikaris-Manias, Symeon, editor. | Politis, Archontis, editor.
Title: Parametric time-frequency domain spatial audio / edited by Ville Pulkki, Symeon Delikaris-Manias, Archontis Politis, Aalto University, Aalto, Finland.
Description: First edition. | Hoboken, NJ, USA : Wiley, 2018. | Includes bibliographical references and index. |
Identifiers: LCCN 2017020532 (print) | LCCN 2017032223 (ebook) | ISBN 9781119252580 (pdf) | ISBN 9781119252610 (epub) | ISBN 9781119252597 (hardback)
Subjects: LCSH: Surround-sound systems–Mathematical models. | Time-domain analysis. | Signal processing. | BISAC: TECHNOLOGY & ENGINEERING / Electronics / General.
Classification: LCC TK7881.83 (ebook) | LCC TK7881.83 .P37 2018 (print) | DDC 621.382/2–dc23
LC record available at https://lccn.loc.gov/2017020532

Cover Design: Wiley
Cover Image: © Vectorig/Gettyimages

Set in 10/12pt WarnockPro by Aptara Inc., New Delhi, India
Printed and bound in Malaysia by Vivar Printing Sdn Bhd

10 9 8 7 6 5 4 3 2 1

Contents

List of Contributors

Ahonen, Jukka
Akukon Ltd, Finland

Alexandridis, Anastasios
Foundation for Research and
Technology-Hellas, Institute of
Computer Science (FORTH-ICS),
Heraklion, Crete, Greece

Alon, David Lou
Department of Electrical and Computer
Engineering, Ben-Gurion University of
the Negev, Israel

Bäckström, Tom
Department of Signal Processing and
Acoustics, Aalto University, Finland

Delikaris-Manias, Symeon
Department of Signal Processing and
Acoustics, Aalto University, Finland

Epain, Nicolas
CARLab, School of Electrical and
Information Engineering, University of
Sydney, Australia

Faller, Christof
Illusonic GmbH, Switzerland
and
École Polytechnique Fédérale de
Lausanne (EPFL), Switzerland

Habets, Emanuël
International Audio Laboratories
Erlangen, Germany

Jin, Craig T.
CARLab, School of Electrical and
Information Engineering, University of
Sydney, Australia

Laitinen, Mikko-Ville
Nokia Technologies, Finland

Mouchtaris, Athanasios
Foundation for Research and
Technology-Hellas, Institute of
Computer Science (FORTH-ICS),
Heraklion, Crete, Greece

Nikunen, Joonas
Department of Signal Processing,
Tampere University of Technology,
Finland

Noohi, Tahereh
CARLab, School of Electrical and
Information Engineering, University of
Sydney, Australia

Pavlidi, Despoina
Foundation for Research and
Technology-Hellas, Institute of
Computer Science (FORTH-ICS),
Heraklion, Crete, Greece

Pertilä, Pasi
Department of Signal Processing,
Tampere University of Technology,
Finland

Pihlajamäki, Tapani
Nokia Technologies, Finland

Politis, Archontis
Department of Signal Processing and
Acoustics, Aalto University, Finland

Pulkki, Ville
Department of Signal Processing and
Acoustics, Aalto University, Finland

Rafaely, Boaz
Department of Electrical and Computer
Engineering, Ben-Gurion University of
the Negev, Israel

Stefanakis, Nikolaos
Foundation for Research and
Technology-Hellas, Institute of
Computer Science (FORTH-ICS),
Heraklion, Crete, Greece

Thiergart, Oliver
International Audio Laboratories
Erlangen, Germany

Vilkamo, Juha
Nokia Technologies, Finland

Virtanen, Tuomas
Department of Signal Processing,
Tampere University of Technology,
Finland

Preface

A plethora of methods for capturing, storing, and reproducing monophonic sound signals has been developed in the history of audio, starting from early mechanical devices, and progressing via analog electronic devices to faithful digital representation. In recent decades there has also been considerable effort to capture and recreate the spatial characteristics of sound scenes to a listener. When reproducing a sound scene, the locations of sound sources and responses of listening spaces should be perceived as in the original conditions, in either faithful replication or with deliberate modification. A vast number of research articles have been published suggesting methods to capture, store, and recreate spatial sound over headphone or loudspeaker listening setups. However, one cannot say that the field has matured yet, as new techniques and paradigms are still actively being published.

Another important task in spatial sound reproduction is the directional filtering of sound, where unwanted sound coming from other directions is attenuated when compared to the sound arriving from the direction of the desired sound source. Such techniques have applications in surround sound, teleconferencing, and head-mounted virtual reality displays.

This book covers a number of techniques that utilize signal-dependent time–frequency domain processing of spatial audio for both tasks: spatial sound reproduction and directional filtering. The application of time–frequency domain techniques in spatial audio is relatively new, as the first attempts were published about 15 years ago. A common property of the techniques is that the sound field is captured with multiple microphones, and its properties are analyzed for each time instance and individually for different frequency bands. These properties can be described by a set of parameters, which are subsequently used in processing to achieve different tasks, such as perceptually motivated reproduction of spatial sound, spatial filtering, or spatial sound synthesis. The techniques are loosely gathered under the title "time–frequency domain *parametric* spatial audio."

The term "parameter" generally denotes any characteristic that can help in defining or classifying a particular system. In spatial audio techniques, the parameter somehow quantifies the properties of the sound field depending on frequency and time. In some techniques described in this book, measures having physical meaning are used, such as the direction of arrival, or the diffuseness of the sound field. Many techniques measure the similarity or dissimilarity of signals from closely located microphones, which also quantifies the spatial attributes of the sound field, although the mapping from parameter value to physical quantities is not necessarily very easy. In all cases, the time- and frequency-dependent parameter directly affects the reproduction of sound, which makes the outputs of the methods depend on the spatial characteristics of the captured sound field. With these techniques, in most cases a significant improvement is obtained with such signal-dependent signal processing compared with more traditional signal-independent processing, when an input with relatively few audio channels is processed.

Signal-dependent processing often relies on implicit assumptions about the properties of the spatial and spectral resolution of the listener, and/or of the sound field. In spatial sound reproduction, the systems should relay sound signals to the ear canals of the listener such that the desired perception of the acoustical surroundings is obtained. The resolution of all perceivable attributes, such as sound spectrum, direction of arrival, or characteristics of reverberation, should be as high as required so that no difference from the original is perceived. On the other hand, the attributes should not be reproduced with an accuracy that is higher than needed, so that the use of computational resources is optimal. Optimally, an authentic reproduction is obtained with a moderate amount of resources, i.e., only a few microphones are needed, the computational requirements are not excessive, and the listening setup consists of only a few electroacoustic transducers.

When the captured acoustical scene deviates from the assumed model, the benefit obtained by the parametric processing may be lost, in addition to potential undesired audible degradations of the audio. An important theme in all the methods presented is how to make them robust to such degradations, by assuming extended and complex models, and/or by handling estimation errors and deviations without detrimental perceptual effects by allowing the result to deviate from reality. Such an optimization requires a deep knowledge of sound field analysis, microphone array processing, statistical signal processing, and spatial hearing. That makes the research topic rich in technological approaches.

The composition of this book was motivated by work on parametric spatial audio at Aalto University. A large number of publications and theses are condensed in this book, aiming to make the core of the work easily accessible. In addition, several chapters are contributed by established international researchers in this topic, offering a wide view of the approaches and solutions in this field.

The first part of the book concerns the analysis of and the synthesis of spatial audio. The first chapter reviews the methods that are commonly used in industrial audio applications when transforming signals to the time–frequency domain. It also provides background knowledge for methods of reproducing sound with controllable spatial attributes, the methods being utilized in several chapters in the book. The two other chapters in this part consider methods for analysis of spatial sound captured with a spherical microphone array: how to decompose the sound field recording into plane waves.

The second part considers systems that consist of a whole sound reproduction chain including capture with a microphone array; time–frequency domain analysis, processing, and synthesis; and often also subjective evaluation of the result. The basic question is how to reproduce a spatial sound scene in such a way that a listener would not notice a difference between original and reproduced occasions. All the methods are parametric in some sense, however, with different assumptions about the sound field and the listener, and with different microphone arrays utilized; the solutions end up being very different.

The third part starts with a review of current signal-dependent spatial filtering approaches. After this, two chapters with new contributions to the field follow. The second chapter discusses a method based on stochastic estimates between higher-order directional patterns, and the third chapter suggests using machine learning and neural networks to perform spatial filtering tasks.

The fourth part extends the theoretic framework to more practical approaches. The first chapter shows a number of commercial devices that utilize parametric time–frequency domain audio techniques. The second chapter discusses the application of the techniques in synthesis of spatial sound for virtual acoustic environments, and the third chapter covers applications in teleconferencing and remote presence.

The reader should possess a good knowledge of the fields of acoustics, audio, psychoacoustics, and digital signal processing; introductions to these fields can be found in other sources.[1] Finally, the working principles of many of the proposed techniques are demonstrated with code examples, written in Matlab®, which focus mostly only on the parametric part of the processing. Tools for time–frequency domain transforms and for linear processing of spatial audio are typically also needed in the implementation of complete audio systems, and the reader might find useful the broad range of tools developed by our research group at Aalto University.[2] In addition, a range of research-related demos are available.[3]

Ville Pulkki, Archontis Politis, and Symeon Delikaris-Manias
Otaniemi, Espoo 2017

---

[1] See, for example, *Communication Acoustics: An Introduction to Speech, Audio and Psychoacoustics*, V. Pulkki and M. Karjalainen, Wiley, 2015.

[2] See http://spa.aalto.fi/en/research/research_groups/communication_acoustics/acoustics_software/.

[3] http://spa.aalto.fi/en/research/research_groups/communication_acoustics/demos/.

About the Companion Website

Don't forget to visit the companion website for this book:

www.wiley.com/go/pulkki/parametrictime-frequency

There you will find valuable material designed to enhance your learning, including:

- Matlab material with selected chapters

Part I

Analysis and Synthesis of Spatial Sound

1

Time–Frequency Processing: Methods and Tools

Juha Vilkamo[1] and Tom Bäckström[2]

[1] *Nokia Technologies, Finland*
[2] *Department of Signal Processing and Acoustics, Aalto University, Finland*

1.1 Introduction

In most audio applications, the purpose is to reproduce sounds for human listening, whereby it is essential to design and optimize systems for perceptual quality. To achieve such optimal quality with given resources, we often use principles in the processing of signals that are motivated by the processes involved in hearing. In the big picture, human hearing processes the sound entering the ears in frequency bands (Moore, 1995). The hearing is thus sensitive to the spectral content of ear canal signals, which changes quickly with time in a complex way. As a result of frequency-band processing, the ear is not particularly sensitive to small differences in weaker sounds in the presence of a stronger masking sound near in frequency and time to the weaker sound (Fastl and Zwicker, 2007). Therefore, a representation of audio signals where we have access to both time and frequency information is a well-motivated choice.

A prerequisite for efficient audio processing methods is a *representation* of the signal that presents features desirable to hearing in an accessible form and also allows high-quality playback of signals. Useful properties of such a representation are, for example, that its coefficients have physically or perceptually relevant interpretations, and that the coefficients can be processed independently from each other. The time–frequency domain is such a domain, and it is commonly used in audio processing (Smith, 2011). Spectral coefficients in this domain explain the signal content in terms of frequency components as a function of time, which is an intuitive and unambiguous physical interpretation. Moreover, time–frequency components are approximately uncorrelated, whereby they can be independently processed and the effect on the output is deterministic. These properties make the spectrum a popular domain for audio processing, and all the techniques discussed in this book utilize it. The first part of this chapter will give an overview of the theory and practice of the tools typically needed in time–frequency processing of audio channels.

The time–frequency domain is also useful when processing the spatial characteristics of sound, for example in microphone array processing. Differences in directions

Parametric Time–Frequency Domain Spatial Audio, First Edition. Edited by Ville Pulkki,
Symeon Delikaris-Manias, and Archontis Politis.
© 2018 John Wiley & Sons Ltd. Published 2018 by John Wiley & Sons Ltd.
Companion Website: www.wiley.com/go/pulkki/parametrictime-frequency

of arrival of wavefronts are visible as differences in time of arrival and amplitude between microphone signals. When the microphone signals are transformed to the time–frequency domain, the differences directly correspond to differences in phase and magnitude in a similar fashion to the way spatial cues used by a human listener are encoded in the ear canal signals (Blauert, 1997). The time–frequency domain differences between microphone channels have proven to be very useful in the capture, analysis, and reproduction of spatial audio, as is shown in the other chapters of this book. The second part of this chapter introduces a few signal processing techniques commonly used, and serves as background information for the reader.

This chapter assumes understanding of basic digital signal processing techniques from the reader, which can be obtained from such basic resources as Oppenheim and Schafer (1975) or Mitra and Kaiser (1993).

1.2 Time–Frequency Processing

1.2.1 Basic Structure

A block diagram of a typical parametric time–frequency processing algorithm is shown in Figure 1.1. The processing involves transforms between the time domain input signal $x_i(t)$, the time–frequency domain signal $x_i(k, n)$, and the time domain output signal $y_j(t)$, where t is the time index in the time domain, and k and n are indexes for the frequency and time frame in the time–frequency domain, respectively; i and j are then the channel indexes in the case of multi-channel input and/or output. Additionally, the processing involves short-time stochastic analysis and parameter-driven processing, where the time domain signal $y(k, n)$ is formed based on the parameters and $x(k, n)$. The parametric data consists of any information describing the frequency band signals, for example the stochastic properties, information based on the audio objects, or user input parameters. In some use cases, such as in parametric spatial audio coding decoders, the stochastic estimation block is not applied, and the processing acts entirely on the parametric data provided in the bit stream.

The parametric processing techniques typically operate on several different sampling rates: The sampling rate F_s of the wide-band signal, the sampling rate F_s/K of the

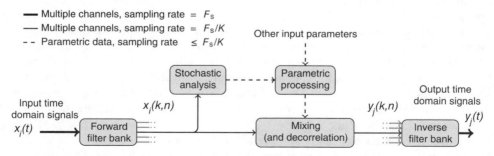

Figure 1.1 Block diagram of a typical parametric time–frequency processing algorithm. The processing operates on three sampling rates: that of the wide-band signal, that of the frequency band signal, and that of the parametric information.

frequency band signals, where K is the downsampling factor, and the sampling rate of the parametric information. Since the samples in the parametric information typically describe the signal properties over time frames, it potentially operates at a sampling rate below F_s/K. The parametric processing can also take place using a varying sampling rate, for example when the frame size adapts with the observed onsets of audio signals. In the following sections, the required background for the processing blocks in Figure 1.1 is discussed in detail.

Audio signals are generally time-varying signals whereby the spectrum is not constant in time. Should we analyze a long segment, then its spectrum would contain a mixture of all the different sounds within that segment. We could then not easily access the independent sounds, but only see their mixture, and the application of efficient processing methods would become difficult. It is therefore important to choose segments of the signal of such a length that we obtain good temporal separation of the audio content. Also, other properties such as constraints on algorithmic delay and requirements on spectral resolution impose demands on the length of analysis windows. It is then clear that while the spectrum or the frequency domain is an efficient domain for audio processing, the time axis also has to be factored in to the representation.

Computationally effective algorithms for time–frequency analysis have enabled their current popular usage. Namely, the basis of most time–frequency algorithms is the fast Fourier transform (FFT), which is a practical implementation of the discrete Fourier transform (DFT). It belongs to the class of super-fast algorithms that have an algorithmic complexity of $\mathcal{O}(N \log N)$, where N is the analysis window length. In practice, the signal is divided into partially overlapping time frames, each of which is processed with an FFT; this is commonly called a short-time Fourier transform (STFT; Smith, 2011). These methods are assumed to be known to the reader, and the rest of the chapter deals with methods based on filter banks, which are more common in practical audio applications.

1.2.2 Uniform Filter Banks

Figure 1.2 illustrates the generic design of uniform filter banks. The block diagram of the forward transform (the left-hand part of the figure) shows a set of band-pass analysis filters with different pass-band frequency regions, followed by downsampling operators, to produce the time–frequency data. The inverse transform (the right-hand side of the figure) is performed with upsampling operations, followed by band-pass synthesis filters. The block diagram is conceptual, and represents equivalent operations to practical

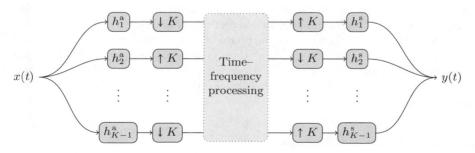

Figure 1.2 General block diagram of a uniform filter bank.

implementations that employ efficient digital signal processing structures, as discussed in Section 1.2.4.

Spatial sound processing techniques in the time–frequency domain typically perform the frequency decompositions with uniform filter banks, where the signal components appear on linear steps in frequency. A uniform grid is practical since it enables the application of efficient signal processing algorithms such as the FFT. From a perceptual viewpoint, *analysis* on a perceptually motivated non-uniform grid can be useful, but time–frequency *processing algorithms* can usually be applied on a uniform grid, since the effect of non-uniform sampling can, for most purposes, be achieved by scaling the magnitudes of spectral components (Bäckström, 2013). When a non-uniform grid is required, a uniform filter bank can be cascaded with filters at the lowest frequency bands to obtain the non-uniform properties (see Section 1.2.7).

Uniform filter banks can be over-, critically, or undersampled. A critically sampled filter bank preserves the overall sampling rate, which means that the sum of the sampling rates of the frequency band signals is the same as that of the input signal. This feature is desirable in coding applications where the objective is to reduce the bit rate by adaptive quantization of the spectral coefficients, whereby an increase in the overall data rate would be counterproductive. Critically sampled time–frequency transforms such as the modified discrete cosine transform are a fundamental part of the well-known MPEG Audio Layer III (mp3; Brandenburg, 1999) and the MPEG Advanced Audio Coding (AAC; ISO/IEC JTC1/SC29/WG11, 1997; Bosi *et al.*, 1997) standards.

Oversampled filter banks, on the other hand, have a higher combined sampling rate than the input signal. A typical oversampled filter bank in this scope is the complex-modulated filter bank, where representation of the signal using the real and the imaginary parts doubles the sampling rate; this is discussed further in Section 1.2.3. Such filter banks are used in processing methods where the objective is signal modification. An intuitive way to consider the sampling rate and the robustness in signal modification is to note that after the transform the frequency bands have a degree of spectral overlap. For robust independent processing of the frequency bands, the partially overlapping spectral data must be expressed by both of the adjacent bands.

Undersampled filter banks are also possible, though since they imply an inherent loss of information, they are generally useful only in analysis applications such as audio segmentation and voice activity detection. Such applications can compromise temporal accuracy in favor of a lower algorithmic complexity, whereby undersampled filter banks become a viable option.

Filter banks can be further categorized according to whether the frequency representation is real or complex valued. In processing applications where the signal is modified on purpose, we usually prefer complex-valued representations, since that gives us direct access to the phase of the signal. Since the phase of signals has a multitude of physical interpretations, complex-valued representations enable the design of algorithms that employ physically motivated techniques. However, since critical sampling is cumbersome with complex-valued filter banks, most audio coding methods rely on real-valued representations.

1.2.3 Prototype Filters and Modulation

The performance of a filter bank depends mainly on analysis and synthesis filters, h_k^a and h_k^s, respectively (see Figure 1.2), which are based on a designed prototype filter

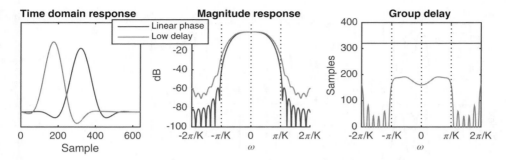

Figure 1.3 Linear-phase and low-delay prototype filters, using $K = 64$. Filter banks using a low-delay prototype filter are useful in applications where the constraints on delay are so stringent that the performance in terms of frequency response must be compromised.

h_0. They are band-pass filters that extract the desired frequency component from the input signal $x(n)$. The characteristics of these filters determine the performance of the filter bank in terms of frequency response. In this section, we discuss a straightforward approach which uses a linear-phase prototype filter. As an alternative, low-delay structures can be obtained using a prototype filter free from the linear-phase property, which requires a different synthesis filter bank structure (Schuller and Smith, 1995; Allamanche *et al.*, 1999). The linear-phase and low-delay prototype filters of the complex-modulated quadrature mirror filter bank type (QMF), applied in several codecs such as Herre *et al.* (2012, 2015) or 3GP (2014), are illustrated in Figure 1.3.

Let $h_0(t)$ be the coefficients of a linear-phase prototype filter and $H_0(e^{i\omega})$ its transfer function, where ω is the angular frequency. The objective in the design of prototype filters is to find a filter $H_0(e^{i\omega})$ such that

$$\begin{cases} |H(e^{i\omega})| & \approx 0 & \text{for } \omega > \pi/K, \\ |H(e^{i\omega})|^2 + |H(e^{i(\omega-\pi/K)})|^2 & \approx 1 & \text{for } 0 \le \omega \le \pi/K, \end{cases} \tag{1.1}$$

where K is the downsampling factor. In other words, the objective is to find a filter whose response at the stop-band is zero, and the response of the pass-band added with a shifted version of itself is unity. A set of pass-bands shifted by π/K then add up to unity (see Figure 1.4(a)). Specific methods for designing a linear-phase prototype filter can be found in, for example, Creusere and Mitra (1995) or Cruz-Roldán *et al.* (2002).

The band-pass filters $h_k(t)$ of uniform filter banks, where k is the frequency band index, are obtained by shifting the spectrum of the prototype filter using modulation. The complex-valued modulation can be expressed by

$$h_k(t) = h_0(t)e^{i\omega_k t + \psi_k}, \tag{1.2}$$

where $\omega_k = \pi(k + \alpha)/K$ is the angular frequency, α is an offset constant, and ψ_k determines the phase of the modulator. For example, defining $\alpha = 0$ means that the band $k = 0$ is centered at the DC frequency $\omega_0 = 0$, whereas defining $\alpha = 0.5$ means a half-band shift. The real-valued modulation is given by

$$h_k(t) = h_0(t)\cos(i\omega_k t + \psi_k) = h_0(t)\frac{e^{i\omega_k t + \psi_k} + e^{-i\omega_k t - \psi_k}}{2}. \tag{1.3}$$

The characteristics of complex- and real-valued modulations are illustrated in Figure 1.4. If the prototype filters are complex modulated, the resulting band-pass

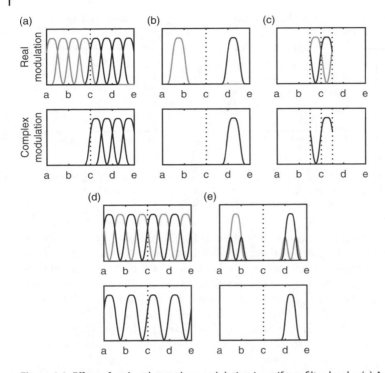

Figure 1.4 Effect of real and complex modulation in uniform filter banks. (a) A set of band-pass filters, (b) individual band-pass filters, (c) their downsampled spectra, (d) the re-upsampled spectra, and (e) the band-pass synthesis-filtered spectra.

spectrum appears either on the positive or the negative frequencies. For real-valued modulators the pass-band spectra appears on both sides, which is also evident from Equation (1.3). Due to the downsampling, spectral information is folded between the new Nyquist frequencies, as shown in Figure 1.4(c). The half-band offset $\alpha = 0.5$ is necessary for real-valued transforms; the spectral overlap of the positive and negative frequencies is thereby minimized and the aliasing components of the real transform are inverses between the adjacent bands. When the frequency bands are not processed independently they cancel each other at the overall output of the filter bank. The complex-valued transform does not have these aliasing components in the first place, which enables the robust independent processing of the bands. The reader is referred to Bosi and Goldberg (2003), Malvar (1992), and Jayant and Noll (1984) for further information on typical time–frequency transforms in audio coding.

1.2.4 A Robust Complex-Modulated Filter Bank, and Comparison with STFT

Our goal, in this section, is to derive the equations of a robust complex-modulated filter bank in a form that can be efficiently implemented using existing FFT algorithms. By default, we use $\alpha = 0$, whereby we obtain a frequency representation with $K + 1$ bands from frequency $\omega_0 = 0$ to $\omega_0 = \pi$, that is, from DC to the Nyquist frequency. The center frequencies of each band are the same as those of the first $(K + 1)$ coefficients of a

$2K$-point FFT. The first and the last bands of this representation are real-valued, and thus the representation has an oversampling factor equal to 2.

As a first step, let us express the equations that correspond to convolving an input signal sequence $x(t)$ with the band-pass filters $h_k(t)$ and performing the downsampling with the factor K. The length of the prototype filter $h_0(t)$ is assumed to be $N_h = 2KC$, where C is a positive integer. An element of the frequency band k is denoted by $X(k, n)$, where n is the downsampled time index. Since applying the downsampling to the band-pass finite impulse response filter output is the same as processing the filter output only every Kth sample, the operations can be expressed as

$$X(k, n) = \sum_{t=0}^{N_h-1} x(t + nK)h_k(t)$$

$$= \sum_{t=0}^{N_h-1} \underbrace{x(t + nK)h_0(t)}_{\text{Processing with prototype filter}} \quad \underbrace{e^{-i\pi kt/K+\psi'_k}}_{\text{Modulation}}, \tag{1.4}$$

where $\psi'_k = \psi_k + \pi k N_h/K$. The notation in Equation (1.4) involves the assumption that the input signal up to $t = (N_h + nK)$ has been accumulated. The first key feature of the equation is that the prototype filter $h_0(t)$ does not depend on the band index k. In other words, for a practical filter bank implementation, we do not need to explicitly derive each of the band-pass filters $h_k(t)$, but only apply $h_0(t)$ to the signal frame. The modulation part in Equation (1.4) accounts for the spectral positions of the band-pass signals.

Let us, then, compare Equation (1.4) with the definition of a $2K$-point STFT which also defines the window length, that is, the case where the prototype filter is of length $N_h = 2K$:

$$X_{\text{STFT}}(k, n) = \sum_{t=0}^{2K-1} x(t + nK)w(t)e^{-i\pi kt/K}, \tag{1.5}$$

where $w(t)$ is the analysis window, which, in a typical configuration, is a square-root Hann window (for a discussion about windowing, see Section 1.2.5). By comparing Equations (1.4) and (1.5), we can readily observe that the STFT is a complex-modulated filter bank with $h_0(t) = w(t)$ and $\psi'_k = 0$. The short prototype filter length of the STFT is a limiting factor for obtaining reasonable stop-band attenuation, as is illustrated in Figure 1.5. In other words, the aliasing effects beyond the adjacent bands are potentially audible at the processed output. Without processing the spectral coefficients, the aliasing effects cancel each other. Thus, the STFT can provide low aliasing when the frequency bands are processed conservatively, such as when applying a spectrally smooth equalizer.

Another important insight from the above derivation is that for the case $\alpha = 0$ and $\psi'_k = 0$, the modulator in Equation (1.4) is cyclic with a period of $2K$ samples. Using the definition $N_h = 2KC$, the equation can be rewritten as

$$X(k, n) = \sum_{t=0}^{2K-1} \left[\sum_{c=0}^{C-1} x(t + (n + 2c)K)h_0(t + 2cK) \right] e^{-i\pi kt/K}, \tag{1.6}$$

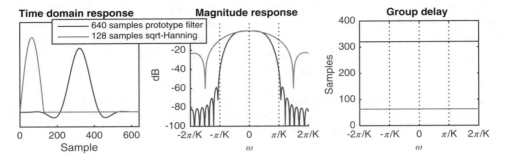

Figure 1.5 Linear-phase prototype filter applied by the MPEG QMF bank, and the square-root Hann window typical in STFT, using $K = 64$.

which is equivalent to performing a $2K$-point FFT on the signal sequence within square brackets. Therefore, even if the applied prototype filter is longer than the transform size, i.e., $C > 1$, we can readily use the existing $2K$-point FFT algorithms to perform the transform itself. The processing block diagram for this approach, equivalent to Equation (1.6), is shown in Figure 1.6. As a result, we obtain an efficient forward transform equivalent to that of the generalized filter bank structure in Figure 1.2.

Similar steps can be taken to also obtain an efficient and robust inverse transform. Let us consider a processed frequency band signal $Y(k, n)$ with $(K + 1)$ bands from the DC frequency to the Nyquist frequency. For formulating the transform, the frequency bands are considered in a $2K$-point conjugate-symmetric form usual for DFTs for real-valued signals,

$$Y(2K - k, n) = Y^*(k, n) \qquad \text{for } k = 1, \dots, (K - 1). \tag{1.7}$$

We now express the equations showing the combined effect after the inverse transform in Figure 1.2 of all frequency bands of $Y(k, n)$ to the output signal $y(t)$. Let us first consider the effect of a single time index n, and denote its corresponding output $\hat{y}_n(t)$. The

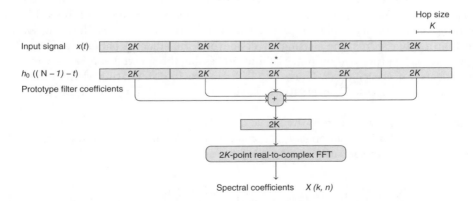

Figure 1.6 Block diagram of a forward transform of a robust filter bank employing the FFT.

upsampling operation determines only the temporal position of the reproduced output. After the band-pass filters $h_k(t)$, and combining the bands, the output is given by

$$\hat{y}_n(t + nK) = \sum_{k=0}^{2K-1} X(k,n)h_k(t) = \sum_{k=0}^{2K-1} X(k,n)h_0(t)e^{i\pi kt/K + \psi_k}$$

$$= h_0(t) \underbrace{\sum_{k=0}^{2K-1} X(k,n)e^{i\pi kt/K + \psi_k}}_{2K\text{-point IDFT when } \psi_k = 0}, \tag{1.8}$$

where it is assumed that $\hat{y}_n(t)$ is zero outside of the range $t = nK, \ldots, (nK + N_h - 1)$. By setting $\psi_k = 0$, we can readily see the equivalence of the processing to using the $2K$-point inverse discrete Fourier transform (IDFT). Since the output of an IDFT is cyclic, Equation (1.8) is equal to the processing in the block diagram in Figure 1.7, which includes a single $2K$-point IFFT operation, repetition of the result as a C-times longer sequence, and performing a point-wise multiplication with the coefficients of the prototype filter $h_0(t)$. If the prototype filter is a $2K$-length square-root Hann window sequence, the expression becomes equivalent to the traditional inverse STFT. For any prototype filter length, the overall output is obtained with overlap-add processing:

$$y(t) = \sum_{n=0}^{\infty} \hat{y}_n(t). \tag{1.9}$$

Similarly to the analysis filter structure, the definition of STFT with $C = 1$ limits the obtained stop-band attenuation accounting for the spectral aliasing components due to the upsampling. A Matlab script realizing the filter bank structure as described in this section is given in Section 1.2.6.

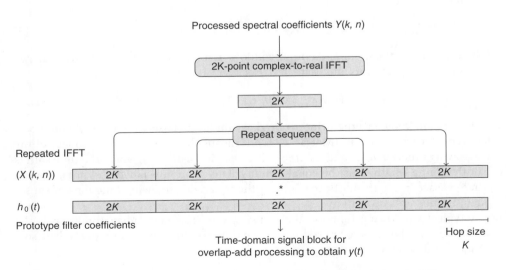

Figure 1.7 Block diagram of an inverse transform of a robust filter bank employing the FFT.

1.2.5 Overlap-Add and Windowing

An important feature of time–frequency transforms in processing applications is that the forward and backward operations themselves, without other processing, do not distort the signal. Such transforms are said to provide *perfect reconstruction*.

In Section 1.2.4 we demonstrated that most filter banks can be implemented using windowing by a prototype filter $h_0(t)$ and application of the discrete Fourier transform. It is well known that the discrete Fourier transform is an orthonormal transform, whereby that part of the transform provides perfect reconstruction. After all, an orthonormal transform is full-rank and, as an added advantage, it can be implemented in a numerically stable manner. Thus, the essential remaining question is whether windowing by $h_0(t)$ and its reverse operation in the inverse transform together provide perfect reconstruction.

Specifically, the question is whether the signal is retained after the processing in Figure 1.2. Observe that the input signal is windowed twice: first with the input window h_0^a, and again after processing with the output window h_0^s. The output of a given frame n is thus

$$x(t)h_0^a(t - nK)h_0^s(t - nK). \tag{1.10}$$

Similarly, the output of the following frame is

$$x(t)h_0^a(t - (n + 1)K)h_0^s(t - (n + 1)K). \tag{1.11}$$

Given that the length of the non-zero part of the frame is $2K$ (assuming $C = 1$), it follows that the overlap between windows is of length K. At the overlap region, $t \in [nK, (n + 1)K - 1]$, the sum of the two outputs should be equal to the original signal,

$$x(t) = x(t)h_0^a(t - nK)h_0^s(t - nK) + x(t)h_0^a(t - (n + 1)K)h_0^s(t - (n + 1)K), \tag{1.12}$$

which means that

$$h_0^a(t)h_0^s(t) + h_0^a(t - K)h_0^s(t - K) = 1. \tag{1.13}$$

Often we choose the input and output windows to be equal, $h_0^a(t) = h_0^s(t)$, whereby we can drop the superscripts and write

$$h_0^2(t) + h_0^2(t + K) = 1 \quad \text{for} \quad t \in [0, K - 1]. \tag{1.14}$$

This relation, known as the Princen–Bradley condition, ensures that we obtain a perfect reconstruction system (Malvar, 1992). It means that the sum of the square of the windowing function $h_0^2(t)$ and its time-shifted version $h_0^2(t + K)$, where the time shift is half the frame length, must be equal to unity. This relation is illustrated in Figure 1.8, where we see in Figure 1.8(b) that adding the squared windowing functions of subsequent frames adds up to unity.

A typical windowing function that satisfies the Princen–Bradley condition is the square root of the Hann or raised-cosine window, which is equivalent to the half-sine window and is defined as

$$h_0(t) = 0.5 \left[1 - \cos\left(\frac{\pi(t + 1/2)}{2K}\right)\right]^{1/2} = \sin\left(\frac{\pi(t + 1/2)}{2K}\right). \tag{1.15}$$

Figure 1.8 (a) Analysis windowing of subsequent windows;
(b) overlap-add at the synthesis stage.

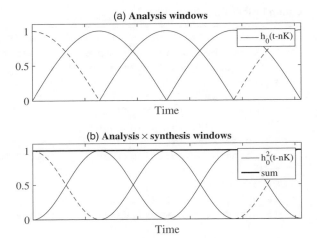

1.2.6   Example Implementation of a Robust Filter Bank in Matlab

An example Matlab implementation[1] of a robust filter bank for time–frequency processing is given in Listings 1.1–1.4. Listing 1.1 consists of an initialization function, including the formulation of the prototype filter coefficients. The cutoff frequencies of the equiripple prototype filters were preformulated for $K = 32, 64, 128, 256$ using the iterative method outlined in Creusere and Mitra (1995). Listings 1.2 and 1.3 respectively provide the forward and the inverse transforms according to the block diagrams in Figures 1.6 and 1.7. Listing 1.4 shows an example script that applies the filter bank to band-pass filter a noise sequence in the transform domain. The resulting spectrum of the script is illustrated in Figure 1.9.

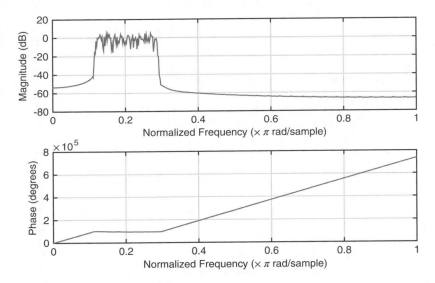

Figure 1.9 Output frequency plot of the Matlab script in Listing 1.1.

---

[1] An open-source C library implementing a filter bank structure of this type is available at https://github.com/jvilkamo/afSTFT/.

Listing 1.1: Initialization for a Filter Bank

```
1   function TF_struct = TF_transform_init(K,numInChans,numOutChans)
2   % Prototype filter frequencies have been pre-formulated ...
        according to Creusere and Mitra.
3   switch K
4       case 32,   f=0.018716082;
5       case 64,   f=0.0093552526;
6       case 128,  f=0.0046768663;
7       case 256,  f=0.0023382376;
8       otherwise, warning('Use K=32, 64, 128, or 256)'); return;
9   end
10  TF_struct.C = 5;
11  TF_struct.N = 2*K*TF_struct.C;
12  TF_struct.prototypeFilter = firceqrip(TF_struct.N-1, f, ...
        [1e-6 1e-5])';
13  TF_struct.K = K;
14  TF_struct.numInChans = numInChans;
15  TF_struct.numOutChans = numOutChans;
16  TF_struct.inBuffer=zeros(TF_struct.N, numInChans);
17  TF_struct.outBuffer=zeros(TF_struct.N, numOutChans);
```

Listing 1.2: Forward Transform

```
1   function [TFdata,TF_struct] = TF_transform_forward(timeData, ...
        TF_struct)
2   if size(timeData,2) ≠ TF_struct.numInChans
3       warning('Wrong number of input channels');return;
4   end
5   K=TF_struct.K;
6   numTFsamples = floor(size(timeData,1)/K);
7   numBands=K+1;
8   TFdata=zeros(numBands,numTFsamples,TF_struct.numInChans);
9   for TFsample = 1:numTFsamples
10      timeDataFrame = timeData((TFsample-1)*K+[1:K],:);
11      TF_struct.inBuffer = [TF_struct.inBuffer(K+1:end,:); ...
            timeDataFrame];
12      prototypeFiltered = TF_struct.inBuffer.*repmat(flipud ...
            (TF_struct.prototypeFilter),[1 TF_struct.numInChans]);
13      folded = prototypeFiltered(1:2*K,:);
14      for c=2:TF_struct.C
15          folded = folded + prototypeFiltered([1:2*K]+2*K*(c-1),:);
16      end
17      transformed = fft(folded);
18      for ch=1:TF_struct.numOutChans
19          TFdata(:,TFsample,ch)=transformed(1:K+1,ch);
20      end
21  end
```

Listing 1.3: Inverse Transform

```
1  function [timeData,TF_struct] = TF_transform_inverse(TFdata, ...
       TF_struct)
2  if size(TFdata,3) ≠ TF_struct.numOutChans
3      warning('Wrong number of output channels');return;
4  end
5  K=TF_struct.K;
6  numTFsamples=size(TFdata,2);
7  numTimeSamples = floor(numTFsamples*K);
8  timeData=zeros(numTimeSamples,TF_struct.numOutChans);
9  FFTdata = zeros(2*K,TF_struct.numOutChans);
10 for TFsample = 1:numTFsamples
11     FFTdata(1:K+1) = squeeze(TFdata(:,TFsample,:));
12     inverseTransformed = ...
           repmat(ifft(FFTdata,'symmetric'),[TF_struct.C, 1])*2*K^2;
13     prototypeFiltered = inverseTransformed.*repmat ...
           (TF_struct.prototypeFilter,[1 TF_struct.numOutChans]);
14     TF_struct.outBuffer = [TF_struct.outBuffer(K+1:end,:); ...
           zeros(K,TF_struct.numOutChans)];
15     TF_struct.outBuffer = TF_struct.outBuffer + prototypeFiltered;
16     timeData([1:K]+(TFsample-1)*K,:) = TF_struct.outBuffer(1:K,:);
17 end
```

Listing 1.4: Example Script Using the Filter Bank

```
1  function TF_transform_example
2  K=64;
3  numInChans=1;
4  numOutChans=1;
5  TF_struct = TF_transform_init(K,numInChans,numOutChans);
6  inAudio = randn(44100,1)/sqrt(44100);
7  [inAudioTF, TF_struct] = TF_transform_forward(inAudio,TF_struct);
8  inAudioTF(1:8,:,:)=0;
9  inAudioTF(20:end,:,:)=0;
10 [outAudio, TF_struct] = TF_transform_inverse(inAudioTF,TF_struct);
11 freqz(outAudio)
```

1.2.7 Cascaded Filters

When a prototype filter with a high stop-band attenuation is applied in a complex-modulated filter bank, the aliasing at the downsampled frequency bands becomes negligible. Therefore, the frequency band signals can be processed and filtered without restriction. In particular, it is typical to apply cascaded filters at the lowest uniform frequency bands to obtain further spectral selectivity to better approximate the perceptual scales. For example, in a typical configuration of the MPEG QMF bank, a 64-band uniform transform operation is applied, of which the first three frequency bands are

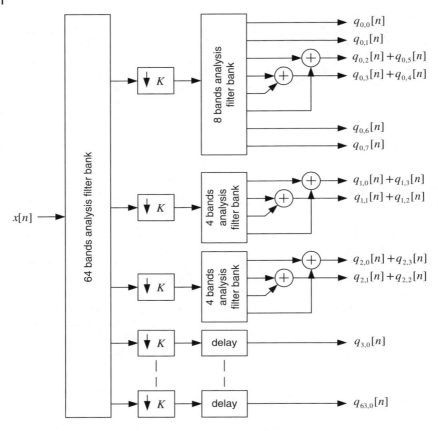

Figure 1.10 Cascaded filters applied in MPEG systems to obtain a higher frequency selectivity at the low frequencies. Adapted from Breebaart *et al.* (2005).

subdivided with linear-phase filters to form in total a 71-band representation (Figure 1.10). The configuration is commonly called the hybrid QMF bank. The filters are designed to be complementary in terms of the amplitude, and not the energy, and no further downsampling is applied. A corresponding delay is applied at the remaining bands. With such a configuration, the synthesis filter bank consists of a simple summation of the hybrid frequency bands, followed by the inverse QMF bank.

1.3 Processing of Spatial Audio

The block diagram of a parametric time–frequency processing algorithm shown in Figure 1.1 also represents most of the parametric spatial audio processing methods discussed in this book. The methods all utilize some kind of stochastic analysis of time–frequency domain differences between microphone signals, which will be discussed in the next section. In the reproduction of spatial audio, the computed time–frequency-dependent parameters are then typically used to control the mixing process, where in some cases decorrelation of signals is used to control the coherence between output channels, as discussed in Section 1.3.2. An optimal method to control the level of decorrelated signal energy is utilized in some of the spatial audio methods discussed in the book; this is described in Section 1.3.3.

1.3.1 Stochastic Estimates

A property in many applications in parametric time–frequency audio processing is that the multiple audio signals are measured by their short-time stochastic properties in frequency bands. Let us define an N_x-channel time–frequency signal vector

$$\mathbf{x}(k, n) = \begin{bmatrix} X_1(k, n) \\ X_2(k, n) \\ \vdots \\ X_{N_x}(k, n) \end{bmatrix}. \tag{1.16}$$

The general expression for the signal covariance matrix is

$$\mathbf{C_x}(k, n) = E[\mathbf{x}(n, k)\mathbf{x}^H(n, k)], \tag{1.17}$$

where $E[\cdot]$ is the expectation operator and $\mathbf{x}^H(n, k)$ is the conjugate transpose of $\mathbf{x}(n, k)$. In adaptive techniques the covariance matrix is estimated using a windowing function over segments of the frequency band signal,

$$\mathbf{C_x}(k, n) = \sum_{n_w=n-N_w/2}^{n+N_w/2-1} w(n_w + N_w/2)\mathbf{x}(n_w, k)\mathbf{x}^H(n_w, k), \tag{1.18}$$

where N_w is the length of the window, typically corresponding to tens of milliseconds. Another typical approach is to use first-order infinite impulse response (IIR) estimation,

$$\mathbf{C_x}(k, n) = \beta\mathbf{C_x}(k, n - 1) + (1 - \beta)\mathbf{x}(n, k)\mathbf{x}(n, k)^H, \tag{1.19}$$

where β is a parameter determining the decay rate of the IIR window. Estimation of the covariance matrix can be performed along an interval in the frequency axis as well, which can be useful at the higher frequency bands in which the uniform frequency resolution is more selective than the perceptual scales. Such a procedure is applied, for example, for reducing the data rate of the parametric spatial information in audio coding, or for reducing the computational complexity of any parameter-driven processing.
Let us study the entries of the covariance matrix,

$$\mathbf{C_x}(k, n) = E[\mathbf{x}(n, k)\mathbf{x}(n, k)^H] = \begin{bmatrix} c_{11} & c_{12} & \cdots & c_{1N_x} \\ c_{21} & c_{22} & & c_{2N_x} \\ \vdots & & \ddots & \vdots \\ c_{N_x 1} & \cdots & & c_{N_x N_x} \end{bmatrix}, \tag{1.20}$$

where

$$c_{ab} = E[X_a(k, n)X_b^*(k, n)]. \tag{1.21}$$

The diagonal values $a = b$ express the estimated channel energies, and the non-diagonal values $a \neq b$ express the cross-correlation of the channels. From these measures, it is possible to obtain the typical measures of the inter-channel level difference,

$$\text{ICLD}_{ab} = 10 \log_{10} \left(\frac{c_{aa}}{c_{bb}} \right), \tag{1.22}$$

the inter-channel phase difference,

$$\text{IPD}_{ab} = \arg(c_{ab}), \tag{1.23}$$

and the inter-channel coherence,

$$\text{ICC}_{ab} = \frac{|c_{ab}|}{\sqrt{c_{aa}c_{bb}}}. \tag{1.24}$$

The channel energies, IPDs, and ICCs provide a set of information equivalent to the covariance matrix, and thus it is only a matter of practicality which representation to apply as part of a parametric processing technique. In techniques that do not account for the phase differences between the channels, the ICC is sometimes considered only by its real part, using an alternative definition:

$$\text{ICC}'_{ab} = \frac{\text{Re}\{c_{ab}\}}{\sqrt{c_{aa}c_{bb}}}. \tag{1.25}$$

The perceptual parametric processing techniques typically analyze and/or synthesize the spatial sound based on the information contained in these stochastic estimates, using either expressions similar to Equations (1.22)–(1.25) or the signal covariance matrix. Since the estimators in Equations (1.18) and (1.19) are essentially low-pass filters, it is meaningful in a computational sense to consider the parameters only sparsely in time. An often-met practical approach is to use a rectangular window estimator with window length N_w corresponding to a time frame in the range of tens of milliseconds, and to apply the estimation every $N_w/2$ time indices. At each of these frames, the parametric data is processed to obtain a rule to mix, process, and/or to gate the frequency band signals. These operations are typically expressible as a matrix of complex-valued mixing weights to process the output frequency band signals based on the input frequency band signals. The weights are formulated for every frame, that is, sparsely in time, and they can be linearly interpolated between the frames prior to applying them to the frequency band signals.

1.3.2 Decorrelation

In many use cases of parametric processing, the input signals consist of a smaller number of prominent independent signals than is required for the output signals. One example is in upmixing, where we want to increase the number of channels—for example, from a two-channel stereophonic signal to a five-channel surround signal. Another example is spatial sound reproduction from a microphone array, where the input signals are highly coherent at low frequencies due to the wavelength with respect to the array size. Signal mixing alone does not increase the number of independent signals, or in the case of array processing, mixing the microphone signals to reduce coherence would result in excessive amplification of the microphone self-noise. Techniques known as *decorre-lating* methods are therefore frequently used to synthesize new incoherent signals from the existing signals.

Ideally, decorrelators produce signals which are incoherent with respect to their inputs and with respect to the outputs of other decorrelators. In addition, the methods should minimize any perceptual modifications of the signal. Various designs of decorrelators have been developed. A simple yet effective method is to apply different delays at different bands and channels (Boueri and Kyriakakis, 2004). Another common class of decorrelators is based on all-pass filters, as applied, for example, in Breebaart *et al.* (2005).

Regardless of the type of decorrelator, its design and tuning are practical engineering tasks. None of the decorrelator structures can preserve the sound quality in a general sense, since there are signal types such as transients and speech for which the phase spectrum has perceptual significance (Laitinen *et al.*, 2013). Typical methods for miti-gating the distortions caused by decorrelation include switching the type of decorrelator applied depending on signal type (Kuntz *et al.*, 2011), and bypassing transients from the decorrelating processes.

In the following sections, we denote the decorrelating operation by

$$\mathbf{x}_D(k, n) = D[\mathbf{x}(k, n)]. \tag{1.26}$$

The objective is that, for the decorrelated signal, $E[\mathbf{x}(n, k)\mathbf{x}_D^H(n, k)] \equiv \mathbf{0}$.

1.3.3 Optimal and Generalized Solution for Spatial Sound Processing Using Covariance Matrices

Although many parametric time–frequency processing techniques can be designed with dedicated signal processing structures, for many tasks it is also an option to use a gener-alized solution as proposed in Vilkamo, Bäckström, and Kuntz (2013). If the task of the parametric processing is defined using expressions that can be applied for a wide range of techniques, it is possible to derive a single solution and implementation that is applica-ble to each of them. Most importantly, we can employ least-squares techniques to opti-mize the audio fidelity for all use cases in scope. For example, the methods described in the following have been applied in spatial sound reproduction from microphone arrays (Chapters 6 and 8 of this book; Vilkamo and Pulkki, 2013; Politis *et al.*, 2015), binaural sound reproduction (Chapters 6 and 8 of this book; Delikaris-Manias *et al.*, 2015), spatial enhancement (Vilkamo and Pulkki, 2015), and stereo-to-surround upmixing (Vilkamo *et al.*, 2013).

Perceptual spatial sound processing is defined here, in a generic and generalized form, as the design of the output signal covariance matrix in frequency bands. The covari-ance matrix contains the perceptually crucial measures of the channel energies, the inter-channel coherences, and the phase differences. It does not, however, contain infor-mation about the actual fine spectral information in the channels, which is addressed separately.

For brevity of notation, let us omit the frequency and the time indices (k, n) and consider a single time–frequency segment in which an application has measured the covariance matrix $\mathbf{C_x} = E[\mathbf{xx}^H]$ of the $N_x \times 1$ input signal $\mathbf{x}$. The $N_y \times 1$ output time–frequency signal is defined as **y**. In the generalized form, the parametric process-ing task is expressed by two matrices:

- a target covariance matrix $\mathbf{C_y} = E[\mathbf{yy}^H]$, which contains the perceptually relevant *spatial information* for the target channels, such as loudspeaker or headphone channels; and
- a prototype matrix **Q**, which determines the relevant *signal information* for the target channels.

The prototype matrix **Q** can be interpreted as an example of the mapping between **x** and **y**, that is, it determines a prototype signal

$$\mathbf{y_Q} = \mathbf{Qx}, \tag{1.27}$$

which is a linear combination or a selection of the input channels. The prototype matrix is designed for each application such that it produces for each output channel the signal content that is most meaningful in a spatial sense. An example of a prototype matrix in the context of microphone array processing is a set of beamformers that are spatially selective towards the directions corresponding to those of the target loudspeaker setup. In the application of spatial enhancement and effects, on the other hand, the prototype matrix is defined to be an identity matrix, since it is desirable that the signal content in the output channels is least affected by the parametric processing.

The method to determine the target covariance matrix $\mathbf{C_y}$ also depends on the task. For example, in spatial sound reproduction based on sound field models, $\mathbf{C_y}$ is deter-mined such that it represents the modeled parameters of the sound field for the target loudspeaker array. This includes placing sound energy in the directions corresponding to the analyzed arriving sound in the original sound field, and spatially spreading the portion of the sound energy that was analyzed as ambience. The adaptive design of the target covariance matrix is exemplified further in the context of the specific use cases in Chapters 6 and 8.

The task for the parametric processing is thus to obtain an output signal **y** with the application-determined target covariance matrix $\mathbf{C_y} = E[\mathbf{yy}^H]$, which is usually different from that of $\mathbf{y_Q}$. The input–output relation for the processing is defined as

$$\mathbf{y} = \mathbf{Mx} + \mathbf{M_D}D[\mathbf{Qx}], \tag{1.28}$$

where **M** is the primary mixing matrix that is solved to produce $\mathbf{C_y}$. However, solving **M** is subject to regularization to ensure robust sound quality. $D[\cdot]$ is a set of decorrelating signal processing operations, and $\mathbf{M_D}$ is a secondary mixing matrix that is designed to process the decorrelated signals to produce a covariance matrix that is complementary to the effect of the regularization of **M**. Assuming incoherence of the decorrelated signals $D[\mathbf{Qx}]$ and the input signals **x**, the expression of the complementary property of the covariance matrices is

$$\mathbf{C_y} = \mathbf{MC_x M}^H + \mathbf{C_D}, \tag{1.29}$$

where $\mathbf{C_D}$ is the covariance matrix for the processed signal $\mathbf{M_D}D[\mathbf{Qx}]$.

Let us temporarily assume that $\mathbf{M_D} = \mathbf{0}$, that is, that the decorrelated signals are not applied. Equation (1.29) simplifies to

$$\mathbf{C_y} = \mathbf{MC_x M}^H. \tag{1.30}$$

The set of solutions for $\mathbf{M}$ fulfilling Equation (1.30) can be found using matrix decompositions. First, let us determine any unitary matrices $\mathbf{P}_x$ and $\mathbf{P}_y$, which satisfy the properties $\mathbf{P}_x\mathbf{P}_x^H = \mathbf{I}$ and $\mathbf{P}_y\mathbf{P}_y^H = \mathbf{I}$, and define a set of decompositions of the covariance matrices:

$$\begin{aligned} \mathbf{C}_x &= \mathbf{K}_x\mathbf{K}_x^H = \mathbf{K}_x\mathbf{P}_x\mathbf{P}_x^H\mathbf{K}_x^H, \\ \mathbf{C}_y &= \mathbf{K}_y\mathbf{K}_y^H = \mathbf{K}_y\mathbf{P}_y\mathbf{P}_y^H\mathbf{K}_y^H. \end{aligned} \tag{1.31}$$

An example of obtaining $\mathbf{K}_x$ and $\mathbf{K}_y$ based on $\mathbf{C}_x$ and $\mathbf{C}_y$ is the Cholesky decomposition. The unitary matrices $\mathbf{P}_x$ and $\mathbf{P}_y$ widen the scope of the decompositions in Equation (1.31) to cover all decompositions fulfilling the defined property. Substituting Equation (1.31) in (1.30), we obtain

$$\mathbf{K}_y\mathbf{P}_y\mathbf{P}_y^H\mathbf{K}_y^H = \mathbf{M}\mathbf{K}_x\mathbf{P}_x\mathbf{P}_x^H\mathbf{K}_x^H\mathbf{M}^H, \tag{1.32}$$

from which we obtain the set of solutions

$$\boxed{\mathbf{M} = \mathbf{K}_y\mathbf{P}_y\mathbf{P}_x^H\mathbf{K}_x^{-1} = \mathbf{K}_y\mathbf{P}\mathbf{K}_x^{-1},} \tag{1.33}$$

where $\mathbf{P} = \mathbf{P}_y\mathbf{P}_x^H$ is any unitary matrix. To recap the steps so far, for any unitary $\mathbf{P}$, the signal $\mathbf{y} = \mathbf{M}\mathbf{x} = \mathbf{K}_y\mathbf{P}\mathbf{K}_x^{-1}\mathbf{x}$ has the target covariance matrix $\mathbf{C}_y$.

As the second step of the derivation, we find a $\mathbf{P}$ that provides the smallest difference in the waveforms of the reproduced signal $\mathbf{y}$ and the prototype signal $\mathbf{y}_Q$. This difference is expressed by an error measure,

$$e = \mathrm{E}[\|\mathbf{G}\mathbf{y}_Q - \mathbf{y}\|^2], \tag{1.34}$$

where $\mathbf{G}$ is a non-negative diagonal matrix that is formulated such that the channel energies of $\mathbf{y}_Q$ match the diagonal of $\mathbf{C}_y$. This normalization ensures that the error measure is weighted with the energies of the reproduced channels. Defining $\mathbf{A} = (\mathbf{G}\mathbf{y}_Q - \hat{\mathbf{y}})$, we can use the matrix trace operator $\mathrm{tr}(\cdot)$ to obtain the error measure in the form

$$e = \mathrm{E}[\mathbf{A}^H\mathbf{A}] = \mathrm{E}[\mathrm{tr}(\mathbf{A}^H\mathbf{A})] = \mathrm{E}[\mathrm{tr}(\mathbf{A}\mathbf{A}^H)], \tag{1.35}$$

where we used the equivalence $\mathrm{tr}(\mathbf{A}_1^H\mathbf{A}_2) = \mathrm{tr}(\mathbf{A}_1\mathbf{A}_2^H)$. Substituting $\mathbf{y}_Q = \mathbf{Q}\mathbf{x}$ and $\mathbf{y} = \mathbf{M}\mathbf{x} = \mathbf{K}_y\mathbf{P}\mathbf{K}_x^{-1}\mathbf{x}$ in Equation (1.35), and substituting the definition of $\mathbf{A}$, we obtain the error measure

$$\begin{aligned} e &= \mathrm{E}\left[\mathrm{tr}\left(\left(\mathbf{G}\mathbf{Q}\mathbf{x} - \mathbf{K}_y\mathbf{P}\mathbf{K}_x^{-1}\mathbf{x}\right)\left(\mathbf{G}\mathbf{Q}\mathbf{x} - \mathbf{K}_y\mathbf{P}\mathbf{K}_x^{-1}\mathbf{x}\right)^H\right)\right] \\ &= \mathrm{tr}(\mathbf{G}\mathbf{Q}\mathbf{C}_x\mathbf{Q}^H\mathbf{G}^H) - 2\mathrm{tr}\left(\mathbf{G}\mathbf{Q}\mathbf{K}_x\mathbf{P}^H\mathbf{K}_y^H\right) + \mathrm{tr}\left(\mathbf{K}_y\mathbf{K}_y^H\right), \end{aligned} \tag{1.36}$$

where we have applied the definition $\mathbf{C}_x = \mathrm{E}[\mathbf{x}\mathbf{x}^H]$, the middle term has been simplified using $\mathbf{C}_x(\mathbf{K}_x^{-1})^H = \mathbf{K}_x$, from Equation (1.31), and the last term has been simplified using the identities $\mathbf{K}_x^{-1}\mathbf{C}_x(\mathbf{K}_x^{-1})^H = \mathbf{I}$ and $\mathbf{P}\mathbf{P}^H = \mathbf{I}$. In Equation (1.36), only the second term depends on the variable $\mathbf{P}$, and we can thus formulate the minimization problem as

$$\begin{aligned} \arg\min_{\mathbf{P}} e &= \arg\max_{P} \mathrm{tr}\left(\mathbf{G}\mathbf{Q}\mathbf{K}_x\mathbf{P}^H\mathbf{K}_y^H\right) \\ &= \arg\max_{P} \mathrm{tr}\left(\mathbf{K}_x^H\mathbf{Q}^H\mathbf{G}^H\mathbf{K}_y\mathbf{P}\right), \end{aligned} \tag{1.37}$$

where we used the relation $\text{tr}(\mathbf{X}^H\mathbf{Y}) = \text{tr}(\mathbf{Y}\mathbf{X}^H)$ (Petersen and Pedersen, 2012). Finally, assuming first that $N_y = N_x$, the optimal solution is found using the fact that, for a non-negative diagonal matrix $\mathbf{S}$ and any unitary matrix $\mathbf{P}_s$, $\text{tr}(\mathbf{S}) \geq \text{tr}(\mathbf{S}\mathbf{P}_s)$, which is a consequence of the Cauchy–Schwartz inequality. Defining a singular value decomposition (SVD) with the relation $\mathbf{U}\mathbf{S}\mathbf{V}^H = \mathbf{K}_x^H\mathbf{Q}^H\mathbf{G}^H\mathbf{K}_y$, where $\mathbf{S}$ is non-negative and diagonal, and $\mathbf{U}$ and $\mathbf{V}$ are unitary, it follows that

$$\text{tr}\left(\mathbf{K}_x^H\mathbf{Q}^H\mathbf{G}^H\mathbf{K}_y\mathbf{P}\right) = \text{tr}(\mathbf{U}\mathbf{S}\mathbf{V}^H\mathbf{P}) = \text{tr}(\mathbf{U}\mathbf{S}\mathbf{V}^H\mathbf{P}\mathbf{U}\mathbf{U}^H) = \text{tr}(\mathbf{S}\mathbf{V}^H\mathbf{P}\mathbf{U}) \leq \text{tr}(\mathbf{S})$$

(1.38)

for any unitary $\mathbf{P}$. The equality holds for

$$\mathbf{P} = \mathbf{V}\mathbf{U}^H, \tag{1.39}$$

which is thus a solution that minimizes the error measure e in Equation (1.34). Furthermore, the optimal solution when $N_x \neq N_y$ is (Vilkamo *et al.*, 2013)

$$\boxed{\mathbf{P} = \mathbf{V}\boldsymbol{\Lambda}\mathbf{U}^H,} \tag{1.40}$$

where $\boldsymbol{\Lambda}$ is an identity matrix appended with zeros to match the dimensions of $\mathbf{P}$.

Regardless of being optimal in a least-squares sense, the aforementioned mixing solution does not yet account for the cases when:

- the matrix $\mathbf{K}_x$ in Equation (1.33) is not invertible;
- the matrix inverse $\mathbf{K}_x^{-1}$ has very large coefficients (when it is ill-conditioned) which typically causes processing artifacts; or
- the number of output channels is larger than the number of input channels.

To account for the first two cases we need to regularize the inverse of matrix $\mathbf{K}_x$. A robust method is to employ the SVD with the relation $\mathbf{K}_x = \mathbf{U}_x\mathbf{S}_x\mathbf{V}_x^H$, whereby regularization of $\mathbf{K}_x$ is equivalent to regularizing $\mathbf{S}_x$ to form the regularized $\mathbf{S}_x'$. We then obtain the regularized matrix as $\mathbf{K}_x'^{-1} = \mathbf{V}_x\mathbf{S}_x'^{-1}\mathbf{U}_x^H$.

A practical regularization is thresholding from below: Let scalars $s_k > 0$ be the diagonal elements of $\mathbf{S}_x$ and set the regularization rule as

$$s_k' := \begin{cases} s_k & \text{for } s_k > \alpha \cdot \max\{s_k\}, \\ \alpha \cdot \max\{s_k\} & \text{for } s_k \leq \alpha \cdot \max\{s_k\}, \end{cases} \tag{1.41}$$

where $\alpha = 0.2$ is a typical value. In other words, we increase the singular values of $\mathbf{K}_x$ until they are at least 0.2 times the largest singular value $\max\{s_k\}$. The regularized inverse $\mathbf{K}_x'^{-1}$ is then substituted for $\mathbf{K}_x^{-1}$ in Equation (1.33), such that we obtain a regularized version $\mathbf{M}$.

Using the regularized $\mathbf{M}$, the mixing solution $\mathbf{M}\mathbf{C}_x\mathbf{M}^H$ no longer yields the target covariance matrix $\mathbf{C}_y$. Consequently, the complementary covariance matrix $\mathbf{C}_D$ in Equation (1.29) becomes non-zero. The secondary mixing matrix $\mathbf{M}_D$ is then designed to process the decorrelated signals to obtain $\mathbf{C}_D$. Mixing the resulting signal with the main signal $\mathbf{M}\mathbf{x}$ as in Equation (1.28) produces the covariance matrix $\mathbf{C}_y$ for the output signal $\mathbf{y}$ by definition. The method to obtain the secondary matrix $\mathbf{M}_D$ is equivalent to the steps to obtain $\mathbf{M}$; however, the matrix $\mathbf{C}_D$ is defined as the target covariance matrix, and an identity matrix as the prototype of matrix $\mathbf{Q}$.

As the main result of the above derivation, we have obtained a generalized technique for parametric spatial sound processing using least-squares optimization, which performs the mixing of the signal and decorrelated signals with respect to three parameter matrices: the measured covariance matrix $\mathbf{C_x}$, the target covariance matrix $\mathbf{C_y}$, and a prototype matrix $\mathbf{Q}$. Application of this technique to specific use cases requires only the design of matrices $\mathbf{C_y}$ and $\mathbf{Q}$, which will be demonstrated in Chapters 6 and 8.

References

3GP (2014) *TS 26.445, EVS Codec Detailed Algorithmic Description; 3GPP Technical Specification (Release 12)*, ETSI, Sophia Antipolis.

Allamanche, E., Geiger, R., Herre, J., and Sporer, T. (1999) MPEG-4 low delay audio coding based on the AAC codec. *Audio Engineering Society Convention 106*, Audio Engineering Society.

Bäckström, T. (2013) Vandermonde factorization of Toeplitz matrices and applications in filtering and warping. *IEEE Transactions on Signal Processing*, **61**(24), 6257–6263.

Blauert, J. (1997) *Spatial Hearing: The Psychophysics of Human Sound Localization* (rev. ed.), MIT Press, Cambridge, MA.

Bosi, M., Brandenburg, K., Quackenbush, *et al.* (1997) ISO/IEC MPEG-2 advanced audio coding. *Journal of the Audio Engineering Society*, **45**(10), 789–814.

Bosi, M. and Goldberg, R.E. (2003) *Introduction to Digital Audio Coding and Standards* Kluwer Academic Publishers, Dordrecht.

Boueri, M. and Kyriakakis, C. (2004) Audio signal decorrelation based on a critical band approach *Audio Engineering Society Convention 117*, Audio Engineering Society.

Brandenburg, K. (1999) MP3 and AAC explained. *Audio Engineering Society 17th International Conference*, Audio Engineering Society.

Breebaart, J., Disch, S., Faller, C., *et al.* (2005) The reference model architecture for MPEG spatial audio coding. *Audio Engineering Society Convention 118*, Audio Engineering Society.

Creusere, C.D. and Mitra, S.K. (1995) A simple method for designing high-quality prototype filters for M-band pseudo QMF banks. *IEEE Transactions on Signal Processing*, **43**(4), 1005–1007.

Cruz-Roldán, F., Amo-López, P., Maldonado-Bascón, S., and Lawson, S.S. (2002) An efficient and simple method for designing prototype filters for cosine-modulated pseudo-QMF banks. *IEEE Signal Processing Letters*, **9**(1), 29–31.

Delikaris-Manias, S., Vilkamo, J., and Pulkki, V. (2015) Parametric binaural rendering using compact microphone arrays. *International Conference on Acoustics, Speech, and Signal Processing*, IEEE.

Fastl, H. and Zwicker, E. (2007) *Psychoacoustics: Facts and Models*. Springer, Berlin.

Herre, J., Hilpert, J., Kuntz, A., and Plogsties, J. (2015) MPEG-H Audio: The new standard for universal spatial/3D audio coding. *Journal of the Audio Engineering Society*, **62**(12), 821–830.

Herre, J., Purnhagen, H., Koppens, J., *et al.* (2012) MPEG spatial audio object coding: The ISO/MPEG standard for efficient coding of interactive audio scenes. *Journal of the Audio Engineering Society*, **60**(9), 655–673.

ISO/IEC JTC1/SC29/WG11 (MPEG) (1997) *MPEG-2 Advanced Audio Coding, AAC*, International Standards Organization.

Jayant, N.S. and Noll, P. (1984) *Digital Coding of Waveforms: Principles and Applications to Speech and Video*, Prentice-Hall, Englewood Cliffs, NJ.

Kuntz, A., Disch, S., Bäckström, T., and Robilliard, J. (2011) The transient steering decorrelator tool in the upcoming MPEG unified speech and audio coding standard, *Audio Engineering Society Convention 131*, Audio Engineering Society.

Laitinen, M.V., Disch, S., and Pulkki, V. (2013) Sensitivity of human hearing to changes in phase spectrum. *Journal of the Audio Engineering Society*, **61**(11), 860–877.

Malvar, H.S. (1992) *Signal Processing with Lapped Transforms*. Artech House, Inc., Norwood, MA.

Mitra, S. and Kaiser, J. (eds.) (1993) *Handbook of Digital Signal Processing*. John Wiley & Sons, Ltd, Chichester.

Moore, B.C.J. (ed.) (1995) *Hearing*. Academic Press, New York.

Oppenheim A.V. and Schafer, R.W. (1975) *Digital Signal Processing*, Prentice-Hall, Englewood-Cliffs, NJ.

Petersen, K.B. and Pedersen, M.S. (2012) *The Matrix Cookbook* vol. 7, Technical University of Denmark.

Politis, A., Vilkamo, J., and Pulkki, V. (2015) Sector-based parametric sound field reproduction in the spherical harmonic domain. *IEEE Journal of Selected Topics in Signal Processing*, **9**(5), 852–866.

Schuller, G. and Smith, M.J. (1995) A new algorithm for efficient low delay filter bank design. *Proc. International Conference on Acoustics, Speech, and Signal Processing*, vol. 2, pp. 1472–1475, IEEE.

Smith. J.O. (2011) *Spectral Audio Signal Processing*, W3K Publishing.

Vilkamo, J., Bäckström, T., and Kuntz, A. (2013) Optimized covariance domain framework for time–frequency processing of spatial audio. *Journal of the Audio Engineering Society*, **61**(6), 403–411.

Vilkamo, J. and Pulkki, V. (2013) Minimization of decorrelator artifacts in directional audio coding by covariance domain rendering. *Journal of the Audio Engineering Society*, **61**(9), 637–646.

Vilkamo, J. and Pulkki, V. (2015) Adaptive optimization of interchannel coherence with stereo and surround audio content. *Journal of the Audio Engineering Society*, **62**(12), 861–869.

2

Spatial Decomposition by Spherical Array Processing

David Lou Alon and Boaz Rafaely

Department of Electrical and Computer Engineering, Ben-Gurion University of the Negev, Israel

2.1 Introduction

Spherical microphone arrays have the advantageous property of rotational symmetry and are, therefore, most suitable for capturing and analyzing three-dimensional sound fields. A fundamental step in the spatial processing of microphone signals is to extract the plane waves that compose the measured sound field in an operation referred to as plane-wave decomposition (PWD; Meyer and Elko, 2002; Abhayapala and Ward, 2002; Gover *et al.*, 2002; Rafaely, 2004). PWD provides a generic spatial representation of sound fields and was previously shown to be useful for various applications, such as beamforming (Meyer and Elko, 2002; Gover *et al.*, 2002), sound field analysis (Abhayapala and Ward, 2002; Park and Rafaely, 2005), and spatial sound field recording (Gerzon, 1973; Abhayapala and Ward, 2002; Bertet *et al.*, 2006; Jin *et al.*, 2014). Chapter 3 of this book presents a PWD formulation that assumes spatial sparsity of the incident sound waves, is signal dependent, and under certain conditions achieves super-resolution with respect to the array capabilities.

An ideal PWD process that accurately estimates the plane-wave density around the microphone array typically requires high spatial resolution over a wide operating bandwidth. For spherical arrays, PWD is usually performed in the spherical harmonics (SH) domain and, in order to achieve high spatial resolution, a PWD with a high SH order is required. However, when using a spherical microphone array system with a practical number of microphones (e.g., a few dozen) the SH order and bandwidth of the PWD are severely limited. At low frequencies, where the wavelength is large compared to the array radius, the high-order coefficients of the sound pressure have low magnitude, therefore limiting the SH order of the PWD that can be estimated in practice. Amplifying the low-magnitude high SH orders also amplifies noise, which leads to a tradeoff between spatial resolution and noise amplification (Meyer and Elko, 2002; Abhayapala and Ward, 2002). At high frequencies, where the wavelength is small, the sound field is represented by high SH orders, which may be higher than the SH orders that can be extracted by the array, due to the limited number of microphones. Using the conventional PWD method

Parametric Time–Frequency Domain Spatial Audio, First Edition. Edited by Ville Pulkki, Symeon Delikaris-Manias, and Archontis Politis.
© 2018 John Wiley & Sons Ltd. Published 2018 by John Wiley & Sons Ltd.
Companion Website: www.wiley.com/go/pulkki/parametrictime-frequency

at high frequencies therefore leads to spatial aliasing that contaminates the PWD estimation (Driscoll and Healy, 1994; Rafaely, 2005).

This chapter presents an overview of the conventional PWD method for spherical microphone arrays, as well as some recent advances that aim to provide improved PWD estimations. The chapter starts with the conventional PWD method and describes the sources of noise amplification and spatial aliasing errors. Then, a quantitative error measure is developed based on the mean squared error of the PWD process. By minimizing the PWD error measure, several optimal PWD methods are developed that achieve robustness to noise, reduced aliasing error, and high spatial resolution over a broad operating frequency. In addition to the analytical development of the optimal PWD methods, simulation examples of a fourth-order array are presented to illustrate, compare, and analyze the performance of different PWD methods.

2.2 Sound Field Measurement by a Spherical Array

Consider the sound pressure around a sphere of radius r at angles $\Omega_q = (\theta_q, \phi_q)$, where θ_q and ϕ_q denote the elevation and azimuth angles in spherical coordinates (Arfken *et al.*, 2012). Assuming that the sound field is composed of plane waves, the pressure $p(k, r, \Omega_q)$ at wave number k can be represented in the SH domain by (Rafaely, 2004)

$$p(k, r, \Omega_q) = \sum_{n=0}^{\infty} \sum_{m=-n}^{n} Y_n^m(\Omega_q) b_n(kr) a_{nm}(k), \tag{2.1}$$

where $Y_n^m(\Omega_q)$ are the SH basis functions of order n and degree m, $b_n(kr)$ are the mode strength functions, also referred to as the *radial* functions (Williams, 1999), and $a_{nm}(k)$ are the spherical Fourier transform coefficients of the spatial density of the plane waves that compose the sound field. The plane waves' spatial density function is defined for every wave number k and direction $\Omega = (\theta, \phi)$ by $a(k, \Omega)$, and the coefficients $a_{nm}(k)$ are referred to in this chapter as the plane-wave coefficients.

By constraining the operating frequency to satisfy $kr \leq N$, the sound pressure around the sphere can be considered an order-limited function up to order N (Ward and Abhayapala, 2001) and Eq. (2.1) can be approximated, with negligible truncation error, using a finite summation:

$$p(k, r, \Omega_q) \approx \sum_{n=0}^{N} \sum_{m=-n}^{n} Y_n^m(\Omega_q) b_n(kr) a_{nm}(k). \tag{2.2}$$

In practice, the sound pressure is measured by Q microphones that not only measure the sound pressure, but also additive noise, such as sensor noise. Nevertheless, the measurement model in Eq. (2.2) and in the next section assumes negligible sensor noise. In Sections 2.4, 2.5, and 2.8, the effect of sensor noise is analyzed.

2.3 Array Processing and Plane-Wave Decomposition

The sound pressure was formulated in the previous section as a continuous function on the surface of a sphere. In this section, the spherical array processing equations are developed. When measured with a spherical microphone array, the sound pressure is spatially sampled at the Q microphone positions. Assuming no sensor noise and negligible truncation error, Eq. (2.2) represents the pressure at each of the array's microphones, and can be reformulated in matrix form as

$$\mathbf{p} = \mathbf{B}\mathbf{a_{nm}}, \tag{2.3}$$

where $\mathbf{p} = [p(k, r_1, \Omega_1), \dots, p(k, r_Q, \Omega_Q)]^T$ is a column vector of length Q holding the sound pressure amplitude sampled by the microphones, located at radius r_q and angles $\Omega_q = (\theta_q, \phi_q)$, with $q = 1, \dots, Q$. $\mathbf{B}$ is a $Q \times (N+1)^2$ matrix as defined in Eq. (2.4), and holds the SH functions $Y_n^m(\Omega_q)$ and the radial functions $b_n(kr_q)$. The latter can be generalized for various array configurations (Rafaely and Balmages, 2007). The $(N+1)^2$-length column vector $\mathbf{a_{nm}} = [a_{00}(k), \dots, a_{NN}(k)]^T$ contains the plane-wave coefficients.

$$\mathbf{B} = \begin{bmatrix} Y_0^0(\Omega_1)b_0(kr_1) & Y_1^{-1}(\Omega_1)b_1(kr_1) & \cdots & Y_N^N(\Omega_1)b_N(kr_1) \\ Y_0^0(\Omega_2)b_0(kr_2) & Y_1^{-1}(\Omega_2)b_1(kr_2) & \cdots & Y_N^N(\Omega_2)b_N(kr_2) \\ \vdots & \vdots & \ddots & \vdots \\ Y_0^0(\Omega_Q)b_0(kr_Q) & Y_1^{-1}(\Omega_Q)b_1(kr_Q) & \cdots & Y_N^N(\Omega_Q)b_N(kr_Q) \end{bmatrix}. \tag{2.4}$$

Equation (2.3) provides a generic measurement model, which is suitable for single-sphere arrays (Meyer and Elko, 2002; Abhayapala and Ward, 2002), for an array composed of multiple concentric spheres (Jin *et al.*, 2006), and for volumetric arrays (Rafaely, 2008; Alon and Rafaely, 2012).

For the case of a single-sphere array, the sound pressure is spatially sampled at Q microphone positions over the sphere surface. Matrix $\mathbf{B}$ can be separated in this case into two matrices, $\mathbf{B} = \mathbf{Y}\,\text{diag}\,(\mathbf{b})$, leading to

$$\mathbf{p} = \mathbf{Y}\,\text{diag}\,(\mathbf{b})\,\mathbf{a_{nm}}, \tag{2.5}$$

where $\mathbf{p} = [p(k, r, \Omega_1), \dots, p(k, r, \Omega_Q)]^T$ holds the sound pressure sampled by the microphones over a constant radius r and angles $\Omega_q = (\theta_q, \phi_q)$. The $Q \times (N+1)^2$ matrix $\mathbf{Y}$ that holds the SH functions is defined in Eq. (2.6), and the $(N+1)^2$-length column vector $\mathbf{b} = [b_0(kr), b_1(kr), b_1(kr), \dots, b_N(kr)]^T$ contains the radial functions.

$$\mathbf{Y} = \begin{bmatrix} Y_0^0(\Omega_1) & Y_1^{-1}(\Omega_1) & \cdots & Y_N^N(\Omega_1) \\ Y_0^0(\Omega_2) & Y_1^{-1}(\Omega_2) & \cdots & Y_N^N(\Omega_2) \\ \vdots & \vdots & \ddots & \vdots \\ Y_0^0(\Omega_Q) & Y_1^{-1}(\Omega_Q) & \cdots & Y_N^N(\Omega_Q) \end{bmatrix}. \tag{2.6}$$

The general measurement model in Eq. (2.3) may be suitable for a wider range of microphone array configurations compared to the single-sphere measurement model in Eq. (2.5). However, for the sake of clarity the formulation in the remainder of this chapter is restricted to arrays with microphones arranged on the surface of a single sphere. Nevertheless, the methods developed here can also be applied to general array configurations using Eq. (2.3).

The measurement model in Eq. (2.5) is an ideal model with a sound field that is assumed to be order limited to order N. Therefore, the number of plane-wave coefficients that are required to accurately represent the sound field in this case is $(N+1)^2$. The first step in spherical array processing, presented in this chapter, is the estimation of the $(N + 1)^2$ plane-wave coefficients from the sound pressure measured by a spherical microphone array. Assuming that the number of microphones satisfies $Q \geq (N + 1)^2$, the SH coefficients of the measured pressure can be calculated by multiplying Eq. (2.5) from the left by $\mathbf{Y}^\dagger = (\mathbf{Y}^H \mathbf{Y})^{-1} \mathbf{Y}^H$, the pseudo-inverse of $\mathbf{Y}$ (Bertet *et al.*, 2006). Then, multiplying again from the left by diag $(\mathbf{b})^{-1}$ leads to an estimation of the plane-wave coefficients, $\mathbf{a_{nm}}$, from the pressure, $\mathbf{p}$, in an operation referred to as conventional PWD:

$$\hat{\mathbf{a}}_{\mathbf{nm}} = (\mathrm{diag}\,(\mathbf{b}))^{-1} \mathbf{Y}^\dagger \mathbf{p}. \tag{2.7}$$

Substituting Eq. (2.5) into Eq. (2.7) leads to the equality

$$\hat{\mathbf{a}}_{\mathbf{nm}} = \mathbf{a}_{\mathbf{nm}}. \tag{2.8}$$

Thus, conventional PWD provides an exact estimation of the true plane-wave coefficients $\mathbf{a_{nm}}$ if the following ideal measurement model assumptions are satisfied: (i) zero measurement noise, (ii) sound pressure limited to order N.

Using the inverse discrete spherical Fourier transform (Rafaely, 2015), the order-N spatial density function, $a_N(k, \Omega)$, can be calculated at any desired direction Ω from the plane-wave coefficients $\mathbf{a_{nm}}$ as follows:

$$a_N(k, \Omega) = \mathbf{y}^\mathrm{T} \mathbf{a_{nm}}, \tag{2.9}$$

where $\mathbf{y} = [Y_0^0(\Omega), Y_1^{-1}(\Omega), \ldots, Y_N^N(\Omega)]^\mathrm{T}$. Moreover, plane-wave coefficients $\mathbf{a_{nm}}$ can be used for different applications with algorithms formulated in the SH domain, such as beamforming (Rafaely *et al.*, 2010), direction of arrival estimation (Nadiri and Rafaely, 2014), and sound reproduction (Ward and Abhayapala, 2001). For example, a general equation for axis-symmetric beamforming in the SH domain can be formulated as

$$y(k, \Omega_l) = \mathbf{y}_l^\mathrm{T}\,\mathrm{diag}\,(\mathbf{d})\mathbf{a_{nm}}, \tag{2.10}$$

where the $(N + 1)^2 \times 1$ beamformer coefficients vector, $\mathbf{d} = [d_0, d_1, d_1, \ldots, d_{NN}]^\mathrm{T}$, controls the spatial and spectral properties of the beam pattern, and $\mathbf{y}_l = [Y_0^0(\Omega_l), Y_1^{-1}(\Omega_l), \ldots, Y_N^N(\Omega_l)]^\mathrm{T}$ controls the beam pattern look direction. Among the common types of beam patterns are the maximum white noise gain (MWNG), with $\mathbf{d}^{\mathrm{MWNG}} = [|b_0(kr)|^2, |b_1(kr)|^2, |b_1(kr)|^2, \ldots, |b_N(kr)|^2]^\mathrm{T}$, and the maximum directivity (MD), with $\mathbf{d}^{\mathrm{MD}} = [1, 1, \ldots, 1]^\mathrm{T}$; the latter also provides the maximal spatial resolution (Rafaely *et al.*, 2010). Using the MD beamformer coefficients $\mathbf{d}^{\mathrm{MD}}$ leads to the equality $y(k, \Omega_l) = a_N(k, \Omega_l)$. Thus, the conventional PWD is also considered to provide the highest spatial resolution, assuming the sound field is limited to order N.

In order to illustrate PWD operation, a simulation example is presented next. The sound field is assumed to be order limited to $N = 4$ and composed of a single, unit-amplitude plane wave arriving from direction $\Omega = (90°, 50°)$. The sound field is measured using an order $N = 4$ array with $Q = 36$ microphones in a nearly uniformly spaced microphone arrangement over the surface of a rigid sphere with radius $r = 4.2$ cm.

Figure 2.1 shows a surface plot of the magnitude of the fourth-order spatial density function, $a_4(k, \Omega)$, calculated using Eq. (2.9), where the gray scale denotes the magnitude of the spatial density function in decibels. The analysis in Figure 2.1(a) is performed at $f = 4.7$ kHz or, equivalently, $kr = 3.6$, which satisfies $kr \leq N$. The peak around $\Omega = (90°, 50°)$ indicates the arrival direction of the plane wave. Figure 2.1(b) shows the

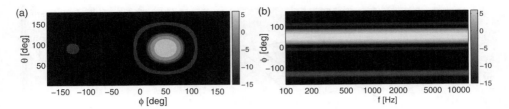

Figure 2.1 Ideal fourth-order spatial density function, $a_4(k, \Omega)$, presented for (a) different directions $\Omega = (\theta, \phi)$ at a selected frequency $f = 4.7\,\text{kHz}$, and for (b) $\Omega = (90°, \phi)$ over all azimuth angles ϕ and over different frequencies.

spatial density function at an elevation angle of $\theta = 90°$ for different azimuth angles and over a wide range of frequencies. The figure shows a single plane wave arriving from an azimuth of around $\phi = 50°$ with a constant magnitude over all frequencies. This is an accurate representation of the fourth-order spatial density function calculated under ideal conditions. However, in reality, the measured signal may be more complex than the representation provided by the ideal model in Eq. (2.5), and errors may occur when using the conventional PWD. The causes for errors in conventional PWD, as well as methods to overcome these errors, will be presented in the next sections.

2.4 Sensitivity to Noise and Standard Regularization Methods

In the previous section the ideal array measurement model was presented. In this section a more practical measurement model, which includes additive sensor noise in addition to the sound pressure measured by the microphones, is considered. The array measurement model is reformulated as follows:

$$\mathbf{p} = \mathbf{Y} \, \text{diag}\,(\mathbf{b})\, \mathbf{a}_{nm} + \mathbf{n}, \tag{2.11}$$

where $\mathbf{n} = [n_1(k), \ldots, n_Q(k)]^T$ is a $Q \times 1$ column vector that holds the additive sensor noise of each of the Q array microphones. The effect of the measurement noise on the estimation of the plane-wave coefficients, $\mathbf{a}_{nm}$, using the conventional PWD is evaluated next by applying the conventional PWD described in Eq. (2.7) to the sound pressure in Eq. (2.11), leading to

$$\hat{\mathbf{a}}_{nm} = \text{diag}\,(\mathbf{b})^{-1}\mathbf{Y}^{\dagger}\mathbf{p} = \mathbf{a}_{nm} + \text{diag}\,(\mathbf{b})^{-1}\mathbf{n}_{nm}, \tag{2.12}$$

where $\mathbf{n}_{nm} = \mathbf{Y}^{\dagger}\mathbf{n} = [n_{00}(k), \ldots, n_{NN}(k)]^T$ represents the $(N+1)^2$ spherical Fourier coefficients of the sensor noise. Equation (2.12) shows that the contribution of the noise coefficients $\mathbf{n}_{nm}$ includes a division by the radial function $b_n(kr)$. Low magnitude of the radial function will lead to a high amplification of the corresponding nth-order coefficients in $\mathbf{n}_{nm}$, which, in turn, may lead to a high error in the estimation of the nth-order plane-wave coefficients in $\hat{\mathbf{a}}_{nm}$. Conventional PWD, therefore, suffers from low robustness to noise at frequencies and SH orders at which the magnitude of the radial function is low.

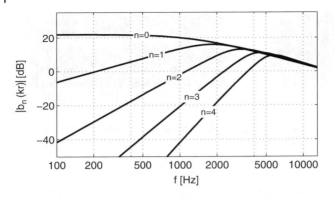

Figure 2.2 Magnitude of the radial function $b_n(k)$ with different orders, for a rigid-sphere array.

In order to illustrate the problem of low robustness to noise, a microphone array with microphones arranged around a rigid sphere is considered. In the case of a rigid-sphere array, the radial function $b_n(kr)$ is given by (Williams, 1999)

$$b_n(kr) = 4\pi i^n \left(j_n(kr) - \frac{j_n'(kr_s)}{h_n^{(2)'}(kr_s)} h_n^{(2)}(kr) \right), \tag{2.13}$$

where $j_n(kr)$ is the spherical Bessel function of the first kind, $h_n^{(2)}(kr)$ is the spherical Hankel function of the second kind, $j_n'(kr)$ and $h_n^{(2)'}(kr)$ are their derivatives, r denotes the radius at which the microphones are positioned, and r_s denotes the rigid-sphere radius. In this example the microphones are mounted on the surface of the rigid sphere and so the two radii are equal: $r = r_s$. Different radial functions $b_n(kr)$ that represent other array configurations have previously been presented (Williams, 1999; Rafaely and Balmages, 2007).

Figure 2.2 shows the magnitude of the radial function, $|b_n(kr)|$, over frequency for different orders n and for an array radius of $r_s = 4.2$ cm. The figure shows that the magnitude of the radial functions is low at low frequencies and for high orders. As explained above, low magnitude of $b_n(kr)$ may cause high amplification of the corresponding noise coefficient $n_{nm}(k)$. As a consequence, conventional PWD is expected to suffer from low robustness to noise at low frequencies and at high orders. Although other array configurations may have different radial functions, low robustness to noise at low frequencies is a common problem in all spherical array configurations, and so the rigid sphere is analyzed as a representative example throughout this chapter.

Several approaches that avoid high noise amplification, and thus achieve improved robustness in the PWD process, have previously been presented. These approaches can be represented in a unified manner using an $(N + 1)^2 \times (N + 1)^2$ transformation matrix $\mathbf{C}$, referred to as the regularization matrix. A regularized PWD method can be formulated by multiplying Eq. (2.12) from the left by the regularization matrix $\mathbf{C}$:

$$\hat{\mathbf{a}}_{nm} = \mathbf{C} \, \text{diag} \, (\mathbf{b})^{-1} \mathbf{Y}^\dagger \mathbf{p} = \mathbf{C} \mathbf{a}_{nm} + \mathbf{C} \, \text{diag} \, (\mathbf{b})^{-1} \mathbf{n}_{nm}. \tag{2.14}$$

Some of the previously presented regularization approaches (Bertet *et al.*, 2006; Moreau, 2006; Spors *et al.*, 2012; Lösler and Zotter, 2015), also assume that each element of $\hat{\mathbf{a}}_{nm}$ is estimated solely from the corresponding element of $\mathbf{a}_{nm}$. The regularization matrix in that case is a diagonal matrix with the regularization function, denoted by $c_n(k)$, on its diagonal:

$$
\mathbf{C} = \begin{bmatrix} c_0(k) & \cdots & 0 \\ \vdots & \ddots & \vdots \\ 0 & \cdots & c_N(k) \end{bmatrix}.
\tag{2.15}
$$

The explicit use of a regularization function has previously been discussed, where it was termed a regularization filter (Bertet *et al.*, 2006) or a limiting filter (Lösler and Zotter, 2015) in the context of high-order Ambisonics, which is equivalent to PWD, but using real-valued SH (Poletti, 2009). The regularized PWD formulation, as in Eq. (2.14), is used next to represent several common PWD methods as special cases by appropriately selecting the regularization functions $c_n(k)$:

1. Conventional PWD with constant regularization function:

$$
c_n^{\text{conv.}}(k) = 1,
\tag{2.16}
$$

   where in this case the regularization matrix is simply an identity matrix $\mathbf{C}^{\text{conv.}} = \mathbf{I}$. Assuming that there is no measurement noise ($\mathbf{n_{nm}} = \mathbf{0}$), substituting this regularization function, $\mathbf{C}^{\text{conv.}}$, in Eq. (2.14) leads to an accurate estimation of the plane-wave coefficients, as in Eq. (2.8). As previously shown in Section 2.3, conventional PWD is also related to the MD beamformer and is, therefore, considered to be the solution providing the highest spatial resolution. However, in the presence of measurement noise the conventional PWD suffers from high noise amplification caused by high values of $1/b_n(kr)$ at low frequencies.

2. Delay-and-sum (DaS) PWD, which uses the standard DaS beamforming coefficients in order to restrain the noise amplification in the PWD process. This regularization method was suggested by Spors *et al.* (2012), and the regularization function can be formulated as

$$
c_n^{\text{DaS}}(k) = \frac{|b_n(kr)|^2}{\beta},
\tag{2.17}
$$

   where $\beta = \frac{1}{(N+1)^2} \sum_{n=0}^{N} (2n+1)|b_n(kr)|^2$ is a normalization parameter. In contrast to conventional PWD, the DaS PWD regularization applies attenuation rather than amplification of the noise coefficients $n_{nm}(k)$ for low magnitudes of $b_n(kr)$. This leads to a robust PWD process, but also to inevitable attenuation of the high-order plane-wave coefficients, which, in turn, leads to a loss of spatial resolution. DaS PWD can be considered to be a normalized version of the MWNG beamformer presented in Section 2.3, using the beamforming coefficients $\mathbf{d}^{\text{MWNG}}$.

3. Plane-wave decomposition with Tikhonov regularization (Tikhonov and Arsenin, 1977), which uses the following regularization function:

$$
c_n^{\text{Tikhonov}}(k) = \frac{|b_n(kr)|^2}{|b_n(kr)|^2 + \lambda^2},
\tag{2.18}
$$

   where λ is a regularization parameter that controls the maximal noise amplification. This regularization function was previously presented in the context of high-order Ambisonics (Bertet *et al.*, 2006; Moreau, 2006; Jin *et al.*, 2014). It is interesting to note that using $\lambda = 0$ leads to the conventional PWD. Furthermore, with $\lambda \neq 0$, and for very low magnitudes of the radial function such that $|b_n(kr)|^2 \ll \lambda^2$, the Tikhonov regularization function becomes proportional to the DaS PWD regularization function. Different approaches for regularization parameter selection have been presented in the past. One of these approaches is the application of Wiener deconvolution filters, which provides the optimal parameter in terms of minimum mean-squared estimation error, assuming the signal to noise ratio (SNR) is known. This approach is widely used in various fields of signal processing, such as restoration of blurred images (Andrews, 1974), seismic signal processing (Berkhout, 1977), and room acoustics (Frey *et al.*, 2013). Among other approaches for parameter selection are the generalized cross validation and the L-curve methods, which are common in acoustic source reconstruction (Kim and Nelson, 2004), and do not require *a priori* information such as the SNR.

The PWD methods presented above differ in their regularization functions, leading to a different tradeoff between the two conflicting features, spatial resolution and robustness to noise. Plane-wave decomposition with Tikhonov regularization provides a compromise between conventional PWD with high spatial resolution and DaS PWD with high robustness to noise. However, it is not clear whether PWD with Tikhonov regularization is optimal, under what conditions it is preferable, and how exactly to select the regularization parameter. In the next section, an optimal PWD method is presented based on the minimization of a mean-squared PWD error measure.

2.5 Optimal Noise-Robust Design

The aim of PWD is to provide an accurate spatial representation of the sound field by estimating the plane-wave coefficients, $\mathbf{a_{nm}}$, from the sound pressure measured by the array microphones, $\mathbf{p}$. In the previous section several PWD methods that provide a compromise between loss of spatial resolution and sensitivity to measurement noise were presented. Both low spatial resolution and measurement noise produce inaccuracies that can be considered as errors in the estimated plane-wave coefficients. In order to quantify these errors, a measure that is based on the mean-squared error of the estimated plane-wave coefficients is presented in this section.

2.5.1 PWD Estimation Error Measure

The estimation error of the plane-wave coefficients, referred to as PWD error and denoted by ϵ_{PWD}, is now derived as the mean-squared error between $\hat{\mathbf{a}}_{\mathbf{nm}}$ computed using Eq. (2.14) and the true $\mathbf{a_{nm}}$. The expression for the PWD error is given by

$$\epsilon_{\mathrm{PWD}} = E[\|\hat{\mathbf{a}}_{\mathbf{nm}} - \mathbf{a_{nm}}\|^2] \tag{2.19}$$
$$= \underbrace{\|(\mathbf{C} - \mathbf{I})\mathbf{a_{nm}}\|^2}_{\text{distortion}} + \underbrace{E[\|\mathbf{C} \operatorname{diag}(\mathbf{b})^{-1}\mathbf{n_{nm}}\|^2]}_{\text{noise}},$$

where $\mathbf{I}$ is an $(N+1)^2 \times (N+1)^2$ identity matrix. It is assumed that the plane-wave spatial density function, $a(k, \Omega)$, and, as a consequence, also $\mathbf{a_{nm}}$, are deterministic. It is also assumed that the sensor noise is random and has a zero mean value, that is, $E[\mathbf{n_{nm}}] = \mathbf{0}$; therefore, the cross terms between the noise and $\mathbf{a_{nm}}$ are zero.

The PWD error in Eq. (2.19) explicitly expresses the contribution of two error components: spatial distortion of the signal and additive measurement noise. It is evident that the distortion error increases as the difference between the regularization matrix $\mathbf{C}$ and the identity matrix $\mathbf{I}$ increases, which leads to increased PWD error. On the other hand, the error also increases as the magnitude of the elements in the matrix $\mathbf{C}$ diag $(\mathbf{b})^{-1}$ increases, leading to higher amplification of the noise, and, as a consequence, to higher noise error. In order to provide a clear derivation of the PWD error measure, ϵ_{PWD}, each of these error components is reformulated separately.

First, the error due to spatial distortion of the signal is considered. In order to develop a general PWD error measure that is independent of the specific measured sound field (i.e. independent of $\mathbf{a}_{nm}$), a diffuse sound field is assumed. Thus, the measured sound field is composed of plane waves with equal magnitude σ_s across all propagation directions, and spatially random phase. The average PWD error measure for a diffuse field can be calculated by substituting the plane-wave coefficients of a single plane wave, $\mathbf{a}_{nm} = \mathbf{y}_0^* \sigma_s = [Y_0^0(\Omega_0), \ldots, Y_N^N(\Omega_0)]^H \sigma_s$, in Eq. (2.19), and then by integrating over all plane-wave arrival directions:

$$\frac{1}{4\pi} \int_{\Omega_0 \in S^2} \|(\mathbf{C} - \mathbf{I})\mathbf{y}_0^* \sigma_s\|^2 d\Omega_0 = \frac{\sigma_s^2}{4\pi} \sum_{l=1}^{(N+1)^2} (\mathbf{c}_l - \mathbf{i}_l) \int_{\Omega_0 \in S^2} \mathbf{y}_0^* \mathbf{y}_0^T d\Omega_0 (\mathbf{c}_l - \mathbf{i}_l)^H \tag{2.20}$$

$$= \frac{\sigma_s^2}{4\pi} \|\mathbf{C} - \mathbf{I}\|_F^2,$$

where $\mathbf{c}_l$ and $\mathbf{i}_l$ represent the lth row of matrix $\mathbf{C}$ and of the identity matrix $\mathbf{I}$ respectively. The SH orthogonality property (Arfken *et al.*, 2012), $\int_{\Omega_0 \in S^2} \mathbf{y}_0^* \mathbf{y}_0^T d\Omega_0 = \mathbf{I}$, was used to solve the integral, and $\| \cdot \|_F$ represents the Frobenius norm.

Second, the noise error is considered separately, by assuming a zero sound field, $\mathbf{a}_{nm} = \mathbf{0}$. It is further assumed that (i) the array measurement noise is composed of sensor noise only, (ii) the array sensors are identical, and (iii) the sensor noise elements composing vector $\mathbf{n}$ in Eq. (2.11) are zero mean, independent and identically distributed (i.i.d.), and with variance denoted by σ_n^2. Thus, the covariance matrix of the noise vector $\mathbf{n}$ is $E[\mathbf{n}\mathbf{n}^H] = \sigma_n^2 \mathbf{I}_Q$, where $\mathbf{I}_Q$ is a $Q \times Q$ identity matrix and the covariance matrix of the noise coefficients $\mathbf{n}_{nm}$ is $E[\mathbf{n}_{nm}\mathbf{n}_{nm}^H] = \mathbf{Y}^\dagger E[\mathbf{n}\mathbf{n}^H](\mathbf{Y}^\dagger)^H = \sigma_n^2(\mathbf{Y}^H\mathbf{Y})^{-1}$. For the special case of a spherical microphone array with the microphones arranged in a nearly uniform sampling scheme, the following relation is satisfied: $\mathbf{Y}^H\mathbf{Y} = \frac{Q}{4\pi}\mathbf{I}$ (Rafaely, 2005; Hardin and Sloane, 1996). This leads to the simplified expression $E[\mathbf{n}_{nm}\mathbf{n}_{nm}^H] = \sigma_n^2 \frac{4\pi}{Q}\mathbf{I}$. Using these assumptions leads to a simplified measurement noise error component expression:

$$E\left[\|\mathbf{C} \text{ diag } (\mathbf{b})^{-1}\mathbf{n}_{nm}\|^2\right] = \sum_{l=1}^{(N+1)^2} \mathbf{c}_l \text{ diag } (\mathbf{b})^{-1} E\left[\mathbf{n}_{nm}\mathbf{n}_{nm}^H\right] \text{ diag } (\mathbf{b}^*)^{-1}\mathbf{c}_l^H$$

$$= \frac{4\pi\sigma_n^2}{Q} \|\mathbf{C} \text{ diag } (\mathbf{b})^{-1}\|_F^2. \tag{2.21}$$

The final step of the PWD error development is to formulate the overall normalized PWD error by incorporating the errors developed in Eqs. (2.20) and (2.21), and normalizing their sum with the error obtained by substituting $\hat{\mathbf{a}}_{mn} = \mathbf{0}$ in Eq. (2.19), that is,

$$
\bar{e}_{\text{PWD}} = \frac{\frac{\sigma_s^2}{4\pi}\|\mathbf{C} - \mathbf{I}\|_F^2 + \frac{4\pi\sigma_n^2}{Q}\|\mathbf{C}\,\text{diag}\,(\mathbf{b})^{-1}\|_F^2}{\frac{\sigma_s^2}{4\pi}\|\mathbf{I}\|_F^2} \tag{2.22}
$$

$$
= \frac{1}{(N+1)^2}\left(\|\mathbf{C} - \mathbf{I}\|_F^2 + (\overline{SNR})^{-1}\|\mathbf{C}\,\text{diag}\,(\mathbf{b})^{-1}\|_F^2\right),
$$

where the parameter $\overline{SNR} = \frac{\sigma_s^2 Q}{\sigma_n^2 (4\pi)^2}$ is a normalized version of the SNR measured at the array input, which is denoted by $SNR = \frac{\sigma_s^2}{\sigma_n^2}$. This expression for the normalized PWD error includes the average contributions to the overall PWD error from the spatial distortion and noise errors. For example, using conventional PWD with $\mathbf{C} = \mathbf{I}$, which means no regularization, leads to zero distortion error. However, using $\mathbf{C} = \mathbf{0}$, which means too much regularisation is imposed, leads to zero measurement noise error. Hence, a regularization matrix $\mathbf{C}$ that balances between the distortion and noise errors is desired.

2.5.2 PWD Error Minimization

An optimal compromise between the spatial distortion and noise errors can be obtained by finding a regularization matrix that minimizes the overall PWD error $\bar{e}_{\text{PWD}}$. This optimal regularization matrix, referred to as the R-PWD regularization matrix, is denoted by $\mathbf{C}^{\text{R-PWD}}$, and is defined as

$$
\mathbf{C}^{\text{R-PWD}} = \arg\min_{\mathbf{C}}(\bar{e}_{\text{PWD}}). \tag{2.23}
$$

In order to solve this minimization problem, Eq. (2.22) is reformulated to include a summation over the $(N + 1)^2$ rows of matrix $\mathbf{C}$:

$$
\bar{e}_{\text{PWD}} = \frac{1}{(N+1)^2}\sum_{l=1}^{(N+1)^2}\left(\|\mathbf{c}_l - \mathbf{i}_l\|^2 + \frac{1}{SNR}\|\mathbf{c}_l\,\text{diag}\,(\mathbf{b})^{-1}\|^2\right). \tag{2.24}
$$

The overall PWD error, $\bar{e}_{\text{PWD}}$, is now represented as a sum of $(N + 1)^2$ independent error terms, and so each term can be minimized separately. The optimal solution for each row is computed by differentiating $\bar{e}_{\text{PWD}}$ with respect to $\mathbf{c}_l^H$ (with $l = 1, \ldots, (N + 1)^2$) and then setting the result to zero:

$$
\mathbf{c}_l - \mathbf{i}_l + \frac{1}{SNR}\mathbf{c}_l\,\text{diag}\,(\mathbf{b})^{-1}\,\text{diag}\,(\mathbf{b}^*)^{-1} = 0. \tag{2.25}
$$

The solutions for the $(N + 1)^2$ rows are then concatenated to form the final R-PWD regularization matrix $\mathbf{C}^{\text{R-PWD}}$,

$$
\mathbf{C}^{\text{R-PWD}} = \left(\mathbf{I} + \frac{1}{SNR}\,\text{diag}\,(\mathbf{b})^{-1}\,\text{diag}\,(\mathbf{b}^*)^{-1}\right)^{-1}, \tag{2.26}
$$

which is a diagonal matrix with the regularization functions on the main diagonal. The regularization function for order n can be reformulated as

$$c_n^{\text{R-PWD}}(k) = \left(\frac{|b_n(kr)|^2}{|b_n(kr)|^2 + \overline{(SNR)}^{-1}} \right), \tag{2.27}$$

which has the form of a Wiener deconvolution filter, and is directly related to the Tikhonov regularization function that was presented in Eq. (2.18). Thus, R-PWD regularization can be viewed as a method for parameter selection of Tikhonov regularization in a minimum mean-squared error sense, similar to the Wiener approach. However, there are several differences between the R-PWD and Tikhonov methods:

1. Unlike Tikhonov regularization, which is a general approach for solving ill-posed (or ill-conditioned) problems, R-PWD regularization was developed as the optimal solution for PWD with spherical microphone arrays and, therefore, it is clear in what sense it is optimal, and under which assumptions: (i) a diffuse sound field, (ii) additive sensor noise, (iii) a nearly uniform sampling scheme, and (iv) an *a priori* known SNR.
2. The R-PWD regularization function depends on the parameter $\overline{SNR}$, which is explicitly dependent on the SNR and on the spherical array configuration. Although the Tikhonov regularization function contains a similar regularization parameter (denoted by λ), the optimal way to set the value of this parameter for spherical microphone arrays is not clear.
3. The R-PWD regularization parameter depends on the SNR, which may vary with frequency, while Tikhonov regularization typically uses a constant regularization parameter.

2.5.3 R-PWD Simulation Study

In order to compare the performance of the R-PWD method with other PWD methods and to evaluate it, an illustrative example is presented. The regularization functions $c_n(k)$ are examined first, assuming the same microphone array configuration as that used in the simulation in Section 2.3: a rigid-sphere microphone array with radius $r = 4.2$ cm and $Q = 36$ microphones arranged in a nearly uniform scheme (Rafaely, 2005). In contrast to the assumption of a noise-free system adopted in Section 2.3, a non-zero level of sensor noise is assumed in this example. Three PWD methods are compared: the conventional PWD, the DaS PWD, and the new R-PWD methods.

Figure 2.3 shows the regularization functions $c_n(k)$ of the three compared PWD methods. Figure 2.3(a) shows the regularization function over frequency for different orders, assuming sensor noise that provides $SNR = 40$ dB. The conventional PWD regularization functions have a constant value $c_n^{\text{conv.}}(k) = 0$ dB across frequency and orders. Figure 2.3(a) illustrates that at higher frequencies the R-PWD regularization functions converge to the conventional PWD regularization functions, that is, to a constant value of $c_n(k) = 0$ dB. However, at lower frequencies the regularization functions using R-PWD are attenuated, compared to the conventional PWD regularization functions, in order to restrain the noise amplification. Additionally, as is evident from comparing Eqs. (2.17) and (2.27), at low frequencies (where the magnitude of $b_n(kr)$ becomes negligible compared with $\overline{(SNR)}^{-1}$) the R-PWD regularization functions have

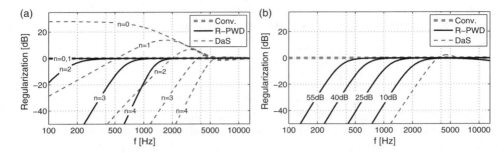

Figure 2.3 Comparison of the regularization functions $c_n(k)$ using conventional PWD, R-PWD, and DaS PWD with (a) a given SNR ($SNR = 40$ dB) and for zeroth- to fourth-order regularization functions, and (b) different SNR values for the third-order R-PWD regularization function $c_3^{\text{R-PWD}}(k)$.

the same slope as the corresponding order of the DaS PWD regularization function, that is, $c_n^{\text{R-PWD}}(k) \propto c_n^{\text{DaS}}(k)$, $kr \ll n$. Thus, for orders higher than $n = 1$, the R-PWD regularization functions have high-pass characteristics with a cut-off frequency defined by the magnitude of $b_n(kr)$ and by the value of $\overline{SNR}$, as shown in Eq. (2.27).

The functions illustrated in Figure 2.3(a) were calculated assuming $SNR = 40$ dB. Changing the SNR value will affect the cut-off frequencies of the R-PWD regularization functions. Figure 2.3(b) shows the third-order R-PWD regularization functions, $c_3^{\text{R-PWD}}(k)$, for different SNR values, compared with $c_3^{\text{conv.}}(k)$ and $c_3^{\text{DaS}}(k)$. This result illustrates that for higher SNR values, the cut-off frequency of the R-PWD regularization function decreases. Thus, as the SNR increases, the value of the R-PWD regularization function at low frequencies will increase, leading to convergence to the conventional PWD regularization function at lower frequencies, which may provide higher spatial resolution. On the other hand, for lower SNR the cut-off frequency of the R-PWD regularization function is obtained at a higher frequency leading to a magnitude that decreases with decreasing SNR, which offers greater noise suppression.

A further qualitative comparison between the performance of the R-PWD, the conventional PWD, and the DaS PWD is performed next, using the same array configuration as in the previous example. The sound field is assumed to be order limited to $N = 4$, and composed of a single unit-amplitude plane wave arriving from a given direction $\Omega = (90°, 50°)$.

Figure 2.4 shows the magnitude of the fourth-order spatial density function, $a_4(k, \Omega)$, which was calculated using the inverse discrete spherical Fourier transform from Eq. (2.9) with the three PWD methods and for $\Omega = (90°, \phi)$ (i.e. for all azimuth angles ϕ), and over a wide frequency range. The gray scale denotes the spatial density function magnitude in decibels. In Figure 2.4(a), the true spatial density function is presented, assuming no sensor noise (as in Figure 2.1(b)). In Figures 2.4(b)–2.4(d) the estimated spatial density function using the three compared PWD methods is presented, assuming sensor noise that generates an $SNR = 40$ dB, as before. It is evident from Figure 2.4(b) that, using conventional PWD, at frequencies below 900 Hz the PWD estimation is completely corrupted. This is due to the sensor noise, which is highly amplified at low frequencies. For the case of DaS PWD, presented in Figure 2.4(c), the PWD has low resolution, which can be considered as spatial distortion error, and no spatial properties

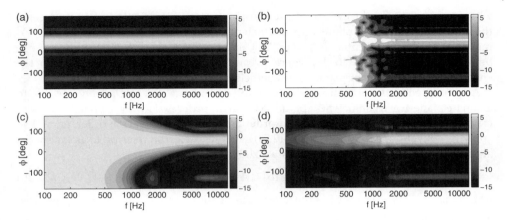

Figure 2.4 Magnitude of the fourth-order spatial density function, $|a_4(k, \Omega)|$, at $\Omega = (90°, \phi)$ for all azimuth angles ϕ over different frequencies: (a) the true spatial density function $a_4(k, \Omega)$, and the estimated spatial density function $\hat{a}_4(k, \Omega)$, using (b) conventional PWD, (c) DaS PWD, (d) R-PWD.

of the sound field can be obtained below 550 Hz. Comparing the R-PWD estimation in Figure 2.4(d) with those of the previously presented methods shows that the R-PWD obtains high spatial resolution with relatively low noise amplification. It is interesting to note that the R-PWD solution attenuates the overall magnitude of the spatial density function at low frequencies. Thus, at very low frequencies, an attenuation of the regularization function in a correct manner leads to a lower overall error. This implies that R-PWD regularization introduces a coloration effect in the form of a high-pass filter on the spatial density function.

2.6 Spatial Aliasing and High Frequency Performance Limit

In the previous section the sound field was assumed to be limited to order N, where N satisfies $Q \geq (N + 1)^2$. It has been shown that this assumption approximately holds up to a frequency that satisfies $kr \leq N$ (Ward and Abhayapala, 2001). However, at higher frequencies the sound field order increases to an order higher than N. In this section the effect of a high-order sound field on the estimated PWD function is presented. The sound field order, which might be higher than the array order N, is denoted by $\tilde{N}$, and is expected to satisfy $(\tilde{N} + 1)^2 \geq Q$. The array measurement model presented in Eq. (2.5) is reformulated as

$$\mathbf{p} = \tilde{\mathbf{Y}} \, \text{diag} \, (\tilde{\mathbf{b}}) \, \tilde{\mathbf{a}}_{\mathbf{nm}}, \tag{2.28}$$

where $\tilde{\mathbf{Y}}$, $\tilde{\mathbf{b}}$, and $\tilde{\mathbf{a}}_{\mathbf{nm}}$ are similar to $\mathbf{Y}$, $\mathbf{b}$, and $\mathbf{a}_{\mathbf{nm}}$ in Eq. (2.5), respectively, only with larger dimensions, corresponding to the sound field order $\tilde{N}$. Thus, the $(\tilde{N} + 1)^2 \times 1$ sound field plane-wave coefficients vector $\tilde{\mathbf{a}}_{\mathbf{nm}}$ can be separated into two shorter vectors:

$$\tilde{\mathbf{a}}_{\mathbf{nm}} = \begin{bmatrix} \mathbf{a}_{\mathbf{nm}} \\ \tilde{\mathbf{a}}_{\Delta} \end{bmatrix}, \tag{2.29}$$

where $\mathbf{a_{nm}}$ represents the desired Nth-order plane-wave coefficients and $\tilde{\mathbf{a}}_\Delta$ represents the undesired high-order plane-wave coefficients. In a similar manner, the $Q \times (\tilde{N} + 1)^2$ matrix $\tilde{\mathbf{Y}}$ can be represented as $\tilde{\mathbf{Y}} = [\mathbf{Y}|\tilde{\mathbf{Y}}_\Delta]$, and the $(\tilde{N} + 1)^2 \times (\tilde{N} + 1)^2$ matrix $\tilde{\mathbf{b}}$ can be represented as $\tilde{\mathbf{b}} = [\mathbf{b}^T|\tilde{\mathbf{b}}_\Delta^T]^T$, where $\tilde{\mathbf{Y}}_\Delta$ and $\tilde{\mathbf{b}}_\Delta$ are introduced to account for the contributions from higher plane-wave coefficient orders in $\tilde{\mathbf{a}}_{nm}$.

As presented in the previous sections, with conventional PWD it is only possible to estimate plane-wave coefficients $\hat{\mathbf{a}}_{nm}$ up to the array order N. However, at high frequencies, for which $kr > N$, the sound field order is higher than the array order, N, and so applying conventional PWD (as in Eq. (2.7)) generates an error, referred to as the spatial aliasing error (Rafaely *et al.*, 2007; Plessas, 2009),

$$\hat{\mathbf{a}}_{nm}^{\text{alias.}} = \text{diag}\,(\mathbf{b})^{-1}(\mathbf{Y})^\dagger\mathbf{p} = \text{diag}\,(\mathbf{b})^{-1}(\mathbf{Y})^\dagger\tilde{\mathbf{Y}}\,\text{diag}\,(\tilde{\mathbf{b}})\tilde{\mathbf{a}}_{nm} = \mathbf{D}\tilde{\mathbf{a}}_{nm}, \tag{2.30}$$

where matrix $\mathbf{D}$ is referred to as the aliasing projection matrix and can be represented as

$$\mathbf{D} = \text{diag}\,(\mathbf{b})^{-1}(\mathbf{Y})^\dagger[\mathbf{Y}|\tilde{\mathbf{Y}}_\Delta]\begin{pmatrix} \text{diag}\,(\mathbf{b}) & \mathbf{0} \\ \mathbf{0} & \text{diag}\,(\tilde{\mathbf{b}}_\Delta) \end{pmatrix}$$

$$= [\mathbf{I}|\Delta_\epsilon], \quad N < \tilde{N}. \tag{2.31}$$

Hence, the aliasing projection matrix $\mathbf{D}$ holds in its right-hand columns, Δ_ϵ, information about the spatial aliasing pattern, that is, how high-order plane-wave coefficients are aliased into the lower estimated orders (Rafaely *et al.*, 2007; Plessas, 2009). The aliasing error matrix Δ_ϵ can be represented as

$$\Delta_\epsilon = \text{diag}\,(\mathbf{b})^{-1}(\mathbf{Y})^\dagger\tilde{\mathbf{Y}}_\Delta\,\text{diag}\,(\tilde{\mathbf{b}}_\Delta). \tag{2.32}$$

The formation of the aliasing error can be explicitly formulated by substituting Eqs. (2.31) and (2.29) in Eq. (2.30); the estimated plane-wave coefficients vector with spatial aliasing is reformulated as

$$\hat{\mathbf{a}}_{nm}^{\text{alias.}} = \mathbf{D}\tilde{\mathbf{a}}_{nm} = [\mathbf{I}|\Delta_\epsilon]\begin{bmatrix} \mathbf{a}_{nm} \\ \tilde{\mathbf{a}}_\Delta \end{bmatrix} = \mathbf{a}_{nm} + \Delta_\epsilon\tilde{\mathbf{a}}_\Delta, \quad N < \tilde{N}. \tag{2.33}$$

Thus, each element in the estimated plane-wave coefficients vector $\hat{\mathbf{a}}_{nm}^{\text{alias.}}$ results from adding the desired contribution from the corresponding coefficient in $\mathbf{a}_{nm}$ to an undesired contribution from the higher-order plane-wave coefficients $\tilde{\mathbf{a}}_\Delta$. The undesired contribution from the coefficients in $\tilde{\mathbf{a}}_\Delta$ that are aliased into the lower estimated coefficients $\hat{\mathbf{a}}_{nm}^{\text{alias.}}$ is referred to as the spatial aliasing error.

It is interesting to note that the aliasing error matrix Δ_ϵ in Eq. (2.32) is composed of two components; the first is the radial functions in $\text{diag}\,(\mathbf{b})$ and $\text{diag}\,(\tilde{\mathbf{b}}_\Delta)$, which depend on frequency and array radius, while the second component is the SH basis functions in $(\mathbf{Y})^\dagger\tilde{\mathbf{Y}}_\Delta$, which depend on the array sampling scheme and are frequency independent. Changing the array sampling scheme will affect both matrices $\mathbf{Y}$ and $\tilde{\mathbf{Y}}_\Delta$ and, as a consequence, will affect the spatial aliasing pattern. As the frequency increases, the magnitude of high-order radial function elements in $\text{diag}\,(\tilde{\mathbf{b}}_\Delta)$ increases and this leads to an increase in the corresponding aliasing error elements. On the other hand, using an array with a smaller radius will reduce the magnitude of high-order radial function elements, leading to a reduction in the corresponding aliasing error elements. Therefore, an array with a smaller radius can be considered to be preferable in terms of reducing the aliasing at high frequencies.

2.7 High Frequency Bandwidth Extension by Aliasing Cancellation

In the previous section, spatial aliasing was introduced for the case where the array order N is lower than the sound field order $\tilde{N}$. In such a case, high-order plane-wave coefficients are aliased into the lower-order coefficients, causing aliasing error. A method to reduce this aliasing error is presented in this section.

2.7.1 Spatial Aliasing Error

In order to minimize the aliasing error, an $(N+1)^2 \times (N+1)^2$ transformation matrix $\mathbf{C}$ is introduced. Equation (2.33) is multiplied from the left by $\mathbf{C}$, leading to

$$\hat{\mathbf{a}}_{nm} = \mathbf{C}\hat{\mathbf{a}}_{nm}^{alias.} = \mathbf{CD}\tilde{\mathbf{a}}_{nm} = \mathbf{Ca}_{nm} + \mathbf{C\Delta}_\epsilon\tilde{\mathbf{a}}_\Delta. \tag{2.34}$$

The rightmost expression in Eq. (2.34) is composed of two components, one with the desired Nth-order plane-wave coefficients and the other with the undesired aliased high-order plane-wave coefficients. This shows the conflict in selecting matrix $\mathbf{C}$; on the one hand, using $\mathbf{C} = \mathbf{I}$ will lead to preserving the desired plane-wave coefficients $\mathbf{a}_{nm}$, but also to significant aliasing error at high frequencies, as explained in the previous section. On the other hand, using $\mathbf{C} = \mathbf{0}$ will lead to zero contribution from the aliasing error, but also to nulling the desired plane-wave coefficients $\mathbf{a}_{nm}$.

The aim of this section is to find the transformation matrix that will minimize the estimation error of the first $(N+1)^2$ plane-wave coefficients, $\mathbf{a}_{nm}$. This matrix, referred to as the *aliasing cancellation* (AC) matrix and denoted by $\mathbf{C}^{AC}$, should satisfy

$$\mathbf{C}^{AC} = \arg\min_{\mathbf{C}} \left(E[\|\hat{\mathbf{a}}_{nm} - \mathbf{a}_{nm}\|^2]\right) = \arg\min_{\mathbf{C}} \left(E[\|\mathbf{CD}\tilde{\mathbf{a}}_{nm} - \mathbf{a}_{nm}\|^2]\right). \tag{2.35}$$

It is important to keep in mind that the true plane-wave coefficients, $\tilde{\mathbf{a}}_{nm}$, are of higher order and include $(\tilde{N}+1)^2$ coefficients. In other words, the relation between the desired Nth-order plane-wave coefficients and the true higher-order plane-wave coefficients is $\mathbf{a}_{nm} = [\mathbf{I}|\mathbf{0}]\tilde{\mathbf{a}}_{nm}$.

Hence, the AC matrix, $\mathbf{C}$, should ideally satisfy

$$\mathbf{CD} = [\mathbf{C}|\mathbf{C\Delta}_\epsilon] \overset{\text{ideally}}{=} [\mathbf{I}|\mathbf{0}]. \tag{2.36}$$

Unfortunately, to calculate each row of the AC matrix in Eq. (2.36) means to exactly solve a linear system with more constraints than degrees of freedom, which might not be possible. Therefore, the objective for the AC matrix is to minimize the overall error due to spatial aliasing. The normalized overall PWD error, $\bar{\epsilon}_{PWD}$, that was presented in Eq. (2.22) is used to evaluate the aliasing error and to develop a method to minimize that aliasing error.

The PWD error measure is developed under the assumption that the spatial density function $a(k, \Omega)$ is deterministic, and so is the plane-wave coefficients vector $\tilde{\mathbf{a}}_{nm}$. In addition, the sensor noise is neglected in this section so the expectation operation in Eq. (2.19) can be removed:

$$\epsilon_{PWD} = \|\hat{\mathbf{a}}_{nm} - \mathbf{a}_{nm}\|^2 = \|(\mathbf{CD} - [\mathbf{I}|\mathbf{0}])\,\tilde{\mathbf{a}}_{nm}\|^2. \tag{2.37}$$

Similar to the steps taken in deriving Eq. (2.20), the development of the overall PWD error begins by assuming that the sound field contains only a single $\tilde{N}$th-order plane wave, arriving from direction Ω_0 with magnitude σ_s. The expression for $\tilde{\mathbf{a}}_{nm}$ in Eq. (2.37) is, therefore, replaced by $\tilde{\mathbf{a}}_{nm} = \tilde{\mathbf{y}}_0^* \sigma_s = [Y_0^0(\Omega_0), \ldots, Y_{\tilde{N}}^{\tilde{N}}(\Omega_0)]^H \sigma_s$, with further normalization with the error obtained for $\hat{\mathbf{a}}_{mn} = \mathbf{0}$. An expression for the normalized single plane-wave error is obtained:

$$\bar{e}_0 = \frac{1}{\|\mathbf{y}_0 \sigma_s\|^2} \|(\mathbf{C}\mathbf{D} - [\mathbf{I}|\mathbf{0}])\tilde{\mathbf{y}}_0^* \sigma_s\|^2 = \frac{4\pi}{(N+1)^2} \sum_{l=1}^{(N+1)^2} \mathbf{e}_l \tilde{\mathbf{y}}_0^* \tilde{\mathbf{y}}_0^T \mathbf{e}_l^H, \tag{2.38}$$

where $\mathbf{e}_l$ is the lth row vector of the error matrix $(\mathbf{C}\mathbf{D} - [\mathbf{I}|\mathbf{0}])$. Next, in order to develop a general expression for the overall PWD error that is independent of the specific measured sound field, it is assumed that the sound is diffuse. This means that plane waves with equal magnitude σ_s arrive from all directions with uniform density. The overall PWD error, $\bar{e}_{\mathrm{PWD}}$, is then computed by integrating, for all plane-wave arrival directions and using the SH orthogonality property, $\int_{\Omega_0 \in S^2} \tilde{\mathbf{y}}_0^* \tilde{\mathbf{y}}_0^T d\Omega_0 = \mathbf{I}$ (Arfken *et al.*, 2012):

$$\bar{e}_{\mathrm{PWD}} = \frac{1}{4\pi} \int_{\Omega_0 \in S^2} \bar{e}_0 d\Omega_0 = \frac{1}{(N+1)^2} \int_{\Omega_0 \in S^2} \sum_{l=1}^{(N+1)^2} \mathbf{e}_l \tilde{\mathbf{y}}_0^* \tilde{\mathbf{y}}_0^T \mathbf{e}_l^H d\Omega_0$$

$$= \frac{1}{(N+1)^2} \sum_{l=1}^{(N+1)^2} \|\mathbf{e}_l\|^2 = \frac{1}{(N+1)^2} \|\mathbf{C}\mathbf{D} - [\mathbf{I}|\mathbf{0}]\|_F^2, \tag{2.39}$$

which is similar to the error measure that has previously been presented (Plessas, 2009) for the special case of $\mathbf{C} = \mathbf{I}$.

The overall PWD error, $\bar{e}_{\mathrm{PWD}}$, is now represented as a sum of $(N+1)^2$ independent error terms, which can, therefore, be minimized separately. The optimal AC solution is solved by differentiating $\bar{e}_{\mathrm{PWD}}$ with respect to $\mathbf{c}_l^H$ (with $l = 1, \ldots, (N+1)^2$) and then setting the result to zero:

$$\mathbf{c}_l \mathbf{D}\mathbf{D}^H - \tilde{\mathbf{i}}_l \mathbf{D}^H = 0. \tag{2.40}$$

The solutions for the $(N+1)^2$ rows are then concatenated to derive the expression for the AC matrix, which is given by

$$\mathbf{C}^{AC} = [\mathbf{I}|\mathbf{0}]\mathbf{D}^H(\mathbf{D}\mathbf{D}^H)^{-1} = (\mathbf{D}\mathbf{D}^H)^{-1}, \tag{2.41}$$

where the last equality is obtained from the assumption that $\mathbf{D} = [\mathbf{I}|\Delta_\epsilon]$, as in Eq. (2.31). Multiplying Eq. (2.30) by the optimal AC matrix $\mathbf{C}^{AC}$ provides an estimation of the $(N+1)^2$ desired plane-wave coefficients $\mathbf{a}_{nm}$ with AC. This procedure, referred to as a AC-PWD, can be described mathematically as follows:

$$\hat{\mathbf{a}}_{nm}^{AC} = \mathbf{C}^{AC} \hat{\mathbf{a}}_{nm}^{\mathrm{alias.}} = (\mathbf{D}\mathbf{D}^H)^{-1} \mathbf{D}\tilde{\mathbf{a}}_{nm}. \tag{2.42}$$

2.7.2 AC-PWD Simulation Study

In order to evaluate the performance of the proposed AC-PWD method, an illustrative example is presented next. The same microphone array configuration that was used

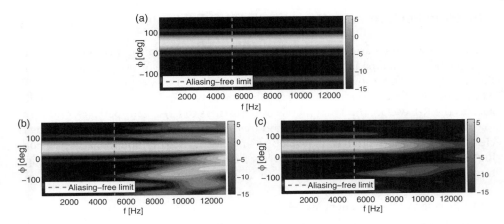

Figure 2.5 Magnitude of the fourth-order spatial density function, $|a_4(k, \Omega)|$, at $\Omega = (90°, \phi)$ for all azimuth angles ϕ over different frequencies with (a) the true spatial density function $a_4(k, \Omega)$, and the estimated spatial density function $\hat{a}_4(k, \Omega)$, with high sound field order $\tilde{N} = 10$, using (b) conventional PWD, (c) AC-PWD.

for the previous simulation examples is used in this section: the array is composed of $Q = 36$ microphones that are arranged in a nearly uniform scheme and mounted on the surface of a rigid sphere of radius $r = 4.2$ cm. At an SH order of $N = 4$, this array provides an aliasing-free bandwidth up to a frequency of $f_{af} = 5.2$ kHz. In the following simulation, we consider a unit-amplitude, order $\tilde{N} = 10$ plane wave with arrival direction $\Omega = (90°, 50°)$. The sound field order, $\tilde{N} = 10$, is chosen to satisfy $kr \leq \tilde{N}$, such that an operating bandwidth up to $f_{max} = 13$ kHz can be analyzed accurately.

Figure 2.5 shows a surface plot of the magnitude of the fourth-order spatial density function over frequency, where the gray scale denotes the magnitude of the spatial density function in decibels. This figure has a linear-scaled x-axis in order to emphasize the differences at high frequencies. The vertical dashed line indicates the aliasing-free frequency limit f_{af}. Figure 2.5(a) presents the true fourth-order spatial density function calculated using Eq. (2.9). This represents the desired aliasing-free result, similar to Figure 2.1. Figures 2.5(b)–(c) present the fourth-order estimated spatial density function, $\hat{a}_4(k, \Omega)$, calculated using the inverse discrete spherical Fourier transform from Eq. (2.9) applied to Eq. (2.34) with the different PWD methods. Figure 2.5(b) presents $\hat{a}_4(k, \Omega)$ using conventional PWD with $\mathbf{C} = \mathbf{I}$. Figure 2.5(c) presents $\hat{a}_4(k, \Omega)$ computed using AC-PWD with $\mathbf{C}^{AC}$.

Comparing Figures 2.5(a) and 2.5(b), it can be seen that as the frequency increases above f_{af}, the aliasing error increases and contaminates the resulting spatial density function. In contrast, a comparison between Figures 2.5(b) and 2.5(c) reveals that the negative effects of the aliasing error are significantly reduced by reduction of the artificial side lobes level, as well as by attenuation of the entire spatial density function level at very high frequencies. This implies that a low-pass filter coloration effect may be noticeable when using AC-PWD. A possible explanation for this may be that, in the framework of a minimum mean-squared error solution, overall attenuation is preferable to an erroneous result at very high frequencies, where the aliasing error is too high and cannot be reduced.

2.8 High Performance Broadband PWD Example

The accuracy of the basic measurement model in Eq. (2.5) was improved in the previous sections by taking into account the contribution of sensor noise, as described in Eq. (2.11), or by incorporating the high sound field orders, as expressed in Eq. (2.28). These improved measurement models enable the development of new PWD methods, such as the R-PWD method that increases robustness to noise and spatial resolution at low frequencies, or the AC-PWD method that reduces aliasing error at high frequencies. The aim of this section is to develop an improved PWD method that achieves both high spatial resolution and robustness at low frequencies, and low aliasing error at high frequencies.

2.8.1 Broadband Measurement Model

To achieve the improved broadband performance, a new generalized measurement model that incorporates the contributions of both additive sensor noise and sound fields with high orders is given by

$$\mathbf{p} = \tilde{\mathbf{Y}} \, \text{diag} \, (\tilde{\mathbf{b}}) \, \tilde{\mathbf{a}}_{\mathbf{nm}} + \mathbf{n}. \tag{2.43}$$

The plane-wave coefficients $\hat{\mathbf{a}}_{\mathbf{nm}}$, up to the array order N, can be estimated from the pressure measured by the array microphones by applying conventional PWD, as in Eq. (2.7):

$$\hat{\mathbf{a}}_{\mathbf{nm}}^{\text{conv.}} = \text{diag} \, (\mathbf{b})^{-1}(\mathbf{Y})^{\dagger}\mathbf{p} = \mathbf{D}\tilde{\mathbf{a}}_{\mathbf{nm}} + \text{diag} \, (\mathbf{b})^{-1}\mathbf{n}_{\mathbf{nm}} \tag{2.44}$$

$$= \mathbf{a}_{\mathbf{nm}} + \Delta_{\epsilon}\tilde{\mathbf{a}}_{\Delta} + \text{diag} \, (\mathbf{b})^{-1}\mathbf{n}_{\mathbf{nm}}.$$

Equation (2.44) shows that the expression for $\hat{\mathbf{a}}_{\mathbf{nm}}^{\text{conv.}}$ includes the desired Nth-order plane-wave coefficient vector $\mathbf{a}_{\mathbf{nm}}$, as well as additional contributions from the undesired aliased high-order plane-wave coefficient vector $\tilde{\mathbf{a}}_{\Delta}$ and from the undesired amplified additive noise vector $\text{diag} \, (\mathbf{b})^{-1}\mathbf{n}_{\mathbf{nm}}$.

2.8.2 Minimizing Broadband PWD Error

In order to reduce both spatial aliasing and noise amplification, a transformation matrix $\mathbf{C}$ is introduced such that

$$\hat{\mathbf{a}}_{\mathbf{nm}} = \mathbf{C}\hat{\mathbf{a}}_{\mathbf{nm}}^{\text{conv.}} = \mathbf{C}\mathbf{a}_{\mathbf{nm}} + \mathbf{C} \left(\Delta_{\epsilon}\tilde{\mathbf{a}}_{\Delta} + \text{diag} \, (\mathbf{b})^{-1}\mathbf{n}_{\mathbf{nm}} \right). \tag{2.45}$$

As shown in the previous sections, the estimated plane-wave coefficients $\hat{\mathbf{a}}_{\mathbf{nm}}$ in Eq. (2.45) express the conflict in selecting matrix $\mathbf{C}$, which is a trade-off between preserving the desired $\mathbf{a}_{\mathbf{nm}}$ while attenuating the undesired contribution of aliased elements and amplified noise. Minimizing both aliasing error at high frequencies and noise amplification error that was shown to be dominant at low frequencies, while preserving the desired true $\mathbf{a}_{\mathbf{nm}}$, will lead to a broadband PWD method, referred to as BB-PWD, with a transformation matrix denoted by $\mathbf{C}^{\text{BB}}$. The BB-PWD transformation matrix $\mathbf{C}^{\text{BB}}$ that minimizes the PWD error, ϵ_{PWD}, should satisfy

$$\mathbf{C}^{\text{BB}} = \arg \min_{\mathbf{C}} (\epsilon_{\text{PWD}}) = \arg \min_{\mathbf{C}} \left(E \left[\|\hat{\mathbf{a}}_{\mathbf{nm}} - \mathbf{a}_{\mathbf{nm}}\|^2 \right] \right). \tag{2.46}$$

The PWD error, denoted by ϵ_{PWD}, is derived for $\hat{\mathbf{a}}_{\text{nm}}$ using Eq. (2.45), in a similar way to the PWD error computation in Eq. (2.19):

$$\epsilon_{\text{PWD}} = E\big[\|\hat{\mathbf{a}}_{\text{nm}} - \mathbf{a}_{\text{nm}}\|^2\big] \tag{2.47}$$
$$= \|(\mathbf{CD} - [\mathbf{I}|\mathbf{0}])\tilde{\mathbf{a}}_{\text{nm}}\|^2 + E\big[\|\mathbf{C} \text{ diag } (\mathbf{b})^{-1}\mathbf{n}_{\text{nm}}\|^2\big].$$

As outlined in Section 2.5, the expression for the PWD error is composed of two components that are related to the error in the plane-wave coefficient estimation and to the error due to the additive sensor noise. The two error components can be reformulated assuming:

- a diffuse sound field, composed of an infinite number of $\tilde{N}$th-order plane waves with amplitude σ_s that arrive from all directions;
- the sensor noise signals are i.i.d. with variance σ_n^2;
- the microphones are arranged in a nearly uniform sampling scheme.

Following the same steps as described for Eqs. (2.19)–(2.22), these assumptions lead to

$$\epsilon_{\text{PWD}} = \sum_{l=1}^{(N+1)^2} \left(\frac{\sigma_s^2}{4\pi}\|\mathbf{c}_l\mathbf{D} - \tilde{\mathbf{i}}_l\|^2 + \frac{4\pi\sigma_n^2}{Q}\|\mathbf{c}_l \text{ diag } (\mathbf{b})^{-1}\|^2 \right). \tag{2.48}$$

The overall PWD error can be expressed in the broadband case as

$$\bar{\epsilon}_{\text{PWD}} = \frac{E[\|\hat{\mathbf{a}}_{\text{nm}} - \mathbf{a}_{\text{nm}}\|^2]}{E[\|\mathbf{a}_{\text{nm}}\|^2]} \tag{2.49}$$
$$= \frac{1}{(N+1)^2} \left(\|\mathbf{CD} - [\mathbf{I}|\mathbf{0}]\|_{\text{F}}^2 + \frac{1}{SNR}\|\mathbf{C} \text{ diag } (\mathbf{b})^{-1}\|_{\text{F}}^2 \right).$$

Minimization of $\bar{\epsilon}_{\text{PWD}}$ is performed separately for each of the $(N+1)^2$ summation elements by differentiating $\bar{\epsilon}_{\text{PWD}}$ with respect to $\mathbf{c}_l^{\text{H}}$ and setting the result to zero:

$$\mathbf{c}_l\mathbf{DD}^{\text{H}} - \tilde{\mathbf{i}}_l\mathbf{D}^{\text{H}} + \frac{1}{SNR}\mathbf{c}_l \text{ diag } (\mathbf{b})^{-1} \text{ diag } (\mathbf{b}^*)^{-1} = 0. \tag{2.50}$$

Solving for each row, $\mathbf{c}_l$, of matrix $\mathbf{C}$ and concatenating the solutions leads to the optimal BB-PWD matrix $\mathbf{C}^{\text{BB}}$:

$$\mathbf{C}^{\text{BB}} = \left(\mathbf{DD}^{\text{H}} + \frac{1}{SNR} \text{ diag } (\mathbf{b})^{-1} \text{ diag } (\mathbf{b}^*)^{-1} \right)^{-1}, \tag{2.51}$$

which can be used to estimate the plane-wave coefficients:

$$\hat{\mathbf{a}}_{\text{nm}}^{\text{BB}} = \mathbf{C}^{\text{BB}}\hat{\mathbf{a}}_{\text{nm}}^{\text{conv.}} = \mathbf{C}^{\text{BB}} \left(\mathbf{D}\tilde{\mathbf{a}}_{\text{nm}} + \text{ diag } (\mathbf{b})^{-1}\mathbf{n}_{\text{nm}} \right). \tag{2.52}$$

It is interesting to note that at low frequencies, where plane-wave coefficients of order higher than N can be neglected, the aliasing projection matrix is reduced to $\mathbf{D} = [\mathbf{I}|\mathbf{0}]$, and the BB-PWD transformation matrix is equal to the R-PWD regularization matrix, $\mathbf{C}^{\text{BB}} = \mathbf{C}^{\text{R-PWD}}$. Furthermore, at high SNR values or at high magnitudes of the radial functions $b_n(kr)$, the BB-PWD transformation matrix converges to the AC-PWD, $\mathbf{C}^{\text{BB}} \approx \mathbf{C}^{\text{AC}}$.

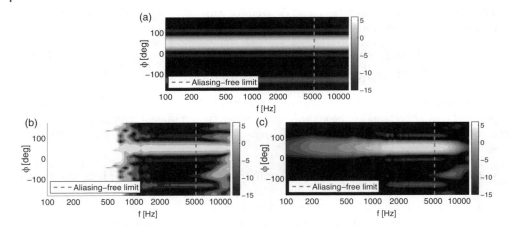

Figure 2.6 Magnitude of the fourth-order spatial density function, $|a_4(k,\Omega)|$, at $\Omega = (90°, \phi)$ for all azimuth angles ϕ over different frequencies with (a) the true spatial density function $a_4(k,\Omega)$, and the estimated spatial density function $\hat{a}_4(k,\Omega)$, using (b) conventional PWD, (c) optimal BB-PWD.

2.8.3 BB-PWD Simulation Study

The performance of the BB-PWD method is analyzed next through a simulation study. The considered array configuration is the same as in Sections 2.5.3 and 2.7.2. The array order $N = 4$ is lower than the sound field order, which is assumed to be composed of a single unit-amplitude plane wave of order $\tilde{N} = 10$, arriving from $\Omega = (90°, 50°)$. The microphone measurements include sensor noise that leads to $SNR = 40\,\text{dB}$.

Figure 2.6 shows the fourth-order spatial density function calculated at different azimuth angles $\Omega = (90°, \phi)$ over a broad frequency range, where the gray scale denotes the magnitude of the spatial density function in decibels. Figure 2.6(a) presents the desired result, the true fourth-order spatial density function, $a_4(k,\Omega)$. Figure 2.6(b) presents the fourth-order estimated spatial density function $\hat{a}_4^{\text{conv.}}(k,\Omega) = \mathbf{y}^T\hat{\mathbf{a}}_{nm}^{\text{conv.}}$ using conventional PWD, as in Eq. (2.44). Figure 2.6(c) presents the fourth-order estimated spatial density function $\hat{a}_4^{\text{BB}}(k,\Omega) = \mathbf{y}^T\hat{\mathbf{a}}_{nm}^{\text{BB}}$ using BB-PWD with matrix $\mathbf{C}^{\text{BB}}$, as described in Eq. (2.52).

In contrast to the desired result in Figure 2.6(a) that shows a constant magnitude over frequency along the plane wave arrival direction $\Omega = (90°, 50°)$, Figure 2.6(b) shows large errors at high frequencies (above $f_{\text{af}} = 5.2\,\text{kHz}$), similar to the aliasing error presented in Figure 2.5(b). In addition, Figure 2.6(b) shows large errors at frequencies lower than 900 Hz, as was previously shown in Figure 2.4(b). Compared to the large errors displayed in Figure 2.6(b), Figure 2.6(c) presents lower errors, obtained by applying BB-PWD, with the result that it is more similar to Figure 2.6(a). The reduced BB-PWD error illustrated in Figure 2.6(c) is similar to the results obtained by R-PWD at low frequencies, as presented in Figure 2.4(d), and to the results obtained at high frequencies by AC-PWD, as presented in Figure 2.5(c). Thus, the new BB-PWD method incorporates the advantages of both the AC-PWD and R-PWD methods and, therefore, offers overall improved broadband performance.

Figure 2.6 gives a qualitative comparison between the conventional PWD and BB-PWD methods by showing the spatial density function in the case of a single plane wave.

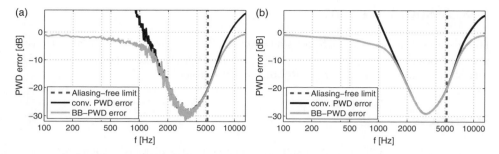

Figure 2.7 The overall PWD error using the conventional PWD and R-PWD methods: (a) the error for a single plane wave, (b) the mean overall PWD error, $\bar{e}_{PWD}$, for a diffuse sound field.

Figure 2.7 shows a more quantitative comparison between these PWD methods using the overall PWD error, $\bar{e}_{PWD}$. Figure 2.7(a) shows the normalized error, $\frac{\|\hat{a}_{nm} - a_{nm}\|^2}{\|a_{nm}^2\|}$, that was obtained for the example of the single plane wave measured with sensor noise that provides $SNR = 40$ dB. It quantifies the improvement achieved by BB-PWD in comparison with the conventional PWD method at both high and low frequencies, as was qualitatively described in the previous example. Figure 2.7(b) shows the overall PWD error, $\bar{e}_{PWD}$, calculated using Eq. (2.49) for the case of a diffuse sound field with averaged noise. Comparing the error obtained by BB-PWD and conventional PWD reveals that below $f = 1.14$ kHz, where the conventional PWD error rises above 0 dB, the BB-PWD yields an error of -5.5 dB. Moreover, at frequencies above 8.86 kHz, where the conventional PWD error rises above 0 dB, BB-PWD achieves an error of -3.5 dB. This reduced error achieved by the BB-PWD method can be considered to be a significant improvement in PWD performance. As previously shown (Alon *et al.*, 2015), such improved PWD performance may lead to a significant improvement in the perceptual evaluation of binaural reproduction signals based on the estimated plane-wave coefficients. It is interesting to note that although the optimal BB-PWD was developed for a diffuse sound field, good performance is also achieved for the case of a single plane wave in Figure 2.7(a). In addition, similar error values are obtained for both the diffuse sound field case and the single plane wave case, Figures 2.7(b) and (a), respectively. Hence, the single plane wave example, as in Figure 2.6, can be considered to be a typical result.

2.9 Summary

This chapter presents a general formulation for the spatial decomposition of a sound field that is measured by a spherical microphone array. This general formulation utilizes a transformation matrix **C** that makes it possible to describe the conventional PWD method and to develop new optimal PWD methods by finding matrices **C** that minimize the PWD error. These include the R-PWD method that provides high robustness to noise and high spatial resolution at low frequencies; the AC-PWD method that provides minimization of the spatial aliasing error at high frequencies; and the BB-PWD method that incorporates both improved features of R-PWD and AC-PWD. Thus, the BB-PWD method yields a wider operating bandwidth with lower error for spherical microphone arrays.

2.10 Acknowledgment

The research leading to these results has received funding from the European Union's Seventh Framework Programme (FP7/2007-2013) under grant agreement no. 609465 as part of the Embodied Audition for RobotS (EARS) project.

References

Abhayapala, T.D. and Ward, D.B. (2002) Theory and design of high order sound field microphones using spherical microphone array. *Proceedings ICASSP*, vol. II, 1949–1952.

Alon, D.L. and Rafaely, B. (2012) Spherical microphone array with optimal aliasing cancellation. *IEEE 27th Convention of Electrical Electronics Engineers in Israel (IEEEI)*, pp. 1–5, Eilat, Israel.

Alon, D.L., Sheaffer, J., and Rafaely, B. (2015) Plane-wave decomposition with aliasing cancellation for binaural sound reproduction. *Audio Engineering Society Convention 139*. Audio Engineering Society.

Andrews, H.C. (1974) Digital image restoration: A survey. *Computer*, 7(5), 36–45.

Arfken, G., Weber, H., and Harris, F. (2012) *Mathematical Methods for Physicists: A Comprehensive Guide*. Elsevier, Amsterdam.

Berkhout, A.J. (1977) Least-squares inverse filtering and wavelet deconvolution. *Geophysics*, 42(7), 1369–1383.

Bertet, S., Daniel, J., and Moreau, S. (2006) 3D sound field recording with higher order Ambisonics: Objective measurements and validation of spherical microphone. *Audio Engineering Society Convention 120*. Audio Engineering Society.

Driscoll, J. and Healy, D. (1994) Computing Fourier transforms and convolutions on the 2-sphere. *Advances in Applied Mathematics*, 15(2), 202–250.

Frey, D., Coelho, V., and Rangayyan, R.M. (2013) Acoustical impulse response functions of music performance halls. *Synthesis Lectures on Speech and Audio Processing*, 9(2), 1–110.

Gerzon, M.A. (1973) Periphony: With-height sound reproduction. *Journal of the Audio Engineering Society*, 21(1), 2–10.

Gover, B., Ryan, J., and Stinson, M. (2002) Microphone array measurement system for analysis of directional and spatial variations of sound fields. *Journal of the Acoustical Society of America*, 112(5), 1980–1991.

Hardin, R.H. and Sloane, N.J.A. (1996) Mclaren's improved snub cube and other new spherical designs in three dimensions. *Discrete Computational Geometry*, 15(4), 429–441.

Jin, C., Epain, N., and Parthy, A. (2014) Design, optimization and evaluation of a dual-radius spherical microphone array. *IEEE/ACM Transactions on Audio, Speech, and Language Processing*, 22(1), 193–204.

Jin, C., Parthy, A., and van Schaik, A. (2006) Optimisation of co-centred rigid and open spherical microphone arrays *Audio Engineering Society Convention 120*. Audio Engineering Society.

Kim, Y. and Nelson, P. (2004) Optimal regularisation for acoustic source reconstruction by inverse methods. *Journal of Sound and Vibration*, 275(3), 463–487.

Lösler, S. and Zotter, F. (2015) Comprehensive radial filter design for practical higher-order ambisonic recording. *Proceedings of the 41. Deutsche Jahrestagung fur Akustik (DAGA 2015)*.

Meyer, J. and Elko, G.W. (2002) A highly scalable spherical microphone array based on an orthonormal decomposition of the sound field. *Proceedings ICASSP*, vol. II, 1781–1784.

Moreau, S. (2006) Étude et réalisation d'outils avancés d'encodage spatial pour la technique de spatialisation sonore Higher Order Ambisonics: microphone 3D et controle de distance. PhD thesis. Université du Maine.

Nadiri, O. and Rafaely, B. (2014) Localization of multiple speakers under high reverberation using a spherical microphone array and the direct-path dominance test. *IEEE/ACM Transactions on Audio, Speech, and Language Processing*, **22**(10), 1494–1505.

Park, M. and Rafaely, B. (2005) Sound-field analysis by plane-wave decomposition using spherical microphone array. *The Journal of the Acoustical Society of America*, **118**(5), 3094–3103.

Plessas, P. (2009) Rigid sphere microphone arrays for spatial recording and holography. Dip. Eng. thesis. Graz University of Technology.

Poletti, M. (2009) Unified description of Ambisonics using real and complex spherical harmonics. *Ambisonics Symposium*, Graz, Austria.

Rafaely, B. (2004) Plane-wave decomposition of the pressure on a sphere by spherical convolution. *Journal of the Acoustical Society of America*, **116**(4), 2149–2157.

Rafaely, B. (2005) Analysis and design of spherical microphone arrays. *IEEE Transactions on Speech and Audio Processing*, **13**(1), 135–143.

Rafaely, B. (2008) The spherical-shell microphone array. *IEEE Transactions on Audio, Speech, and Language Processing*, **16**(4), 740–747.

Rafaely, B. (2015) *Fundamentals of Spherical Array Processing*, Springer Topics in Signal Processing, vol. 8, Springer, New York.

Rafaely, B. and Balmages, I. (2007) Open-sphere designs for spherical microphone arrays. *IEEE Transactions on Audio, Speech and Language Processing*, **15**, 727–732.

Rafaely, B., Peled, Y., Agmon, M., Khaykin, D., and Fisher, E. (2010) Spherical microphone array beamforming, in *Speech Processing in Modern Communications: Challenges and Perspectives*, eds. I. Cohen, J. Benesty, and S. Gannot, pp. 281–305. Springer-Verlag, Berlin.

Rafaely, B., Weiss, B., and Bachmat, E. (2007) Spatial aliasing in spherical microphone arrays. *IEEE Transactions on Signal Processing*, **55**(3), 1003–1010.

Spors, S., Wierstorf, H., and Geier, M. (2012) Comparison of modal versus delay-and-sum beamforming in the context of data-based binaural synthesis. *Audio Engineering Society Convention 132*. Audio Engineering Society.

Tikhonov, A.N. and Arsenin, V.I. (1977) *Solutions of Ill-Posed Problems*. V.H. Winston, Washington, DC.

Ward, D.B. and Abhayapala, T.D. (2001) Reproduction of a plane-wave sound field using an array of loudspeakers. *IEEE Transactions on Speech and Audio Processing*, **9**(6), 697–707.

Williams, E.G. (1999) *Fourier Acoustics: Sound Radiation and Nearfield Acoustic Holography*. Academic Press, London, UK.

3

Sound Field Analysis Using Sparse Recovery

Craig T. Jin, Nicolas Epain, and Tahereh Noohi

CARLab, School of Electrical and Information Engineering, University of Sydney, Australia

3.1 Introduction

The analysis or decomposition of a sound field into plane-wave source signals and their directions has many uses, such as in sound source tracking, direction of arrival estimation, source separation, sound field reproduction, and architectural acoustics. The history of microphone array processing is replete with numerous methods for sound field analysis. For example, there are simple beamforming methods that can provide an energy map of the sound field (Gover *et al.*, 2002; Park and Rafaely, 2005). There are numerous time difference of arrival (TDOA) methods based on the cross-correlation of microphone signals (for a detailed review, refer to Benesty *et al.*, 2008). There are also more elaborate techniques such as multiple signal classification (MUSIC; Schmidt, 1986) and the estimating signal parameters via rotational invariance technique (ESPRIT; Roy and Kailath, 1989), which have recently been applied to spherical microphone arrays (see EB-MUSIC: Rafaely *et al.*, 2010; EB-ESPRIT: Sun *et al.*, 2011). There are also direction-of-arrival techniques based on multi-dimensional maximum-likelihood estimation which have recently been proposed for spherical microphone arrays (Tervo and Politis, 2015). The previous chapter of this book presents a PWD operation based on robust signal-independent beamforming principles, where the resolution capability relies only on the properties of the array. In this chapter, we describe a sparse recovery method for sound field decomposition that provides increased spatial resolution compared to other methods.

The sparse recovery methods described here were initially developed using spherical microphone arrays (SMAs), but have general applicability to any microphone array. The use of SMAs relates to our interest in sound field reproduction and the fact that SMAs provide a uniform panoramic view of the sound field. With respect to sound field reproduction, the sparse recovery methods described here provide a means to identify audio objects (i.e., sound sources) and to *upscale* a sound scene to higher spherical harmonic order (Wabnitz *et al.*, 2011, 2012; Wu *et al.*, 2012). In this chapter, we will take a fairly descriptive approach to the use of the sparse recovery method for sound field decomposition and focus on the technique and its application.

Parametric Time–Frequency Domain Spatial Audio, First Edition. Edited by Ville Pulkki, Symeon Delikaris-Manias, and Archontis Politis.
© 2018 John Wiley & Sons Ltd. Published 2018 by John Wiley & Sons Ltd.
Companion Website: www.wiley.com/go/pulkki/parametrictime-frequency

3.2 The Plane-Wave Decomposition Problem

We begin with the classic plane-wave decomposition problem. Consider a microphone array for which all sound sources are sufficiently far from the microphone array that we can model the sound field as a sum of N plane waves incoming from many directions in space. Assume that the sound field is observed via a set of K ($K \ll N$) signals, which might, for instance, consist of K microphone signals or K spherical Fourier transform (SFT) signals. Further, assume that a short-time Fourier transform is applied, so that signals are expressed in the time–frequency domain.

We represent the plane-wave signals using the notation of a three-dimensional array: $\mathcal{X}[n,t,f]$, with different plane waves, indexed by n, running along the first dimension; time, t, running along the second dimension; and frequency, f, running along the third dimension. Similarly, we represent the *observation* signals using the notation of a three-dimensional array: $\mathcal{B}[k,t,f]$, with different observations, indexed by k, running along the first dimension; time, t, running along the second dimension; and frequency, f, running along the third dimension. Because analysis is often performed for each frequency separately, we consider a slice of the plane-wave array for a fixed, given frequency f. The slice is represented by the plane-wave matrix $\mathbf{X}(f)$:

$$\mathbf{X}(f) = \mathcal{X}[\cdot,\cdot,f], \tag{3.1}$$

where the matrix element $\mathbf{X}_{n,t}(f)$ is given by $\mathcal{X}[n,t,f]$. Similarly, we consider the observation matrix, $\mathbf{B}(f)$, corresponding to a slice of the observation array for a fixed, given frequency f:

$$\mathbf{B}(f) = \mathcal{B}[\cdot,\cdot,f], \tag{3.2}$$

where the matrix element $\mathbf{B}_{k,t}(f)$ is given by $\mathcal{B}[k,t,f]$.

We assume that the relationship between the observation matrix, $\mathbf{B}(f)$, and the plane-wave matrix, $\mathbf{X}(f)$, is known and specified by the transformation matrix, $\mathbf{A}(f)$:

$$\mathbf{A}(f)\mathbf{X}(f) = \mathbf{B}(f). \tag{3.3}$$

The transformation matrix, $\mathbf{A}(f)$, is the matrix expressing the contribution of the different plane waves to the observation signals:

$$\mathbf{A}(f) = [\mathbf{a}(\Omega_1,f), \mathbf{a}(\Omega_2,f), \ldots, \mathbf{a}(\Omega_N,f)],$$
$$\mathbf{a}(\Omega_n,f) = [a_1(\Omega_n,f), a_2(\Omega_n,f), \ldots, a_K(\Omega_n,f)]^{\mathrm{T}}, \tag{3.4}$$

where $a_k(\Omega_n,f)$ represents the complex-valued gain between the plane wave, with frequency f, incoming from direction Ω_n and the kth observation signal. The vector $\mathbf{a}(\Omega_n,f)$ is commonly referred to as an *array manifold* vector.

Using the above formalism, we describe the plane-wave decomposition problem as solving Equation (3.3) for the unknown plane-wave matrix, $\mathbf{X}(f)$, given both the observation matrix, $\mathbf{B}(f)$, and the transformation matrix, $\mathbf{A}(f)$. In other words, we want to determine the plane-wave signals given the observation signals and knowledge of the array manifold vectors. One difficulty is that the set of *possible* plane-wave signal directions is infinite. In order to achieve high-resolution localization, one includes a large number of array manifold vectors into the matrix, $\mathbf{A}(f)$, one for each direction considered. Each possible direction considered corresponds to a column of matrix $\mathbf{A}(f)$ with the implication that $\mathbf{A}(f)$ becomes a very wide matrix leading to an under-determined problem. In the ensuing discussion, we refer to $\mathbf{A}(f)$ as a dictionary – a plane-wave dictionary – because we are expressing the observation signals as a sum of plane-wave contributions that are *defined* by $\mathbf{A}(f)$.

3.2.1 Sparse Plane-Wave Decomposition

The trade-off between increasing the resolution of the plane-wave dictionary and having an increasingly under-determined problem leads to consideration of sparse recovery methods. To begin, assume that the rank of $\mathbf{A}$ is K, which is generally true if the directions are distributed sufficiently evenly over the sphere. The plane-wave decomposition problem is then under-determined because there is an infinite number of valid and possible solutions to Equation (3.3). A classic way to circumvent this problem is to choose the solution, $\mathbf{X}(f)$, with the least energy, that is, the smallest Frobenius norm (or ℓ_2 norm when there is only one time sample). This solution is known as the least-norm solution and is given by:

$$\bar{\mathbf{X}}(f) = \mathbf{A}^H(f)(\mathbf{A}(f)\mathbf{A}^H(f))^{-1}\mathbf{B}(f). \tag{3.5}$$

The matrix $\mathbf{A}^H(\mathbf{A}\mathbf{A}^H)^{-1}$ is often referred to as the Moore–Penrose pseudo-inverse of $\mathbf{A}$. The trouble with the least-norm solution is that it tends to distribute the energy evenly across plane-wave directions. This physical assumption is generally wrong and leads to an undesirable spatial blurring of the acoustic image.

A better alternative to the least-norm solution is to consider the *sparsest* solution, that is, the solution with the fewest plane waves that still explains the observations. Mathematically, this solution can be defined as:

$$\hat{\mathbf{X}}(f) = \underset{\mathbf{X}(f)}{\arg\min}\{\|\mathbf{X}(f)\|_{0,2} : \mathbf{A}(f)\mathbf{X}(f) = \mathbf{B}(f)\}, \tag{3.6}$$

where $\|\cdot\|_{0,2}$ denotes the $\ell_{0,2}$ norm. The $\ell_{0,2}$ norm of matrix $\mathbf{X}$ is defined as the total number of plane-wave signals for which the corresponding signal energy is non-zero, that is, the total number of indices n for which:

$$\sqrt{\sum_{t=1}^{T} |\mathbf{X}_{n,t}(f)|^2} \neq 0. \tag{3.7}$$

This solution is advantageous for a few reasons. First, it provides a sharper acoustic image than the least-norm solution because it finds the smallest number of plane waves that explain the observations, that is, the solution is spatially sparse. Also, it is reasonable to apply the principle of Occam's razor and explain the observations in terms of the smallest number of simultaneous sources.

3.2.2 The Iteratively Reweighted Least-Squares Algorithm

In practice it is extremely difficult to solve Equation (3.6) for the least $\ell_{0,2}$ norm solution. Instead, one solves the problem for the least $\ell_{p,2}$ norm solution, where $0 < p \leq 1$. Mathematically, this also promotes sparsity. The least $\ell_{p,2}$ norm solution can be found using the iteratively reweighted least-squares algorithm (IRLS; Daubechies *et al.*, 2010), also referred to as the focal under-determined system solver (FOCUSS; Cotter *et al.*, 2005).

The idea behind the IRLS algorithm is that the $\ell_{p,2}$ norm of the solution can be expressed as a weighted $\ell_{2,2}$ norm (the Frobenius norm). We have:

$$\|\mathbf{X}(f)\|_{p,2}^{p} = \sum_{n=1}^{N} \left(\sum_{t=1}^{T} |\mathbf{X}_{n,t}(f)|^2 \right)^{\frac{p}{2}}$$

$$= \sum_{n=1}^{N} \left(\sum_{t=1}^{T} |\mathbf{X}_{n,t}(f)|^2 \right)^{\frac{p-2}{2}} \left(\sum_{t=1}^{T} |\mathbf{X}_{n,t}(f)|^2 \right)$$

$$= \left\| \mathbf{W}^{-\frac{1}{2}}(f)\mathbf{X}(f) \right\|_2^2, \tag{3.8}$$

where $\mathbf{W}(f)$ is the diagonal matrix given by:

$$\mathbf{W}(f) = \text{diag}(w_1(f), w_2(f), \ldots, w_N(f)),$$

$$w_n(f) = \left(\sum_{t=1}^{T} |\mathbf{X}_{n,t}(f)|^2 \right)^{\frac{2-p}{2}}. \tag{3.9}$$

Therefore, finding the solution with the least $\ell_{p,2}$ norm is equivalent to solving a weighted least-norm problem in which the weights depend on the solution. So long as there is little danger of confusion, in the ensuing discussion we omit the frequency dependence of the variables for the purpose of notational simplicity.

The IRLS algorithm solves the following weighted least-norm problem:

$$\text{minimize } \left\| \mathbf{W}^{-\frac{1}{2}}\mathbf{X} \right\|_2^2 \text{ subject to } \mathbf{AX} = \mathbf{B}. \tag{3.10}$$

For fixed $\mathbf{W}$, this problem has a closed-form solution, $\mathbf{X}_{\mathbf{W}}$, given by:

$$\mathbf{X}_{\mathbf{W}} = \mathbf{WA}^{\mathrm{H}}(\mathbf{AWA}^{\mathrm{H}})^{-1}\mathbf{B}. \tag{3.11}$$

Note that this result can be easily demonstrated using the method of Lagrange multipliers. Because the weights in Equation (3.9) depend on matrix $\mathbf{X}$, the IRLS algorithm calculates the solution iteratively, as summarized in Algorithm 3.1.

The IRLS algorithm is presented here for completeness. Please note that details that remain unexplained here will be described later in the chapter. At this stage, the important points are as follows. The weights are first all initialized to one. The algorithm then repeats the following two steps in alternation until convergence: (i) update the solution $\mathbf{X}$ assuming a fixed weight matrix $\mathbf{W}$ (line 7 of the algorithm); and (ii) update the weight matrix given the solution $\mathbf{X}$ (line 12 of the algorithm). Note that the term ϵ in the algorithm is a regularization term that prevents the weights from being equal to 0 (Daubechies *et al.*, 2010). Also, in order to improve the convergence of the algorithm, we run the first ten iterations with $p = 1$ and then switch to $p = 0.8$, as suggested by Daubechies *et al.* (2010).

In practice, it is also important to note that when the number of time samples T is larger than the number of observation signals, K, principal component analysis (PCA) may be used to accelerate the computations, as proposed by Malioutov *et al.* (2005). Indeed, if $T > K$, the vectors of observation signals corresponding to the different time instants are not linearly independent and the observation signals can be replaced by their PCA representation, which is a $K \times K$ matrix. Mathematically, the singular value decomposition (SVD) of the observation signal matrix $\mathbf{B}$ is given by:

$$\mathbf{B} = \mathbf{U}\boldsymbol{\Sigma}\mathbf{V}^{\mathrm{H}}, \tag{3.12}$$

where $\mathbf{U}$ is the $K \times K$ matrix of singular vectors, $\boldsymbol{\Sigma}$ is the $K \times K$ diagonal matrix of the singular values, and $\mathbf{V}$ is the $T \times K$ matrix of the K principal components. To accelerate the computations, we replace matrix $\mathbf{B}$ by its PCA representation $\tilde{\mathbf{B}}$, given by:

$$\tilde{\mathbf{B}} = \mathbf{U}\boldsymbol{\Sigma}. \tag{3.13}$$

We then solve the sparse plane-wave decomposition problem to obtain $\tilde{\mathbf{X}}$, from which the source signals, $\mathbf{X}$, are obtained as:

$$\mathbf{X} = \tilde{\mathbf{X}}\mathbf{V}^{\mathrm{H}}. \tag{3.14}$$

Algorithm 3.1 The regularized IRLS algorithm for sparse plane-wave decomposition.

1: **Inputs:** plane-wave dictionary $\mathbf{A}$, observation signals $\mathbf{B}$, relative energy of noise β

2: **Outputs:** plane-wave signals $\mathbf{X}$

3: **Initialization:**

4: $\qquad \mathbf{W} = \mathbf{I}_N$

5: **Until convergence, do:**

6: $\qquad \gamma \leftarrow \frac{1}{N}\frac{\beta}{1-\beta}\mathrm{tr}(\mathbf{AWA}^{\mathrm{H}})$

7: $\qquad \mathbf{X} \leftarrow \mathbf{WA}^{\mathrm{H}}(\mathbf{AWA}^{\mathrm{H}} + \gamma\mathbf{I})^{-1}\mathbf{B}$

8: $\qquad \text{for } n = \{1, 2, \dots, N\}, e_i \leftarrow \sum_{t=1}^{T} |\mathbf{X}_{n,t}|^2$

9: $\qquad e_{\max} \leftarrow \max\{e_i, i = 1, \dots, N\}$

10: $\qquad \epsilon \leftarrow \min\left(\epsilon, \frac{e_{\max}}{N^2}\right)$

11: $\qquad \text{for } i = \{1, 2, \dots, N\}, w_i \leftarrow (e_i + \epsilon)^{\frac{2-p}{2}}$

12: $\qquad \mathbf{W} \leftarrow \mathrm{diag}(w_1, w_2, \dots, w_N)$

3.3 Bayesian Approach to Plane-Wave Decomposition

In non-sparse sound conditions, it is critical to appropriately regularize the matrix inversion in Equation (3.11). In this regard, we now turn our attention to the Bayesian approach to the plane-wave decomposition problem to gain more insight. From a Bayesian viewpoint, we modify Equation (3.3) to explicitly consider the noise. In other words, we model the observation signals, $\mathbf{B}(f)$, as a linear mixture of plane-wave signals, $\mathbf{X}(f)$, multiplied by the plane-wave dictionary matrix, $\mathbf{A}(f)$, but now explicitly polluted by some noise, modeled as $\mathbf{N}(f)$:

$$\mathbf{B}(f) = \mathbf{A}(f)\mathbf{X}(f) + \mathbf{N}(f). \tag{3.15}$$

The noise matrix $\mathbf{N}(f)$ is defined similarly to $\mathbf{B}(f)$ so that the matrix element $\mathbf{N}_{k,t}(f)$ is defined by the three-dimensional array $\mathcal{N}[k, t, f]$, with observations, indexed by k, running along the first dimension; time, t, running along the second dimension; and frequency, f, running along the third dimension. As previously, we omit the frequency dependence of the variables in the following discussion for the purpose of notational simplicity.

In the Bayesian framework the probabilities of signals are considered, and so we assume that the plane-wave signals constitute a zero-mean multivariate Gaussian distribution. The probability, $P(\mathbf{X})$, of $\mathbf{X}$ taking a particular value can be expressed as (Hyvärinen *et al.*, 2001, p. 32):

$$P(\mathbf{X}) \propto e^{-\frac{1}{2}\text{tr}[\mathbf{X}^H \mathbf{C}_{xx}^{-1} \mathbf{X}]}, \tag{3.16}$$

where $\mathbf{C}_{xx} = \text{E}\{\mathbf{X}^H \mathbf{X}\}$ denotes the covariance matrix of the plane-wave signals, $\text{E}\{\cdot\}$ is the statistical expectation operator, $\mathbf{X}^H$ is the Hermitian transpose of $\mathbf{X}$, and $\text{tr}[\cdot]$ is the matrix trace operator. Assuming that the noise signals also constitute a zero-mean multivariate Gaussian distribution, the probability, $P(\mathbf{B}|\mathbf{X})$, of the observation signals, $\mathbf{B}$, *given* a particular set of plane-wave signals, $\mathbf{X}$, becomes:

$$P(\mathbf{B}|\mathbf{X}) \propto e^{-\frac{1}{2}\text{tr}[(\mathbf{AX}-\mathbf{B})^H \mathbf{C}_{nn}^{-1}(\mathbf{AX}-\mathbf{B})]}, \tag{3.17}$$

where $\mathbf{C}_{nn}$ denotes the covariance matrix of the noise signals.

Within the Bayesian framework, the unknown plane-wave signals of interest are described by the probability $P(\mathbf{X}|\mathbf{B})$, which is the probability of the plane-wave signals $\mathbf{X}$ given a particular set of observation signals $\mathbf{B}$. According to Bayes' rule (Hyvärinen *et al.*, 2001, p. 30), $P(\mathbf{X}|\mathbf{B})$ is given by:

$$P(\mathbf{X}|\mathbf{B}) = \frac{P(\mathbf{B}|\mathbf{X})\, P(\mathbf{X})}{P(\mathbf{B})}, \tag{3.18}$$

where we refer to $P(\mathbf{X}|\mathbf{B})$ as the posterior probability of the plane-wave signals; $P(\mathbf{X})$ as the prior probability of the plane-wave signals; $P(\mathbf{B}|\mathbf{X})$ as the likelihood; and $P(\mathbf{B})$ as the evidence. Assuming that the evidence is described by a uniform distribution, it becomes an inconsequential normalization factor and we have:

$$P(\mathbf{X}|\mathbf{B}) \propto P(\mathbf{B}|\mathbf{X})\, P(\mathbf{X})$$
$$\propto e^{-\frac{1}{2}\text{tr}[(\mathbf{AX}-\mathbf{B})^H \mathbf{C}_{nn}^{-1}(\mathbf{AX}-\mathbf{B})+\mathbf{X}^H \mathbf{C}_{xx}^{-1}\mathbf{X}]}. \tag{3.19}$$

The most likely set of plane-wave signals given the observation signals is found by maximizing the probability $P(\mathbf{X}|\mathbf{B})$, which is equivalent to maximizing the *log-likelihood* function, $LL(\mathbf{X})$, given by:

$$LL(\mathbf{X}) = \ln(P(\mathbf{X}|\mathbf{B}))$$
$$= K - \frac{1}{2}\text{tr}\left[(\mathbf{AX} - \mathbf{B})^H \mathbf{C}_{nn}^{-1}(\mathbf{AX} - \mathbf{B}) + \mathbf{X}^H \mathbf{C}_{xx}^{-1}\mathbf{X}\right], \tag{3.20}$$

where K is some constant. This in turn leads to the following minimization problem to determine the expected plane-wave signals, $\hat{\mathbf{X}}$:

$$\hat{\mathbf{X}} = \underset{\mathbf{X}}{\arg\min}\, \text{tr}\left[(\mathbf{AX} - \mathbf{B})^H \mathbf{C}_{nn}^{-1}(\mathbf{AX} - \mathbf{B}) + \mathbf{X}^H \mathbf{C}_{xx}^{-1}\mathbf{X}\right]$$
$$= \underset{\mathbf{X}}{\arg\min}\, \left[\|\mathbf{C}_{nn}^{-\frac{1}{2}}(\mathbf{AX} - \mathbf{B})\|_2^2 + \|\mathbf{C}_{xx}^{-\frac{1}{2}}\mathbf{X}\|_2^2\right]. \tag{3.21}$$

The cost function, $J(\mathbf{X})$, for this minimization problem is given by $J(\mathbf{X}) = \|\mathbf{C}_{nn}^{-\frac{1}{2}}(\mathbf{AX} - \mathbf{B})\|_2^2 + \|\mathbf{C}_{xx}^{-\frac{1}{2}}\mathbf{X}\|_2^2$. Its minimization consists of solving a weighted damped least-squares problem (Menke, 2012, p. 58): the first term is the weighted norm of the error, while the second term is the weighted norm of the solution, which can be described as a regularization or damping term. It can be shown (see the appendix for a proof) that the solution $\hat{\mathbf{X}}$ is given by:

$$\hat{\mathbf{X}} = \mathbf{C}_{xx}\mathbf{A}^H(\mathbf{AC}_{xx}\mathbf{A}^H + \mathbf{C}_{nn})^{-1}\mathbf{B}. \tag{3.22}$$

The Bayesian solution given in Equation (3.22) is remarkably similar to Equation (3.11) for the IRLS update. The two equations become even more similar if we assume that the plane-wave signals are mutually uncorrelated with each other and that the noise signals are also mutually uncorrelated and have equal power, λ. In this case, the Bayesian solution for $\hat{\mathbf{X}}$ takes the form:

$$\hat{\mathbf{X}} = \mathbf{WA}^H(\mathbf{AWA}^H + \lambda\mathbf{I})^{-1}\mathbf{B}, \tag{3.23}$$

where $\mathbf{W}$ is a diagonal matrix, the non-zero entries of which are the plane-wave signal powers, and λ is the power of each noise signal. Equation (3.23) can be recognized as the Tikhonov-regularized form of the update solution used in the IRLS algorithm. In other words, by applying Tikhonov regularization to the IRLS update equation, one obtains an optimal solution assuming that the source powers are correctly specified by $\mathbf{W}$ and the white noise power is correctly specified by λ. This viewpoint provides an alternative means to understand the IRLS algorithm – and that is as a sparse-recovery expectation maximization (EM) algorithm (Dempster *et al.*, 1977). In the $\mathcal{M}$ step, the new plane-wave signals $\mathbf{X}$ are calculated given the model parameters $\mathbf{W}$ and λ, while in the $\mathcal{E}$ step the model parameters are updated given the new plane-wave signals. The relationship between the IRLS and EM frameworks was first reported by Dempster *et al.* (1980), although the details relating the IRLS framework to sparse-recovery problems had not yet been formulated at that time. More recently, the connection between the Bayesian and compressive sensing approaches was also described by Gerstoft *et al.* (2015).

3.4 Calculating the IRLS Noise-Power Regularization Parameter

When the noise-power regularization parameter is accurately estimated, experience has shown that the Tikhonov-regularized form of the IRLS update equation greatly improves the quality and accuracy of the IRLS solution in non-sparse sound conditions. This significant observation is later demonstrated with Figure 3.1 in Section 3.5. Therefore, we now focus on how to estimate the noise-power regularization parameter, λ, for the IRLS algorithm. We begin with some nomenclature and denote the average power of the observation signals plus noise (the total signal) as p_{total}, the average power of the noise signals as p_{noise}, and the average power of the observation signals due to the sources only (no noise) as p_{sig}. If we assume that the source signals and noise signals are mutually uncorrelated, we then have:

$$p_{total} = p_{sig} + p_{noise}. \tag{3.24}$$

The noise-power regularization parameter, λ, should then be set to the value of p_{noise}.

As it is difficult to directly estimate the value of the noise power, p_{noise}, we start by assuming that the relative noise power, $\beta = p_{\text{noise}}/p_{\text{total}}$, is either known or can be estimated. According to the Bayesian statistical model developed in the previous section, the matrix $\mathbf{AWA}^H$ represents the covariance matrix of the observation signals *without* noise. We therefore have (recall that N indicates the total number of plane-wave sources):

$$p_{\text{sig}} = \frac{\text{tr}[\mathbf{AWA}^H]}{N}. \tag{3.25}$$

We can use the fact that

$$\frac{p_{\text{noise}}}{p_{\text{sig}}} = \frac{p_{\text{noise}}}{p_{\text{total}} - p_{\text{noise}}} = \frac{\beta}{1 - \beta} \tag{3.26}$$

to conclude that λ can be estimated as:

$$\lambda = p_{\text{noise}} = \frac{p_{\text{noise}}}{p_{\text{sig}}} p_{\text{sig}} = \frac{\beta}{1 - \beta} \left(\frac{\text{tr}[\mathbf{AWA}^H]}{N} \right). \tag{3.27}$$

This assumes, of course, that the relative noise power, β, can be determined. In practice, β can be estimated using calculations of *diffuseness*, as detailed by Epain and Jin (2016). We briefly review the method for calculating β in the next section.

3.4.1 Estimation of the Relative Noise Power

To consider the estimation of the relative noise power, β, we start with a scenario in which the sound field consists of N active sound sources, where N is less than the number of observation signals. We also assume that the observation signals are polluted by perfectly uncorrelated Gaussian white noise with variance λ. Using the same assumptions and notations as described in Section 3.3, the covariance matrix of the observation signals, $\mathbf{C_{bb}} = \mathbf{B}^H\mathbf{B}$, is then given by:

$$\begin{aligned} \mathbf{C_{bb}} &= \mathbf{AC_{xx}A}^H + \mathbf{C_{nn}} \\ &= \mathbf{AC_{xx}A}^H + \lambda\mathbf{I}. \end{aligned} \tag{3.28}$$

In addition, we shall assume that the observation signals are in the spherical Fourier transform domain and comprise $K = (L + 1)^2$ order-L spherical harmonic (SH) signals.

The eigenvalue spectrum of the covariance matrix, $\mathbf{C_{bb}}$, provides a means to estimate the relative noise power, β. The eigenvalues of $\mathbf{C_{bb}}$, $(\sigma_1, \sigma_2, \ldots, \sigma_K)$, sorted in decreasing order, are given by (Strang, 2006, p.306):

$$\begin{cases} \sigma_i = \lambda + \omega_i & \text{for } 1 \leq i \leq N, \\ \sigma_i = \lambda & \text{for } N < i \leq K, \end{cases} \tag{3.29}$$

where ω_i denotes the ith eigenvalue of the matrix $\mathbf{AC_{xx}A}^H$. We use the COMEDIE (covariance matrix eigenvalue diffuseness estimation) diffuseness estimator (Epain and Jin, 2016) to estimate the relative noise power:

$$\hat{\beta} = 1 - \frac{\gamma}{\gamma_0}, \tag{3.30}$$

where:

- γ is the deviation of the eigenvalues of the SH signal covariance matrix from their mean, that is:

$$\gamma = \frac{1}{\langle \sigma \rangle} \sum_{i=1}^{K} |\sigma_i - \langle \sigma \rangle|,$$

$$\text{where } \langle \sigma \rangle = \frac{1}{K} \sum_{i=1}^{K} \sigma_i; \tag{3.31}$$

- γ_0 is a normalization constant given by:

$$\gamma_0 = 2[K - \bar{N}]; \tag{3.32}$$

- $\bar{N}$ is a fixed constant often set to 1, but ideally equal to the number of sources.

A simple understanding of the COMEDIE diffuseness estimator can be derived from going through the calculations. To simplify matters, we assume that all source signals have the same power, so that $\omega_i = \omega$ for $i \in [1, \ldots, N]$. In this case, the mean of the eigenvalues, $\langle \sigma \rangle$, is given by:

$$\langle \sigma \rangle = \lambda + \frac{N\omega}{K}, \tag{3.33}$$

and we find that:

$$\sum_{i=1}^{K} |\sigma_i - \langle \sigma \rangle| = \sum_{i=1}^{N} \left| \omega - \frac{N\omega}{K} \right| + \sum_{N+1}^{K} \left| -\frac{N\omega}{K} \right|$$

$$= N \frac{(K-N)\omega}{K} + (K-N)\frac{N\omega}{K}$$

$$= 2(K-N)\frac{N\omega}{K}. \tag{3.34}$$

The normalized deviation, γ/γ_0, of the eigenvalues is given by:

$$\frac{\gamma}{\gamma_0} = \frac{\frac{N\omega}{K}}{\lambda + \frac{N\omega}{K}} \frac{K-N}{K-\bar{N}}, \tag{3.35}$$

and when $\bar{N} = N$,

$$\frac{\gamma}{\gamma_0} = \frac{P_{\text{sig}}}{P_{\text{total}}}. \tag{3.36}$$

The estimated relative noise power, $\hat{\beta}$, is thus given by:

$$\hat{\beta} = \frac{\lambda}{\lambda + \frac{N\omega}{K}} + \frac{N - \bar{N}}{K - \bar{N}} \frac{N\omega}{K}, \tag{3.37}$$

and can be compared with the true relative noise power, β:

$$\beta = \frac{\lambda}{\lambda + \text{tr}[\mathbf{A}\mathbf{C}_{\text{xx}}\mathbf{A}^{\text{H}}]/K}. \tag{3.38}$$

One sees that $\hat{\beta} = \beta$ when $\bar{N} = N$. So long as $K \gg N$, the exact value of $\bar{N}$ is not critical. It is extremely important to appreciate that N refers to the number of true sources and not the number of reflections. The sound environment can be reverberant and the methods described here continue to work. The relationship between the relative noise power and the concept of diffuseness in the multi-source case requires significantly more elaboration than would be appropriate here. For more details and explanation regarding the COMEDIE diffuseness estimator, please refer to Epain and Jin (2016).

3.5 Numerical Simulations

In order to study the behavior of the IRLS sparse-recovery algorithm, we set up numerical simulations in which we varied the following three parameters: (i) the SH order of the observation signals, L; (ii) the number of sound sources; and (iii) the signal to noise ratio (SNR) of the observation signals. For each parameter combination, we randomly generated 250 sets of source and measurement noise signals. Both the source signals and noise signals consisted of 1024 samples of uncorrelated Gaussian white noise. The source signal powers were picked uniformly at random in the interval $[-20, 0]$ dB and the source directions were picked uniformly at random over the surface of the sphere.

For each set of observation signals $\mathbf{B}$, we performed a sparse plane-wave decomposition using the IRLS algorithm with both the non-regularized update equation and the regularized update equation. In the case of the regularized IRLS, the relative noise level, β, was estimated using the algorithm described in Epain and Jin (2016). In every case, the spatial analysis was performed using a dictionary of 3074 plane-wave directions corresponding to the vertices of a Lebedev grid. For each set of signals and each method, we calculated the mismatch between the acoustic energy map corresponding to the actual source directions and powers and the acoustic energy map obtained with the sparse recovery analysis. The mismatch, E, between two acoustic energy maps, map 1 and map 2, is given by:

$$E = \frac{K_{11} + K_{22} - 2K_{12}}{K_{11} + K_{22}}, \tag{3.39}$$

where K_{ij} is given by:

$$K_{ij} = \sum_{m=1}^{M} \sum_{n=1}^{N} \sqrt{\rho_m^{(i)} \rho_n^{(j)}} \, k\left(\Omega_m^{(i)}, \Omega_n^{(j)}\right), \tag{3.40}$$

where $\Omega_m^{(i)}$ and $\rho_m^{(i)}$ denote the mth direction in map i and the corresponding power value, respectively. The function $k(\cdot, \cdot)$ is a spatial kernel function defined as:

$$k\left(\Omega_m^{(i)}, \Omega_n^{(j)}\right) = \max\left(1 - \frac{\angle(\Omega_m^{(i)}, \Omega_n^{(j)})}{\pi/12}, 0\right), \tag{3.41}$$

where $\angle(\Omega_m^{(i)}, \Omega_n^{(j)})$ denotes the angle between directions $\Omega_m^{(i)}$ and $\Omega_n^{(j)}$.

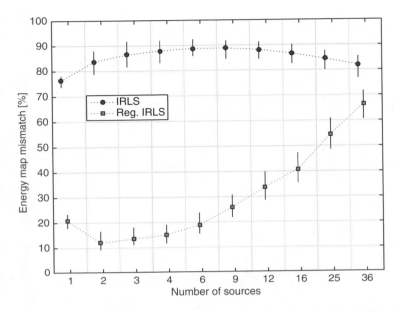

Figure 3.1 Comparison of the accuracy of two different SR-based sound field analysis methods, as a function of the number of sound sources. The signal to noise ratio of the SH signals is 0 dB. The markers and bars represent the median and interquartile range of the energy map mismatch, respectively.

We begin by comparing the accuracy of the sparse recovery analysis results obtained using either the regularized or non-regularized IRLS. This is illustrated in Figure 3.1 for SH order $L = 3$ and an SNR of 0 dB. Due to the high level of noise, the non-regularized IRLS applied to the observation signals fails completely and yields energy map mismatch values close to 100% even when there are only a few sound sources. Using the regularized IRLS dramatically improves the accuracy of the analysis. In the following, we only consider results obtained using the regularized IRLS.

We now compare the results obtained for different SH orders, as a function of the SNR and number of sound sources. This is illustrated in Figure 3.2. The accuracy of the analysis is almost equal for SNRs varying from $+\infty$ to 0 dB regardless of the SH order, which indicates that our approach makes possible the use of sparse recovery in the presence of noise. Comparing the results obtained for SH order 1 to 4, we see that higher-order SH signals can be accurately analyzed in the presence of more sound sources. This is expected because higher-order SH signals provide a higher angular resolution, hence more sources can be resolved. In summary, our approach yields reasonable results (mismatch lower than 15%–20%) when the number of sound sources is less than about half the number of observation signals, $(L + 1)^2/2$, for SNRs greater than 0 dB.

3.6   Experiment: Echoic Sound Scene Analysis

In this section we apply the regularized IRLS to the analysis of signals recorded in an actual room with an SMA. The measurement setup is shown in Figure 3.3. The

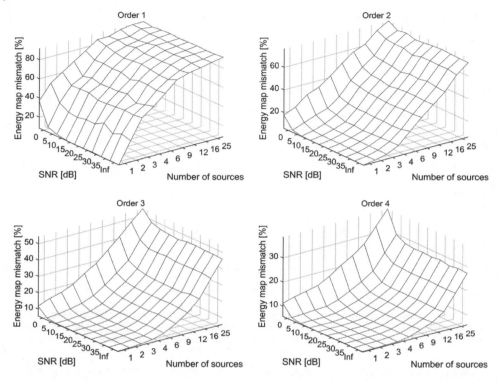

Figure 3.2 Average energy map mismatch resulting from the analysis of SH signals using regularized IRLS, as a function of the number of sound sources and SNR, for SH orders 1–4.

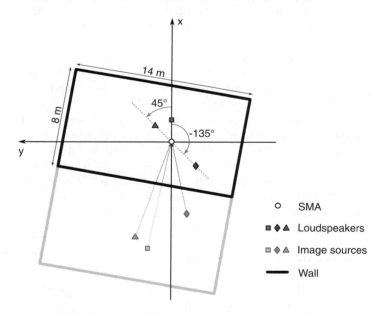

Figure 3.3 Bird's-eye view diagram of the experiment described in Section 3.6.

microphone array was the dual-concentric SMA described in Jin *et al.* (2014), which consists of 64 microphones located on the surface of two spheres: 32 microphones are distributed evenly on a rigid sphere of radius 28 mm and the other 32 are distributed on an open sphere of radius 95.2 mm. Impulse responses were measured using an NTI Audio Talkbox (NTI Audio, 2016) loudspeaker which was moved to the three following locations relative to the SMA (distance, azimuth): (1.8 m, 0°), (1.8 m, 45°), and (2.7 m, −135°). In the three locations the height of the loudspeaker's membrane center matched that of the SMA center, 1.5 m, and therefore the elevation of the speaker relative to the SMA was 0° for the three positions. Also, the loudspeaker pointed directly at the SMA in each location. The room in which the impulse responses were measured is an office space with dimensions $14 \times 8 \times 3$ m, approximately. The room's measured T60 reverberation time is about 0.5 s, and the direct-to-reverberant ratio (DRR) was estimated to 7 dB when the speaker was located 1.8 m from the SMA and 4.5 dB when the speaker was located 2.7 m from the SMA.

The impulse responses measured with the SMA for the three speaker locations were convolved with three anechoic speech signals to generate echoic microphone signals. The speech signals were 4 s long. Two consisted of English sentences spoken by male talkers, and the third consisted of a Japanese sentence spoken by a female talker. Also, the speech signals were equalized in amplitude such that their root-mean-square (RMS) level was equal prior to being convolved with the impulse responses. The presence of microphone self noise was simulated by adding Gaussian white noise to the microphone signals, such that the SNR of the microphone signals was −30 dB. Then, the noisy microphone signals were convolved with SH encoding filters to obtain SH signals up to order $L = 3$. The SH encoding filters were calculated using the method described in Jin *et al.* (2014). As explained there, due to the geometry of the SMA, clean order-3 SH signals can be obtained only over the frequency band ranging from 900 to 10 000 Hz. Thus a band-pass filter was applied to the SH signals to remove signals below 900 Hz and above 10 000 Hz.

We analyzed the SH signals using three different spatial sound field analysis techniques. First, we applied spherical beamforming and steered the beam in a large number of directions over the sphere. We then calculated the power of the beamformer output for every steering direction to obtain a map of the incoming sound energy. Second, we applied the MUSIC algorithm (Schmidt, 1986) to the SH signals. The MUSIC spectrum for direction Ω_n is given by:

$$\mu(\Omega_n) = \left(\sum_{i=4}^{K} \mathbf{a}(\Omega_n)^{\mathrm{T}} \mathbf{u}_i \right)^{-1}, \tag{3.42}$$

where $\mathbf{u}_i$ is the ith eigenvector of the SH signal covariance matrix. Note that the calculation of the MUSIC spectrum requires knowledge of the number of sources. In the formula above, only the eigenvectors $\mathbf{u}_i$ corresponding to the noise subspace are summed. The summation starts with $i = 4$ because there are three sound sources. Lastly, we performed a sparse plane-wave decomposition of the band-passed SH signals by decomposing the signals over a dictionary of 3074 plane-wave directions using the regularized IRLS algorithm. We then calculated the power of the plane-wave signals, thus obtaining a map of the incoming acoustic energy.

In the first instance, we analyzed the scene using the band-pass SH signals in their entirety, that is, analyzing the four seconds of signals at once. The results are shown in Figure 3.4. Although the map resulting from the spherical beamforming technique presents higher power values in the directions where the speakers are located, its resolution is very low. In particular, sources S_1 and S_2, located at azimuths 0° and 45°, respectively, are not resolved by the spherical beamformer. Compared to the spherical beamforming map, the map obtained using the MUSIC algorithm has a higher resolution and presents a clear peak in each of the three speaker directions. However, the amplitude of these peaks does not reflect the power of the sources accurately. Given that the speech signals have equal powers, and given the distance at which the speakers are located, the power received at the SMA from source S_3 (located at azimuth $-135°$) should be approximately 3.5 dB lower than that received from sources S_1 and S_2. By contrast, the peak observed in the MUSIC spectrum in the direction of S_3 is about 12 dB lower than that observed in the directions of S_1 and S_2, which is understandable given that the MUSIC spectrum is not designed to correctly indicate source power. Still referring to Figure 3.4, the map obtained using the sparse recovery method presents five sharp energy peaks. The three highest peaks are within a few degrees of the loudspeaker directions S_1, S_2, and S_3. Interestingly, the two remaining peaks correspond to S_1', the reflection of source S_1 on the floor, and S_1'', the reflection of S_1 on the wall located at the back of the SMA (see Figure 3.3). Regarding the power of the sources, the peak corresponding to S_3 is about 6 dB lower than the peaks observed in the directions S_1 and S_2, which is relatively close to the expected 3.5 dB difference. In summary, the sparse recovery algorithm made it possible to detect the three speaker locations and power with excellent accuracy, as well as two of the reflections occurring on the wall.

In a second experiment we analyzed a frame of 1024 samples of SH signals. We chose a time frame where the three sources were active simultaneously. The results of the analysis are shown in Figure 3.5. Overall, the same observations can be made as for the analysis of the whole 4 s of signals. In particular, the resolution of the MUSIC spectrum is significantly higher than that of the map obtained using spherical beamforming. On the other hand, although it is clearly visible in the spherical beamforming map that the power of S_2, the source located at azimuth 45°, is higher than that of the two other sources, this is not reflected in the amplitude of the peaks observed in the MUSIC spectrum map. Considering the map obtained using our sparse recovery method, we again observe energy peaks within a few degrees of each speaker location. Also, the amplitude of the peak corresponding to source S_2 is clearly higher than those corresponding to S_1 and S_3, which is consistent with the results obtained using spherical beamforming. Regarding the other energy peaks, the sparse recovery analysis detected S_3', which is S_3's floor reflection; S_1'', which is the reflection of S_1 on the back wall; and S_2' and S_2'', which are S_2's floor and ceiling reflections, respectively. The fact that more reflections corresponding to S_2 are detected can be explained by the fact that S_2 is louder than the two other sources over the duration of the time frame. Note that, as was the case when analyzing the entire SH signals, the localization accuracy for the source reflections is lower than for the sources themselves. In particular, S_2' seems shifted to the right compared to its expected location, and the elevation of the peaks corresponding to S_1'' and S_3' are lower than expected.

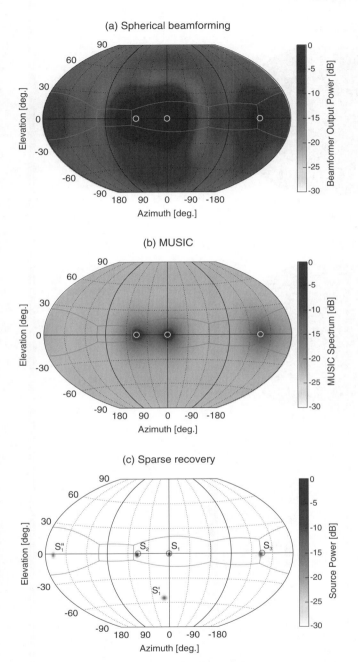

Figure 3.4 Comparison of the sound field energy maps obtained by analyzing 4 s of recorded SH signals with: (a) spherical beamforming, (b) MUSIC, and (c) the sparse recovery method presented in this paper. The light, solid lines (blue in the online version) indicate the wall edges and the circles indicate the loudspeaker positions.

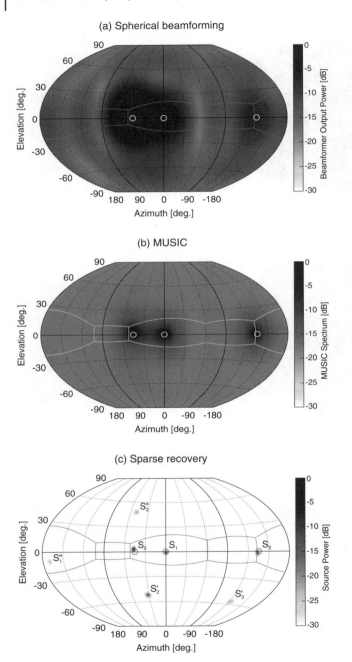

Figure 3.5 Comparison of the sound field energy maps obtained by analyzing 1024 samples of recorded SH signals with: (a) spherical beamforming, (b) MUSIC, and (c) the sparse recovery method presented in this paper. The light, solid lines (blue in the online version) indicate the wall edges and the circles indicate the loudspeaker positions.

3.7 Conclusions

Plane-wave decomposition is a powerful tool for sound field analysis, and this chapter describes the implementation and capabilities of a regularized IRLS sparse-recovery algorithm for plane-wave decomposition. We have argued that the sparse-recovery approach is advantageous because one tries to explain the sound field in terms of the fewest possible number of acoustic sources. Without being given any other information, this is a reasonable approach to pursue in terms of applying the principle of Occam's razor to acoustic scene analysis. We have shown that the IRLS algorithm provides an intuitive (although at times mathematically laborious) expectation-maximization approach to finding plane-wave sources. One first updates the expected values for the plane-wave source powers and noise powers, and then determines which plane-wave source configuration has the maximum likelihood – all the while promoting a sparse solution. The sparse-recovery solutions provide excellent headroom in terms of thresholding and searching for acoustic sources. That is to say, many source localization algorithms are exquisitely sensitive to thresholds applied while searching for acoustic sources. Sparse-recovery solutions, on the other hand, are robust and insensitive to any thresholds applied. In an effort to keep this chapter to a reasonable length and focus, we have not described more recent advances in our sparse-recovery research. These advances take the form of a denoising method that can help separate direct and ambient signals (Epain and Jin, 2013), and also a spatial priming technique (Noohi *et al.*, 2015) that greatly enhances the robustness and ability to work with a large number of sources, even when the number of acoustic sources and observation signals are approximately equal. We hope this chapter has highlighted some of the characteristics and advantages of the sparse-recovery approach and encourages others to explore this area in more detail.

Appendix

In this appendix we derive Equation (3.22). The cost function, $J(\mathbf{X})$, is given by:

$$
\begin{aligned}
J(\mathbf{X}) &= \operatorname{tr}\left[(\mathbf{AX} - \mathbf{B})^H \mathbf{C}_{nn}^{-1}(\mathbf{AX} - \mathbf{B}) + \mathbf{X}^H \mathbf{C}_{xx}^{-1}\mathbf{X}\right] \\
&= \operatorname{tr}\left[\mathbf{X}^H \mathbf{A}^H \mathbf{C}_{nn}^{-1}\mathbf{AX} - \mathbf{X}^H \mathbf{A}^H \mathbf{C}_{nn}^{-1}\mathbf{B} - \mathbf{B}^H \mathbf{C}_{nn}^{-1}\mathbf{AX}\right] + \operatorname{tr}\left[\mathbf{B}^H \mathbf{C}_{nn}^{-1}\mathbf{B} + \mathbf{X}^H \mathbf{C}_{xx}^{-1}\mathbf{X}\right].
\end{aligned}
$$
(3.43)

Note that $J(\mathbf{X})$ is a quadratic form and thus has one global minimum. The partial derivative of the cost function with respect to $\mathbf{X}$ is given by:

$$
\frac{\partial J}{\partial \mathbf{X}} = 2\mathbf{A}^H \mathbf{C}_{nn}^{-1}\mathbf{AX} - 2\mathbf{A}^H \mathbf{C}_{nn}^{-1}\mathbf{B} + 2\mathbf{C}_{xx}^{-1}\mathbf{X}.
$$
(3.44)

The cost function's minimum corresponds to the point where the partial derivative is equal to zero. Equating the partial derivative to zero, we obtain:

$$
\frac{\partial J}{\partial \mathbf{X}} = 0 \iff \left(\mathbf{A}^H \mathbf{C}_{nn}^{-1}\mathbf{A} + \mathbf{C}_{xx}^{-1}\right)\mathbf{X} = \mathbf{A}^H \mathbf{C}_{nn}^{-1}\mathbf{B}.
$$
(3.45)

Thus, the solution $\hat{X}$ is given by:

$$\hat{X} = \left(A^H C_{nn}^{-1} A + C_{xx}^{-1}\right)^{-1} A^H C_{nn}^{-1} B = MB,$$

where $M = \left(A^H C_{nn}^{-1} A + C_{xx}^{-1}\right)^{-1} A^H C_{nn}^{-1}.$ (3.46)

The expression for matrix M can be modified as follows:

$$
\begin{aligned}
M &= M(AC_{xx}A^H + C_{nn})(AC_{xx}A^H + C_{nn})^{-1} \\
&= \left(A^H C_{nn}^{-1} A + C_{xx}^{-1}\right)^{-1} \left(A^H C_{nn}^{-1} AC_{xx}A^H + A^H\right) \times (AC_{xx}A^H + C_{nn})^{-1} \\
&= \left(A^H C_{nn}^{-1} A + C_{xx}^{-1}\right)^{-1} \left(A^H C_{nn}^{-1} A + C_{xx}^{-1}\right) C_{xx}A^H \times (AC_{xx}A^H + C_{nn})^{-1} \\
&= C_{xx}A^H(AC_{xx}A^H + C_{nn})^{-1}.
\end{aligned}
$$ (3.47)

Therefore, we have:

$$
\begin{aligned}
\hat{X} &= MB \\
&= C_{xx}A^H(AC_{xx}A^H + C_{nn})^{-1}B.
\end{aligned}
$$ (3.48)

References

Benesty, J., Chen, J., and Huang, Y. (2008) *Microphone Array Signal Processing*, Springer, New York, chapter 9.

Cotter, S., Rao, B., Engan, K., and Kreutz-Delgado, K. (2005) Sparse solutions to linear inverse problems with multiple measurement vectors. *IEEE Transactions on Signal Processing*, **53**(7), 2477–2488.

Daubechies, I., DeVore, R., Fornasier, M., and Güntürk, C.S. (2010) Iteratively reweighted least squares minimization for sparse recovery. *Communications on Pure and Applied Mathematics*, **63**(1), 1–38.

Dempster, A., Laird, N., and Rubin, D. (1977) Maximum likelihood from incomplete data via the EM algorithm. *Journal of the Royal Statistical Society*, **39**(1), 1–38.

Dempster, A., Laird, N., and Rubin, D. (1980) Iteratively reweighted least squares for linear regression when errors are normal/independent distributed. *Multivariate Analysis*, **V**, 35–57.

Epain, N. and Jin, C. (2013) Super-resolution sound field imaging with sub-space pre-processing. *Proceedings of the 2013 International Conference on Acoustics, Speech, and Signal Processing*, pp. 350–354, Vancouver, Canada.

Epain, N. and Jin, C. (2016) Spherical harmonic signal covariance and sound field diffuseness. *IEEE/ACM Transactions on Audio, Speech, and Language Processing*, **24**(10), 1796–1807.

Gerstoft, P., Xenaki, A., and Mecklenbräuker, C.F. (2015) Multiple and single snapshot compressive beamforming. *The Journal of the Acoustical Society of America*, **138**(4), 2003–2014.

Gover, B., Ryan, J., and Stinson, M. (2002) Microphone array measurement system for analysis of directional and spatial variations of sound fields. *Journal of the Acoustical Society of America*, **112**(5), 1980–1991.

Hyvärinen, A., Karhunen, J., and Oja, E. (2001) *Independent Component Analysis*. Wiley Interscience, Chichester.

Jin, C., Epain, N., and Parthy, A. (2014) Design, optimization and evaluation of a dual-radius spherical microphone array. *IEEE/ACM Transactions on Audio, Speech, and Language Processing*, **22**(1), 193–204.

Malioutov, D., Çetin, M., and Willsky, A.S. (2005) A sparse signal reconstruction perspective for source localization with sensor arrays. *IEEE Transactions on Signal Processing*, **53**(8), 3010–3022.

Menke, W. (2012) *Geophysical Data Analysis: Discrete Inverse Theory*. Academic Press, New York.

Noohi, T., Epain, N., and Jin, C. (2015) Super-resolution acoustic imaging using sparse recovery with spatial priming. *Proceedings of the 2015 International Conference on Acoustics, Speech, and Signal Processing*, pp. 2414–2418, Brisbane, Australia.

NTI Audio (2016) *TalkBox – Acoustic Generator*, http://www.nti-audio.com/products/talkbox.aspx (accessed April 7, 2016).

Park, M. and Rafaely, B. (2005) Sound-field analysis by plane-wave decomposition using spherical microphone array. *Journal of the Acoustical Society of America*, **5**(118), 3094–3103.

Rafaely, B., Peled, Y., Agmon, M., Khaykin, D., and Fisher, E. (2010) Spherical microphone array beamforming, in *Speech Processing in Modern Communication: Challenges and Perspectives* (eds I. Cohen, J. Benesty, and S. Gannot), Springer, New York.

Roy, R. and Kailath, T. (1989) ESPRIT: Estimation of signal parameters via rotational invariance techniques. *IEEE Transactions on Acoustics, Speech, and Signal Processing*, **37**(7), 984–995.

Schmidt, R. (1986) Multiple emitter location and signal parameter estimation. *IEEE Transactions on Antennas and Propagation*, **34**(3), 276–280.

Strang, G. (2006) *Linear Algebra and its Applications*, 4th edition, Thomson Brooks/Cole, New York.

Sun, H., Teutsch, H., Mabande, E., and Kellermann, W. (2011) Robust localization of multiple sources in reverberant environments using EB-ESPRIT with spherical microphone arrays. *Proceedings of the 2011 ICASSP*, Prague, Czech Republic.

Tervo, S. and Politis, A. (2015) Direction of arrival estimation of reflections from room impulse responses using a spherical microphone array. *IEEE/ACM Transactions on Audio, Speech, and Language Processing*, **23**(10), 1539–1551.

Wabnitz, A., Epain, N., and Jin, C. (2012) A frequency-domain algorithm to upscale ambisonic sound scenes. *Proceedings of the 2012 ICASSP*, Kyoto, Japan.

Wabnitz, A., Epain, N., McEwan, A., and Jin, C. (2011) Upscaling ambisonic sound scenes using compressed sensing techniques. *Proceedings of WASPAA*, New Paltz, NY.

Wu, P., Epain, N., and Jin, C. (2012) A dereverberation algorithm for spherical microphone arrays using compressed sensing techniques. *Proceedings of the 2012 ICASSP*, Kyoto, Japan.

Part II

Reproduction of Spatial Sound

4

Overview of Time–Frequency Domain Parametric Spatial Audio Techniques

Archontis Politis, Symeon Delikaris-Manias, and Ville Pulkki

Department of Signal Processing and Acoustics, Aalto University, Finland

4.1 Introduction

Spatial sound reproduction is mainly concerned with delivering appropriate spatial cues using a reproduction system, loudspeakers or headphones, in order to provide a convincing, immersive, and engaging aural experience, close to the intentions of the producer for a synthetic sound scene, or to the original experience for a recorded one. Spatial sound content is commonly produced with a certain reproduction setup in mind, with the present standard being discrete surround multichannel systems such as the 5.1 and 7.1 surround setups. Hence, the content already contains all the relevant spatial cues created by the producer for the corresponding setup. In this case, a state-of-the-art reproduction method has to address the issues of (i) compression, meaning how to reduce the bandwidth of the content for delivery and transmission without affecting its perceptual impression, and (ii) up-mixing or down-mixing, meaning to recreate the spatial sound scene in target systems other than the originally intended one, in the best way possible. In the case of recorded sound scenes, there is the additional complexity of (iii) capturing the scene with a microphone array in such a way that it makes delivery of the spatial cues at reproduction possible. If the reproduction method relies on time-dependent parameters extracted from the signals in order to achieve the above tasks, then it is termed parametric. This is in contrast to performing, for example, up-mixing through a signal-independent (static) matrixing of the input signals. Parametric spatial compression, up-mixing, and reproduction are all interrelated, as the extracted parameters are commonly appropriate for all of the above tasks.

State-of-the-art parametric methods should be *flexible*, able to handle a variety of input formats, either multichannel content or recordings coming from a variety of recording setups. Additionally, the method should be able to handle a variety of target reproduction setups, 2D discrete surround layouts, headphone audio, or large emerging 3D speaker layouts (Ono *et al.*, 2013), generating the target signals from the same content. The method should also be *effective*, in the sense that it should utilize the reproduction system in the best way possible to deliver the appropriate spatial audio cues,

Parametric Time–Frequency Domain Spatial Audio, First Edition. Edited by Ville Pulkki,
Symeon Delikaris-Manias, and Archontis Politis.
© 2018 John Wiley & Sons Ltd. Published 2018 by John Wiley & Sons Ltd.
Companion Website: www.wiley.com/go/pulkki/parametrictime-frequency

while not affecting non-spatial attributes such as spectral balance. Finally, the method should be *efficient*, since many potential applications require real-time operation, such as binaural rendering for head-tracked headphone reproduction, rendering for virtual interactive environments and virtual reality, and telepresence/teleconferencing.

Traditionally, sound scene recording and reproduction, as practiced in music recording, restricts the delivery from microphone signals to loudspeaker signals to one-to-one mapping by extending stereophonic recording principles to multichannel arrangements. The principles of these approaches are outlined, for example, by Williams (2004) and Theile (2001). This approach does not, however, scale well with emerging multi-loudspeaker setups, and it cannot deliver audio for personalized headphone-based audio. Ideally, a compact microphone array with a small number of microphones and appropriate processing should be able to deliver the audio for any arrangement. Some non-parametric approaches attempt this by beamforming to the loudspeaker directions (Ono *et al.*, 2013; Hur *et al.*, 2014), or constructing beamformers that resemble head-related transfer functions (HRTFs; Chen *et al.*, 1992; Sakamoto *et al.*, 2010), as shown in Figure 4.1(a). However, they require both a description of the recording and reproduction setup. More flexible non-parametric methods, such as Ambisonics (Gerzon, 1973; Poletti, 2005; Moreau *et al.*, 2006; Zotter and Frank, 2012), separate the mapping from the array signals to the playback signals into two stages, encoding and decoding, as shown in Figure 4.1(b). This is conducted by expressing both recording and reproduction signals in some transform domain in which their spatial properties have a common representation. Non-ambisonic methods that perform this separation have also been proposed by Farina *et al.* (2013) and Hur *et al.* (2011). Non-parametric methods can suffer from low spatial resolution with small practical arrays, a fact which affects their perceptual performance (Solvang, 2008; Avni *et al.*, 2013; Yang and Xie, 2014; Stitt *et al.*, 2014).

Parametric approaches consider the recording setup and the reproduction setup, but they additionally impose some assumptions or structure on the sound field and its spatial properties, and they try to estimate that structure from the recorded signals, as illustrated in Figure 4.1(c). Commonly, the estimation of parameters is perceptually motivated. Various different approaches exist on which assumptions or sound field model to use; an overview of recent research is presented in the next part of the book. In terms of flexibility and scalability, parametric methods can surpass non-parametric ones. The extracted spatial parameters allow the whole arsenal of audio engineering tools for perceptual sound spatialization to be used optimally for a variety of target setups, such as appropriate panning functions, HRTFs, and reverberation algorithms. For the same reason, it is possible that parametric approaches offer a greater degree of realism or transparency with regards to the original sound scene, and potentially higher perceptual quality. However, compared to non-parametric approaches, more care should be given to avoiding artifacts related to inaccurate estimation of parameters and to the application of time–frequency processing. Artifacts of non-parametric approaches are mainly related to spectrum coloration in the output, while artifacts from parametric approaches can be more severe since the spectrum modification is time-variant. Related problems have been addressed before in speech enhancement and multichannel spatial audio coding (SAC), and these are areas worth looking at for effective solutions. These solutions are mostly based on smoothing the estimated parameters over time and/or frequency (Boashash, 2015). In terms of efficiency, parametric approaches are

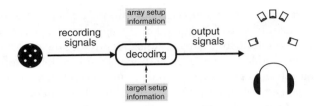

(a) Non-parametric: Direct approaches.

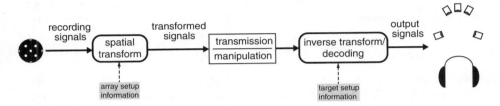

(b) Non-parametric: Transform-based approaches with separated encoding/decoding.

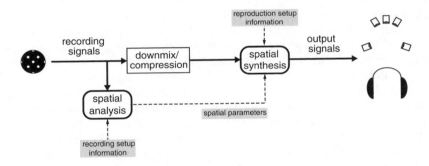

(c) Parametric approaches.

Figure 4.1 Basic processing blocks of parametric and non-parametric approaches to spatial sound reproduction. Adapted from Politis 2016.

computationally more demanding than non-parametric ones, which essentially avoid the estimation stage and realize a static set of filters or gains mapping the inputs to the outputs. However, similarly to speech enhancement and SAC, real-time implementations with no perceivable latency are feasible (Benesty and Huang, 2013), and various recent parametric methods have been demonstrated to work in real time.

4.2   Parametric Processing Overview

Parametric approaches vary widely in their scope and the assumptions that they make on the type of input, the spatial structure of the signals, and the target output system. In general, the various methods may combine elements from statistical signal processing, psychoacoustics, array processing, source separation, and adaptive filtering. Most of the assumptions and constraints are determined by the application itself. A teleconference system with spatialized sound, for example, aims to reproduce the talkers around a table at the receiving room, and hence a two-dimensional sound model is more likely appropriate, utilizing a linear or circular array, and with the objective to suppress reverberant sound that may reduce intelligibility. On the other hand, a general system for music recording and reproduction of sound scenes that can then adapt to multiple reproduction setups would most likely utilize a three-dimensional array, such as a spherical one, with a design that does not dismiss reverberant components but tries to recreate them appropriately at the output. A distinction is made throughout this chapter between systems that do coding and up-mixing, recording and reproduction, and enhancement.

4.2.1 Analysis Principles

Analysis of the parameters depends on the application, the constraints on the type of microphone array or input format that should be used, quality or efficiency constraints, and whether the focus is on reproduction, enhancement, or separation of the sound scene components. Most common array types used in the literature are uniform linear arrays, circular arrays, or spherical arrays, with the first two able to handle 2D analysis, and the latter suitable for full 3D analysis. Linear arrays are a natural choice, for example, in immersive communication systems where the array is integrated on the edge of a screen (Beracoechea *et al.*, 2006; Thiergart *et al.*, 2011). Circular arrays are suitable for teleconferencing/telepresence applications with the speakers arranged around the array, and for general sound reproduction constrained to two dimensions (Ahonen *et al.*, 2012; Alexandridis *et al.*, 2013). Spherical microphone arrays have been developed mostly in the context of non-parametric sound scene analysis and reproduction, especially in the context of ambisonic recording and reproduction (Gerzon, 1973; Moreau *et al.*, 2006). Recently, many parametric acoustic analysis and enhancement methods have been adapted for them (Teutsch, 2007). Non-spherical compact arrays of a few microphones have also been used with portability in mind (Cobos *et al.*, 2010; Nikunen and Virtanen, 2014). Large planar or arbitrary omnidirectional arrays are less common, even though they seem suitable for auralization of soundscapes and for moving virtually inside the recording during reproduction (Gallo *et al.*, 2007; Niwa *et al.*, 2008; Thiergart *et al.*, 2013; Verron *et al.*, 2013).

The core assumptions of each method that differentiate it from others are the way it treats the spatial dependencies between the input channels, and how it uses these to compute the parameters. Since the methods themselves are grouped and presented according to these assumptions in the following subsections, we will mention briefly here some of the most common principles. All the methods with no exception operate in a time–frequency transform domain, such as the short-time Fourier transform or a filterbank. Many methods in enhancement, up-mixing, and reproduction of recordings impose a model of one dominant directional source and some diffuse/ambient sound at each time–frequency tile. To separate these two signals the whole arsenal of microphone array estimation methods can be utilized. The directional component is usually associated with a direction of arrival (DoA), which can be found with efficient techniques based on properties such as cross-correlation or acoustic intensity. Descriptions of these methods can be found in microphone array processing textbooks (Brandstein and Ward,

2013). If the application requires estimation of multiple directional sources per time–frequency tile, then more advanced estimation techniques can be used – steered beam-forming methods such as steered response power (SRP), or subspace-based ones such as multiple signal classification (MUSIC) and estimation of signal parameters via rotational invariance techniques (ESPRIT) – at an increased computational cost.

An alternative to processing the microphone signals is to express them in a transform domain which decouples the spatial properties of the sound scene from the array in use. Typically this is done through the spherical harmonic transform, which projects the sound scene signals into an orthogonal set of directional functions on the sphere, the spherical harmonics (SH). For a thorough overview of spherical acoustic processing the reader is referred to Chapter 2 and the references therein. Parametric methods that operate in the spherical harmonic domain share common input signals with non-parametric ambisonic techniques and can benefit by integrating ambisonic processing (Vilkamo *et al.*, 2009; Politis *et al.*, 2015; Herre *et al.*, 2015). The simplest SH representation of a sound scene requires the first four terms of the expansion, resulting in only four signals, and is known in spatial sound literature as B-format. Even though B-format is characterized by a low spatial resolution for effective non-parametric reproduction, it has been used with success by certain parametric techniques (Pulkki, 2007; Berge and Barrett, 2010) offering high perceptual quality and efficiency, especially when compared with non-parametric techniques. Methods that assume input from coincident directional microphones can also be included in this category (Faller, 2008), as any number of standard directional patterns can be generated from B-format signals at the same spot (Gerzon, 1975a).

4.2.2 Synthesis Principles

In the analysis stage, directional parameters that correspond to meaningful spatial properties of the sound scene are extracted from the array recordings. These parameters should then be used to achieve the required task, whether it is up-mixing, reproduction, or spatial enhancement. For example, estimated DoAs can adapt beamformers for the separation of directional signals. Additionally, the DoAs can be used to respatialize directional components at the target setup, or enhance components from certain directions only. Other commonly estimated parameters are power ratios between directional components and diffuse or ambient sound. Such parameters, apart from their intuitive meaning as a mixing value between directional and diffuse components, can be used to construct a separation filter between these components from the recordings. The separated ambient sound can then be suppressed from the mix, for enhancement of the directional components, or distributed appropriately at the output channels for up-mixing and reproduction applications.

An advantage of parametric approaches for sound scene reproduction is the fact that they can employ effective spatialization techniques at synthesis, making the rendering target independent. For example, after signals for the directional components have been estimated, their loudspeaker rendering can be done with amplitude panning for 2D setups, or with vector base amplitude panning (VBAP) for arbitrary 3D setups. Since the DoAs are updated constantly, the same happens with the panning gains. Binaural synthesis for headphones is similarly performed, but HRTFs are used instead of amplitude panning gains.

Regarding the ambient or diffuse component, high-quality rendering can be achieved again by employing audio engineering techniques such as artifical reverberation or decorrelation. Commonly, a single ambient signal is extracted from the recordings, which, if directly distributed to the channels of the target setup, results in a spatially correlated sound that violates the diffuse sound assumptions and reduces the perceived quality of reproduction. Instead, if statistically independent copies of the ambient signal are distributed to the outputs, high-quality diffuse reproduction can be achieved that is both enveloping and spatially ambiguous. In parametric reproduction and up-mixing, various kinds of decorrelators have been used for this task, such as exponentially decaying random sequences (Hawksford and Harris, 2002), random sub-band delays with time constraints (Laitinen *et al.*, 2011), phase randomization (Pulkki and Merimaa, 2006), or modeled reverberation tails (Faller, 2006b). It is possible that instead of completely uncorrelated diffuse output signals, a specific degree of coherence is desired between them. This is the case of rendering ambient signals for headphones, where even in completely diffuse sound the binaural signals exhibit a sinc-like frequency-dependent coherence, from fully coherent at low frequencies to fully incoherent at higher ones.

More complex spatialization systems, for which recording has proven difficult, such as wave field synthesis (WFS) (Berkhout *et al.*, 1993; Ahrens, 2012), can also benefit from the flexibility of the parametric approach. If an estimate of the source signals has been obtained, along with their DoAs, then rendering at a WFS system using the appropriate driving functions is possible, as proposed by Beracoechea *et al.* (2006) and Cobos *et al.* (2012). A related concept is up-mixing of ambisonic signals, from lower-order to higher-order ones, which again are difficult or impossible to obtain at all frequencies with a practical array. This can be achieved simply by re-encoding the extracted components to ambisonic signals of a higher order, which can then take full advantage of large loudspeaker arrays or HRTFs. Alternatively, compressed sensing techniques seem suitable for the same task, as shown by Wabnitz *et al.* (2011).

4.2.3 Spatial Audio Coding and Up-Mixing

Spatial audio coding refers to methods that aim to compress existing multichannel audio content using a parametric approach in order to reduce bandwidth and processing requirements, as shown in Figure 4.2(a). Ideally, a SAC system should be able to recreate the multichannel content without loss of perceived quality, and ideally with zero perceptual differences (transparency). In contrast to SAC, up-mixing refers to decomposing the audio channels of multichannel content in order to redistribute them to a different target system; a classic example is stereo to multichannel up-mixing, or stereo to binaural conversion. Even though at first look the aims are different, both SAC and up-mixing strive for the same goal: process the recording in such a way that when recreated at the target system, the perceptual properties of the original are preserved. Even though there are SAC approaches that concentrate mostly on compression and efficiency, other methods focus on extracting perceptual spatial parameters that are also suitable for the up-mixing task.

Well-established SAC methods are the MPEG Surround codec specification (Breebaart *et al.*, 2005a; Herre *et al.*, 2008), binaural cue coding (BCC) by Baumgarte and Faller (2003), and parametric stereo (PS) by Breebaart *et al.* (2005b). A comprehensive overview of these approaches is presented in Breebaart and Faller (2007). Binaural cue

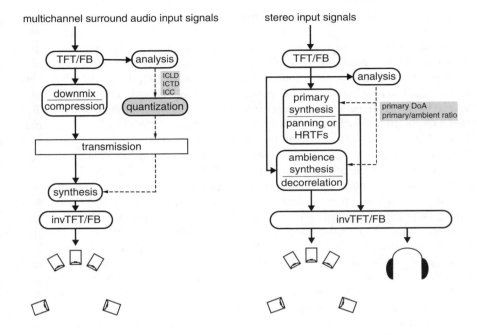

(a) Surround spatial audio coding example.   (b) Stereo up-mixing example.

Figure 4.2 Typical processing blocks of spatial audio coding and up-mixing systems. Source: Politis 2016. Reproduced with permission of Archontis Politis.

coding and PS compute as main spatial parameters the inter-channel level differences (ICLD), inter-channel time differences (ICTD), and inter-channel coherences (ICC). Originally, BCC used only ICLD and ICTD, but was later extended to also use ICC. Extraction of phase, level differences, and coherences between the channels are assumed to capture the statistical signal dependencies that need to be preserved and recreated, either in coding or up-mixing, and they are ubiquitous in most parametric methods. For example, major binaural directional cues can be described by such a set of parameters (Plogsties *et al.*, 2000). These parameters can all be extracted from the covariance matrix of the input channels, as detailed in Section 1.3.1. MPEG Surround starts from similar principles, but it uses a different approach to coding. Instead of extracting the aforementioned parameters, it uses a two-to-three and three-to-two tree structure of consecutive down-mixes of the channels, along with associated metadata.

Up-mixing shares many common parts with coding, such as the time–frequency transform with auditory-like resolution, and extraction of the inter-channel dependencies. However, since the signals are not intended to be reproduced by the same setup, an estimation of their spatial properties is required using some additional assumptions. The most commonly used assumption is that of a single dominant directional source plus diffuse ambient sound at each sub-band, as shown in Figure 4.2(b). The directional component is strongly correlated between channels, while the diffuse component is fully or partially uncorrelated with the directional component and between channels.

Additionally, a spatialization model is assumed for the directional component, which results in a certain correlation between the channels and permits estimation of its DoA by inverting the process. The model can be, for example, the well-known panning laws used in sound mixing. Even though these assumptions may seem crude at first (for example, for more than a single panned source at a certain moment), they hold reasonably well in practice. The reasons for this are twofold: first, the spectra of multiple broadband sources can be rather sparse in the frequency domain, so that only a single one dominates a certain narrow frequency band, a fact that has been termed "W-disjoint orthogonality" (Rickard and Yilmaz, 2002). Secondly, even if there are multiple sources active at a narrow frequency band, the auditory system can be dominated by a single localization cue corresponding to a weighted average of them (Faller and Merimaa, 2004).

Up-mixing methods differ in how they separate the directional and diffuse components, an approach also termed direct–diffuse decomposition, or primary–ambient extraction. Most methods construct a minimum least-squares estimator of the component signals, taking into account the inter-channel dependencies and the assumed component dependencies (Avendano and Jot, 2004; Faller, 2006a; Walther and Faller, 2011; Thompson *et al.*, 2012). Another popular alternative has been the principal component analysis (PCA) of the input signals, mostly stereo, which naturally results in components that are decorrelated between themselves, and achieves effective primary ambient decomposition under certain conditions (Irwan and Aarts, 2002; Driessen and Li, 2005; Briand *et al.*, 2006; Merimaa *et al.*, 2007; He *et al.*, 2014). A related concept has been termed "up-mixing in a unified-transform domain" by Short *et al.* (2007). A comparison between least-squares and PCA-based techniques is presented in He *et al.* (2014). Apart from the channel-dependent spatial parameters, a few techniques extract some global spatial descriptors that are used for the decomposition. Such examples are the intensity/diffuseness used in DirAC, or the ambisonic-inspired velocity and energy vectors used in spatial audio scene coding (SASC) by Goodwin and Jot (2008). Both sets of descriptors are related, as shown by Merimaa (2007). Recent proposals aim to apply blind source separation (BSS) principles to the decomposition problem, by estimating statistically independent components in the channel signals. One such approach is non-negative matrix (NMF) or tensor (NTF) factorization, a popular method in BSS tasks, and applied to up-mixing by Nikunen *et al.* (2011).

Headphone reproduction of multichannel content can be performed directly by spatializing the input setup to the target setup using HRTFs, an approach, however, that has been found to produce unconvincing and low-quality results, with coloration of the material, in-head localization, and no sense of reverberation. Spatial audio coding and up-mixing techniques seem more suitable for this task, with most of the major methods implementing a binaural version, such as MPEG Surround (Breebaart *et al.*, 2006), PS (Faller and Breebaart, 2011), SASC (Goodwin and Jot, 2007), and DirAC (Laitinen and Pulkki, 2009). Direct–diffuse decomposition has been shown to be quite effective, rendering the direct component with HRTFs and diffusing the ambient component appropriately at the two ears, fixing most of the aforementioned issues.

4.2.4 Spatial Sound Recording and Reproduction

In the parametric approach to recording and reproduction of spatial sound, the signals from a microphone array are used to analyze the spatial parameters, and the parameters then affect the process of synthesis, as shown in Figure 4.1(c). Contrary to SAC and up-mixing, reproduction of recorded sound scenes relies strongly on a description of the capture setup, meaning the geometry of the array, the directivity patterns of the microphones, their orientation, and any scattering effects due to any baffles used to increase directionality or separation between channels. All these are usually modeled or measured in the directional array response, termed "steering vector" in the array processing literature. Knowledge of the steering vector means knowledge of the correlations between the recorded channels for a directional source incident from some DoA, and for diffuse sound with equal power from all directions.

Similar to up-mixing, many reproduction methods share the same assumption of estimating one dominant directional cue and one cue for ratios between directional and ambient components. One way to do that is through an energetic analysis of the sound field, through estimation of the acoustic energy and intensity at the origin of the microphone array. Acoustic intensity, and more specifically its active component (Jacobsen, 2007), seems suitable for estimating a dominant DoA since it shows the direction of mean energy transfer, which is generally believed to correlate with perceptual localization. After normalization with the total acoustic energy, intensity also provides a measure of coherent versus incoherent energy in the recording. This index, termed "diffuseness" in the literature, can be used for separation of the coherent and incoherent parts directly as a time–frequency mask. Hence, intensity can provide the main components for parametric spatial sound processing of a recorded sound scene, being a ratio between the coherent and incoherent sound energy, and the DoA of the coherent part. Intensity can be efficiently estimated from the four-channel B-format signals, which in turn can be computed from a variety of near-coincident recording arrangements. The most prominent method using this approach is directional audio coding (DirAC; Merimaa and Pulkki, 2005; Pulkki, 2007). Another early related attempt by Hurtado-Huyssen and Polack (2005) proposed to reproduce the intensity oscillation in the recording as a spatially distributed sound source. Other intensity-based methods have been presented by Günel *et al.* (2008), where the directional statistics of the intensity vector are used as spatial filters to separate directional components in the sound scene.

A more traditional array processing view on estimation of the directional and diffuse components of the sound scene is taken by other techniques, considering the captured sound scene in each time–frequency tile as the combination of a number of point sources and a uniformly distributed diffuse field. DirAC itself can be seen in that light, under the sound field model of a single plane wave and diffuse sound, in which case the time-averaged intensity vector will point opposite to the DoA of the plane wave, and the diffuseness will be inversely related to the direct-to-diffuse energy ratio. This view is followed, for example, in the work of Thiergart *et al.* (2011). Similar models are considered by Faller (2008) for coincident recording front ends for SAC, by Gallo *et al.* (2007) for capturing and manipulating spatial soundscapes using large-scale distributed omnidirectional arrays, and by Thiergart *et al.* (2013) for scene recording with a movable listening spot. The methods vary on the number of directional components that are extracted, and whether a diffuse component is assumed. The method of Cobos *et al.* (2010) considers only a single plane-wave component per time–frequency tile, with its DoA estimated from the phase differences of a compact four-microphone array and its amplitude taken from one of the microphones. Even this minimal approach is found to perform well for

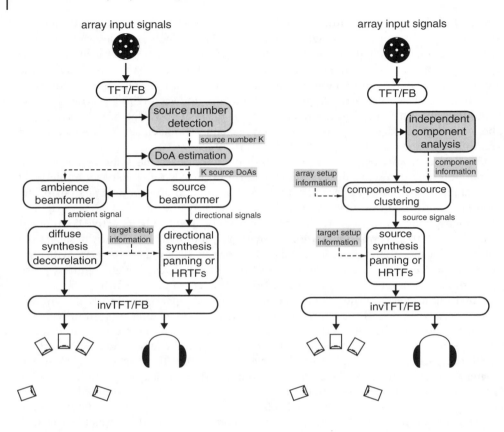

(a) Spatial filtering example.　　　　(b) Source separation example.

Figure 4.3 Basic processing examples for reproduction based on spatial filtering and source separation. Source: Politis 2016. Reproduced with permission of Archontis Politis.

binaural reproduction. That work is extended with a diffuse sound model and applied to reproduction on large WFS systems (Cobos *et al.*, 2012). The HARPEX method assumes two active plane waves per time–frequency tile, and solves their directions and amplitudes by a matrix decomposition of the sub-band B-format signals (Berge and Barrett, 2010). Extraction of two plane waves from B-format is also studied and augmented with a diffuse component by Thiergart and Habets (2012). Extraction of multiple directional sources using adaptive superdirective beamformers has been conducted in Alexandridis *et al.* (2013), by Beracoechea *et al.* (2006), and by Hur *et al.* (2011). A general parametric spatial filtering framework for the estimation of a number of plane waves of interest plus diffuse sound, suitable for both enhancement and sound reproduction, is presented by Thiergart *et al.* (2014), detailed in Chapter 7, and illustrated in Figure 4.3(a).

An alternative to the spatial filtering approach is the BSS approach to sound scene decomposition, up-mixing, and reproduction. Even though most work has focused on up-mixing with channel-based content, there have been a few studies considering multichannel recordings. Two such examples are the work by Niwa *et al.* (2008) using independent component analysis (ICA), and by Nikunen and Virtanen (2014) using NMF. In a sense, BSS seems suitable for the decomposition task, since it aims to minimize statistical dependence between source components, which would most likely correspond to different physical sources. However, recovering source signals of high quality is still a difficult task. Chapter 9 presents approaches to these problems and shows effective solutions when information from the array setup is used in order to cluster the separated components into sensible source signals. The basic principles of these approaches are shown in Figure 4.3(b).

Some recent approaches aim to directly estimate the spatial statistics of the sound scene, without the inclusion of a sound-field model, which can then be used to distribute the recordings appropriately to the loudspeakers. One such example is the spatial covariance matrix method of Hagiwara *et al.* (2013), which requires, however, apart from the scene recording, an additional reference recording at the reproduction setup in order to match the statistics of the two adaptively. A more efficient approach, presented in detail in Chapter 8, is where both reference signals and ideal signal statistics for the target setup are approximated by beamforming operations, and these signals are further enhanced by the adaptive mixing solution detailed in Section 1.3.3. The technique has been developed for either loudspeakers (Vilkamo and Delikaris-Manias, 2015) or headphones (Delikaris-Manias *et al.*, 2015). This type of parametrization accounts for multiple incoherent or mutually dependent plane waves, providing fundamental robustness for varying types of sound fields. Finally, a method based on sparse plane-wave decomposition, employing a compressed sensing solution, is presented by Wabnitz *et al.* (2011); it is used primarily to upscale from a low-order representation of the sound scene, such as a B-format recording, to a higher-order one, such as a high-order Ambisonics recording. However, the method, which is suitable for estimating multiple directional components in a single time–frequency tile, can be used for various sound reproduction and enhancement tasks. An overview of these compressed sensing approaches is given in Chapter 3.

4.2.5 Auralization of Measured Room Acoustics and Spatial Rendering of Room Impulse Responses

Auralization of measured impulse responses (IRs) refers to the method of recording the acoustic IR of an enclosure of acoustical interest, and then listening to the result of its convolution with a test signal. This technique is of major interest in the study of auditorium acoustics and their perceptual effects (Vorländer, 2007; Lokki *et al.*, 2012; Kuusinen and Lokki, 2015). Apart from temporal and spectral effects, the complex spatial effects of the auralized space on sound sources should be preserved as much as possible. Appropriately distributing the captured room impulse response (RIR) from a set of microphones to loudspeakers or headphones is referred to here as rendering of spatial impulse responses (SIRs).

Auralization can be performed in a non-parametric way, by distributing the captured SIR to loudspeakers through some means of inversion or decoding matrix, for example using the ambisonic approach (Gerzon, 1975b). However, RIRs exhibit a clear structure that can be exploited to perform estimation of its components, suitable for a parametric approach to analysis/synthesis of SIRs. For example, RIRs are usually separated into three parts: (i) an early part consisting of the direct sound and early reflections, separated in time; (ii) a denser middle part of reflections with progressively higher echo density; and (iii) a late reverberant part with diffuse behavior. Even though some of the techniques mentioned in the previous sections on parametric recording and reproduction can be applied without modification to recorded RIRs, better results should be expected if this specific structure or parametrization is taken into account.

There is a wealth of literature on localization of early reflections and visualization of RIRs using, for example, cross-correlation between microphones (Yamasaki and Itow, 1989; Tervo *et al.*, 2011, 2013), sound intensity (Yamasaki and Itow, 1989; Merimaa *et al.*, 2001), plane-wave decomposition beamforming (Rafaely *et al.*, 2007; Donovan *et al.*, 2008), subspace estimation methods (Sun *et al.*, 2012; Huleihel and Rafaely, 2013; Tervo and Politis, 2015; Bianchi *et al.*, 2015), and compressed sensing (Mignot *et al.*, 2013). However, not many methods have approached the reproduction task yet. Two established ones are spatial impulse response rendering (SIRR; Merimaa and Pulkki, 2005; Pulkki and Merimaa, 2006) and the spatial decomposition method (SDM; Tervo *et al.*, 2013). The former is the precursor of DirAC and follows the same principles of spatializing the omnidirectional RIR from the B-format recordings, based on the direction of the intensity vector and the diffuseness in frequency bands; it has been used in a variety of auralization tasks (Pätynen and Lokki, 2011; Lokki *et al.*, 2012; Laird *et al.*, 2014). The more recent SDM extracts directional information for a single reflection at each time frame from arbitrary array geometries, assuming a broadband reflection model. An omnidirectional signal is then spatialized according to the estimated DoA. This method omits the frequency dependency of SIRR and the diffuse estimation, and uses a much higher temporal resolution; it has been used in a series of studies on SIR visualization and auralization (Pätynen *et al.*, 2013; Tervo *et al.*, 2014; Kuusinen and Lokki, 2015; Tahvanainen *et al.*, 2015). Binaural rendering of SIRs is possible by either SIRR or SDM. An alternative method for generating binaural RIRs from B-format signals, close in spirit to SIRR, has been proposed by Menzer and Faller (2008). Finally, some recent proposals focus solely on the spatial reproduction of the diffuse reverberation, for loudspeakers (Oksanen *et al.*, 2013; Romblom *et al.*, 2016) or headphones (Menzer and Faller, 2010).

References

Ahonen, J., Del Galdo, G., Kuech, F., and Pulkki, V. (2012) Directional analysis with microphone array mounted on rigid cylinder for directional audio coding. *Journal of the Audio Engineering Society*, **60**(5), 311–324.

Ahrens, J. (2012) *Analytic Methods of Sound Field Synthesis*. Springer, New York.

Alexandridis, A., Griffin, A., and Mouchtaris, A. (2013) Capturing and reproducing spatial audio based on a circular microphone array. *Journal of Electrical and Computer Engineering*, **2013**, 1–16.

Avendano, C. and Jot, J.M. (2004) A frequency-domain approach to multichannel upmix. *Journal of the Audio Engineering Society*, **52**(7/8), 740–749.

Avni, A., Ahrens, J., Geier, M., Spors, S., Wierstorf, H., and Rafaely, B. (2013) Spatial perception of sound fields recorded by spherical microphone arrays with varying spatial resolution. *The Journal of the Acoustical Society of America*, **133**(5), 2711–2721.

Baumgarte, F. and Faller, C. (2003) Binaural cue coding – part I: Psychoacoustic fundamentals and design principles. *IEEE Transactions on Speech and Audio Processing*, **11**(6), 509–519.

Benesty, J. and Huang, Y. (2013) *Adaptive Signal Processing: Applications to Real-World Problems*. Springer Science & Business Media, New York.

Beracoechea, J.A., Casajus, J., García, L., Ortiz, L., and Torres-Guijarro, S. (2006) Implementation of immersive audio applications using robust adaptive beamforming and wave field synthesis. *120th Convention of the AES*, Paris, France.

Berge, S. and Barrett, N. (2010) High angular resolution planewave expansion. *2nd International Symposium on Ambisonics and Spherical Acoustics*, Paris, France.

Berkhout, A.J., de Vries, D., and Vogel, P. (1993) Acoustic control by wave field synthesis. *Journal of the Acoustical Society of America*, **93**(5), 2764–2778.

Bianchi, L., Verdi, M., Antonacci, F., Sarti, A., and Tubaro, S. (2015) High resolution imaging of acoustic reflections with spherical microphone arrays. *2015 IEEE Workshop on Applications of Signal Processing to Audio and Acoustics (WASPAA)*, pp. 1–5. IEEE.

Boashash, B. (2015) *Time–Frequency Signal Analysis and Processing: A Comprehensive Reference*. Academic Press, New York.

Brandstein, M. and Ward, D. (2013) *Microphone Arrays: Signal Processing Techniques and Applications*. Springer, New York.

Breebaart, J., Disch, S., Faller, C., *et al.* (2005a) MPEG spatial audio coding/MPEG surround: Overview and current status. *119th Convention of the AES*, New York.

Breebaart, J. and Faller, C. (2007) *Spatial Audio Processing: MPEG Surround and Other Applications*. John Wiley & Sons Ltd, Chichester.

Breebaart, J., Herre, J., Jin, C., Kjörling, K., Koppens, J., Plogsties, J., and Villemoes, L. (2006) Multi-channel goes mobile: MPEG surround binaural rendering. *29th International Conference of the AES: Audio for Mobile and Handheld Devices*, Seoul, Korea.

Breebaart, J., van de Par, S., Kohlrausch, A., and Schuijers, E. (2005b) Parametric coding of stereo audio. *EURASIP Journal on Applied Signal Processing*, **2005**, 1305–1322.

Briand, M., Martin, N., and Virette, D. (2006) Parametric representation of multichannel audio based on principal component analysis. *120th Convention of the AES*, Paris, France.

Chen, J., Van Veen, B.D., and Hecox, K.E. (1992) External ear transfer function modeling: A beamforming approach. *The Journal of the Acoustical Society of America*, **92**(4), 1933–1944.

Cobos, M., Lopez, J.J., and Spors, S. (2010) A sparsity-based approach to 3D binaural sound synthesis using time–frequency array processing. *EURASIP Journal on Advances in Signal Processing*, **2010**(1), 1–13.

Cobos, M., Spors, S., Ahrens, J., and Lopez, J.J. (2012) On the use of small microphone arrays for wave field synthesis auralization. *45th International Conference of the AES: Applications of Time–Frequency Processing in Audio*, Helsinki, Finland.

Delikaris-Manias, S., Vilkamo, J., and Pulkki, V. (2015) Parametric binaural rendering utilizing compact microphone arrays. *2015 IEEE International Conference on Acoustics, Speech and Signal Processing (ICASSP)*, pp. 629–633. IEEE.

Donovan, A.O., Duraiswami, R., and Zotkin, D. (2008) Imaging concert hall acoustics using visual and audio cameras. *IEEE International Conference on Acoustics, Speech, and Signal Processing (ICASSP)*, Las Vegas, NV.

Driessen, P.F. and Li, Y. (2005) An unsupervised adaptive filtering approach of 2-to-5 channel upmix. *119th Convention of the AES*, New York.

Faller, C. (2006a) Multiple-loudspeaker playback of stereo signals. *Journal of the Audio Engineering Society*, **54**(11), 1051–1064.

Faller, C. (2006b) Parametric multichannel audio coding: Synthesis of coherence cues. *IEEE Transactions on Audio, Speech, and Language Processing*, **14**(1), 299–310.

Faller, C. (2008) Microphone front-ends for spatial audio coders. *125th Convention of the AES*, San Francisco, CA.

Faller, C. and Breebaart, J. (2011) Binaural reproduction of stereo signals using upmixing and diffuse rendering. *131st Convention of the AES*, New York.

Faller, C. and Merimaa, J. (2004) Source localization in complex listening situations: Selection of binaural cues based on interaural coherence. *The Journal of the Acoustical Society of America*, **116**(5), 3075–3089.

Farina, A., Amendola, A., Chiesi, L., Capra, A., and Campanini, S. (2013) Spatial PCM sampling: A new method for sound recording and playback. *52nd International Conference of the AES: Sound Field Control*, Guildford, UK.

Gallo, E., Tsingos, N., and Lemaitre, G. (2007) 3D-audio matting, postediting, and rerendering from field recordings. *EURASIP Journal on Advances in Signal Processing*, **2007**(1), 1–16.

Gerzon, M.A. (1973) Periphony: With-height sound reproduction. *Journal of the Audio Engineering Society*, **21**(1), 2–10.

Gerzon, M.A. (1975a) The design of precisely coincident microphone arrays for stereo and surround sound. *50th Convention of the AES*, London, UK.

Gerzon, M.A. (1975b) Recording concert hall acoustics for posterity. *Journal of the Audio Engineering Society*, **23**(7), 569–571.

Goodwin, M. and Jot, J.M. (2007) Binaural 3-D audio rendering based on spatial audio scene coding. *123rd Convention of the AES*, New York.

Goodwin, M. and Jot, J.M. (2008) Spatial audio scene coding. *125th Convention of the AES*, San Francisco, CA, USA.

Günel, B., Hacihabiboglu, H., and Kondoz, A.M. (2008) Acoustic source separation of convolutive mixtures based on intensity vector statistics. *IEEE Transactions on Audio, Speech, and Language Processing*, **16**(4), 748–756.

Hagiwara, H., Takahashi, Y., and Miyoshi, K. (2013) Wave front reconstruction using the spatial covariance matrices method. *Journal of the Audio Engineering Society*, **60**(12), 1038–1050.

Hawksford, M.J. and Harris, N. (2002) Diffuse signal processing and acoustic source characterization for applications in synthetic loudspeaker arrays. *Audio Engineering Society Convention 112*. Audio Engineering Society.

He, J., Tan, E.L., and Gan, W.S. (2014) Linear estimation based primary-ambient extraction for stereo audio signals. *IEEE/ACM Transactions on Audio, Speech, and Language Processing*, **22**(2), 505–517.

Herre, J., Hilpert, J., Kuntz, A., and Plogsties, J. (2015) MPEG-H Audio – The new standard for universal spatial/3D audio coding. *Journal of the Audio Engineering Society*, **62**(12), 821–830.

Herre, J., Kjörling, K., Breebaart, J., *et al.* (2008) MPEG Surround: The ISO/MPEG standard for efficient and compatible multichannel audio coding. *Journal of the Audio Engineering Society*, **56**(11), 932–955.

Huleihel, N. and Rafaely, B. (2013) Spherical array processing for acoustic analysis using room impulse responses and time-domain smoothing. *The Journal of the Acoustical Society of America*, **133**(6), 3995–4007.

Hur, Y., Abel, J.S., Park, Y.-c., and Youn, D.H. (2011) Techniques for synthetic reconfiguration of microphone arrays. *Journal of the Audio Engineering Society*, **59**(6), 404–418.

Hur, Y., Abel, J.S., Park, Y.-c., and Youn, D.H. (2014) A bank of beamformers implementing a constant-amplitude panning law. *The Journal of the Acoustical Society of America*, **136**(3), EL212–EL217.

Hurtado-Huyssen, A. and Polack, J.D. (2005) Acoustic intensity in multichannel rendering systems. *119th Convention of the AES*, New York.

Irwan, R. and Aarts, R.M. (2002) Two-to-five channel sound processing. *Journal of the Audio Engineering Society*, **50**(11), 914–926.

Jacobsen, F. (2007) Sound intensity, in *Springer Handbook of Acoustics* (ed. Rossing TD), chapter 25, pp. 1053–1075. Springer, New York.

Kuusinen, A. and Lokki, T. (2015) Investigation of auditory distance perception and preferences in concert halls by using virtual acoustics. *The Journal of the Acoustical Society of America*, **138**(5), 3148–3159.

Laird, I., Murphy, D., and Chapman, P. (2014) Comparison of spatial audio techniques for use in stage acoustic laboratory experiments. *EAA Joint Symposium on Auralization and Ambisonics*, Berlin, Germany.

Laitinen, M.V., Kuech, F., Disch, S., and Pulkki, V. (2011) Reproducing applause-type signals with directional audio coding. *Journal of the Audio Engineering Society*, **59**(1/2), 29–43.

Laitinen, M.V. and Pulkki, V. (2009) Binaural reproduction for directional audio coding. *IEEE Workshop on Applications of Signal Processing to Audio and Acoustics (WASPAA)*, New Paltz, NY, USA.

Lokki, T., Pätynen, J., Kuusinen, A., and Tervo, S. (2012) Disentangling preference ratings of concert hall acoustics using subjective sensory profiles. *The Journal of the Acoustical Society of America*, **132**(5), 3148–3161.

Menzer, F. and Faller, C. (2008) Obtaining binaural room impulse responses from B-format impulse responses. *125th Convention of the AES*, San Francisco, CA, USA.

Menzer, F. and Faller, C. (2010) Investigations on an early-reflection-free model for BRIRs. *Journal of the Audio Engineering Society*, **58**(9), 709–723.

Merimaa, J. (2007) Energetic sound field analysis of stereo and multichannel loudspeaker reproduction. *123rd Convention of the AES*, New York.

Merimaa, J., Goodwin, M.M., and Jot, J.M. (2007) Correlation-based ambience extraction from stereo recordings, *123rd Convention of the AES*, New York.

Merimaa, J., Lokki, T., Peltonen, T., and Karjalainen, M. (2001) Measurements, analysis, and visualization of directional room responses. *111th Convention of the AES*, New York.

Merimaa, J. and Pulkki, V. (2005) Spatial impulse response rendering I: Analysis and synthesis. *Journal of the Audio Engineering Society*, **53**(12), 1115–1127.

Mignot, R., Daudet, L., and Ollivier, F. (2013) Room reverberation reconstruction: Interpolation of the early part using compressed sensing. *IEEE Transactions on Audio, Speech, and Language Processing*, **21**(11), 2301–2312.

Moreau, S., Bertet, S., and Daniel, J. (2006) 3D sound field recording with higher order Ambisonics – objective measurements and validation of spherical microphone. *120th Convention of the AES*, Paris, France.

Nikunen, J. and Virtanen, T. (2014) Direction of arrival based spatial covariance model for blind sound source separation. *IEEE/ACM Transactions on Audio, Speech, and Language Processing*, **22**(3), 727–739.

Nikunen, J., Virtanen, T., and Vilermo, M. (2011) Multichannel audio upmixing based on non-negative tensor factorization representation. *IEEE Workshop on Applications of Signal Processing to Audio and Acoustics (WASPAA)*, New Paltz, NY, USA.

Niwa, K., Nishino, T., and Takeda, K. (2008) Encoding large array signals into a 3D sound field representation for selective listening point audio based on blind source separation. *IEEE International Conference on Acoustics, Speech, and Signal Processing (ICASSP)*, Las Vegas, NV, USA.

Oksanen, S., Parker, J., Politis, A., and Valimaki, V. (2013) A directional diffuse reverberation model for excavated tunnels in rock. *IEEE International Conference on Acoustics, Speech, and Signal Processing (ICASSP)*, Vancouver, Canada.

Ono, K., Nishiguchi, T., Matsui, K., and Hamasaki, K. (2013) Portable spherical microphone for Super Hi-Vision 22.2 multichannel audio. *135th Convention of the AES*, New York.

Pätynen, J. and Lokki, T. (2011) Evaluation of concert hall auralization with virtual symphony orchestra. *Building Acoustics*, **18**(3/4), 349–366.

Pätynen, J., Tervo, S., and Lokki, T. (2013) Analysis of concert hall acoustics via visualizations of time–frequency and spatiotemporal responses. *The Journal of the Acoustical Society of America*, **133**(2), 842–857.

Plogsties, J., Minnaar, P., Olesen, S.K., Christensen, F., and Møller, H. (2000) Audibility of all-pass components in head-related transfer functions. *Audio Engineering Society Convention 108*. Audio Engineering Society.

Poletti, M.A. (2005) Three-dimensional surround sound systems based on spherical harmonics. *Journal of the Audio Engineering Society*, **53**(11), 1004–1025.

Politis, A. (2016) Microphone array processing for parametric spatial audio techniques. Doctoral dissertation, School of Electrical Engineering, Department of Signal Processing and Acoustics, Espoo, Finland.

Politis, A., Vilkamo, J., and Pulkki, V. (2015) Sector-based parametric sound field reproduction in the spherical harmonic domain. *IEEE Journal of Selected Topics in Signal Processing*, **9**(5), 852–866.

Pulkki, V. (2007) Spatial sound reproduction with Directional Audio Coding. *Journal of the Audio Engineering Society*, **55**(6), 503–516.

Pulkki, V. and Merimaa, J. (2006) Spatial impulse response rendering II: Reproduction of diffuse sound and listening tests. *Journal of the Audio Engineering Society*, **54**(1/2), 3–20.

Rafaely, B., Balmages, I., and Eger, L. (2007) High-resolution plane-wave decomposition in an auditorium using a dual-radius scanning spherical microphone array. *The Journal of the Acoustical Society of America*, **122**(5), 2661–2668.

Rickard, S. and Yilmaz, Ö. (2002) On the approximate W-disjoint orthogonality of speech. *2002 IEEE International Conference on Acoustics, Speech, and Signal Processing (ICASSP)*, vol. 1, pp. I–529. IEEE.

Romblom, D., Depalle, P., Guastavino, C., and King, R. (2016) Diffuse field modeling using physically inspired decorrelation filters and B-format microphones: Part I. Algorithm. *Journal of the Audio Engineering Society*, **64**(4), 177–193.

Sakamoto, S., Kodama, J., Hongo, S., Okamoto, T., Iwaya, Y., and Suzuki, Y. (2010) A 3D sound-space recording system using spherical microphone array with 252ch microphones. *20th International Congress on Acoustics (ICA)*, Sydney, Australia.

Short, K.M., Garcia, R.A., and Daniels, M.L. (2007) Multichannel audio processing using a unified-domain representation. *Journal of the Audio Engineering Society*, **55**(3), 156–165.

Solvang, A. (2008) Spectral impairment of two-dimensional higher order Ambisonics. *Journal of the Audio Engineering Society*, **56**(4), 267–279.

Stitt, P., Bertet, S., and van Walstijn, M. (2014) Off-centre localisation performance of Ambisonics and HOA for large and small loudspeaker array radii. *Acta Acustica united with Acustica*, **100**(5), 937–944.

Sun, H., Mabande, E., Kowalczyk, K., and Kellermann, W. (2012) Localization of distinct reflections in rooms using spherical microphone array eigenbeam processing. *The Journal of the Acoustical Society of America*, **131**(4), 2828–2840.

Tahvanainen, H., Pätynen, J., and Lokki, T. (2015) Studies on the perception of bass in four concert halls. *Psychomusicology: Music, Mind, and Brain*, **25**(3), 294–305.

Tervo, S., Korhonen, T., and Lokki, T. (2011) Estimation of reflections from impulse responses. *Building Acoustics*, **18**(1–2), 159–174.

Tervo, S., Laukkanen, P., Pätynen, J., and Lokki, T. (2014) Preferences of critical listening environments among sound engineers. *Journal of the Audio Engineering Society*, **62**(5), 300–314.

Tervo, S., Pätynen, J., Kuusinen, A., and Lokki, T. (2013) Spatial decomposition method for room impulse responses. *Journal of the Audio Engineering Society*, **61**(1/2), 17–28.

Tervo, S. and Politis, A. (2015) Direction of arrival estimation of reflections from room impulse responses using a spherical microphone array. *IEEE/ACM Transactions on Audio, Speech and Language Processing (TASLP)*, **23**(10), 1539–1551.

Teutsch, H. (2007) *Modal Array Signal Processing: Principles and Applications of Acoustic Wavefield Decomposition*. Springer, Berlin.

Theile, G. (2001) Natural 5.1 music recording based on psychoacoustic principles. *19th International Conference of the AES: Surround Sound – Techniques, Technology, and Perception*, Bavaria, Germany.

Thiergart, O., Del Galdo, G., Taseska, M., and Habets, E.A.P. (2013) Geometry-based spatial sound acquisition using distributed microphone arrays. *IEEE Transactions on Audio, Speech, and Language Processing*, **21**(12), 2583–2594.

Thiergart, O. and Habets, E.A.P. (2012) Robust direction-of-arrival estimation of two simultaneous plane waves from a B-format signal. *IEEE 27th Convention of Electrical and Electronics Engineers in Israel (IEEEI)*, Eilat, Israel.

Thiergart, O., Kallinger, M., Del Galdo, G., and Kuech, F. (2011) Parametric spatial sound processing using linear microphone arrays, in *Microelectronic Systems: Circuits, Systems and Applications* (ed. Heuberger, A., Elst, G., and Hanke, R.), chapter 30, pp. 321–329. Springer, Berlin.

Thiergart, O., Taseska, M., and Habets, E.A.P. (2014) An informed parametric spatial filter based on instantaneous direction-of-arrival estimates. *IEEE/ACM Transactions on Audio, Speech, and Language Processing*, **22**(12), 2182–2196.

Thompson, J., Smith, B., Warner, A., and Jot, J.M. (2012) Direct–diffuse decomposition of multichannel signals using a system of pairwise correlations. *133rd Convention of the AES*, San Francisco, CA, USA.

Verron, C., Gauthier, P.A., Langlois, J., and Guastavino, C. (2013) Spectral and spatial multichannel analysis/synthesis of interior aircraft sounds. *IEEE Transactions on Audio, Speech, and Language Processing*, **21**(7), 1317–1329.

Vilkamo, J. and Delikaris-Manias, S. (2015) Perceptual reproduction of spatial sound using loudspeaker-signal-domain parametrization. *IEEE/ACM Transactions on Audio, Speech, and Language Processing*, **23**(10), 1660–1669.

Vilkamo, J., Lokki, T., and Pulkki, V. (2009) Directional audio coding: Virtual microphone-based synthesis and subjective evaluation. *Journal of the Audio Engineering Society*, **57**(9), 709–724.

Vorländer, M. (2007) *Auralization: Fundamentals of Acoustics, Modelling, Simulation, Algorithms and Acoustic Virtual Reality*. Springer, Berlin.

Wabnitz, A., Epain, N., McEwan, A., and Jin, C. (2011) Upscaling ambisonic sound scenes using compressed sensing techniques. *IEEE Workshop on Applications of Signal Processing to Audio and Acoustics (WASPAA)*, New Paltz, NY, USA.

Walther, A. and Faller, C. (2011) Direct–ambient decomposition and upmix of surround signals. *IEEE Workshop on Applications of Signal Processing to Audio and Acoustics (WASPAA)*, New Paltz, NY, USA.

Williams, M. (2004) Multichannel sound recording using 3, 4 and 5 channels arrays for front sound stage coverage. *117th Convention of the AES*, San Francisco, CA, USA.

Yamasaki, Y. and Itow, T. (1989) Measurement of spatial information in sound fields by closely located four point microphone method. *Journal of the Acoustical Society of Japan*, **10**(2), 101–110.

Yang, L. and Xie, B. (2014) Subjective evaluation on the timbre of horizontal Ambisonics reproduction. *International Conference on Audio, Language and Image Processing (ICALIP)*, Shanghai, China.

Zotter, F. and Frank, M. (2012) All-round ambisonic panning and decoding. *Journal of the Audio Engineering Society*, **60**(10), 807–820.

5

First-Order Directional Audio Coding (DirAC)

Ville Pulkki,[1] Archontis Politis,[1] Mikko-Ville Laitinen,[2]
Juha Vilkamo,[2] and Jukka Ahonen[3]

[1] *Department of Signal Processing and Acoustics, Aalto University, Finland*
[2] *Nokia Technologies, Finland*
[3] *Akukon Ltd, Finland*

Directional audio coding (DirAC; Pulkki, 2007) is a non-linear time–frequency domain method to reproduce spatial sound. The sound field affects the processing instantaneously in auditory frequency bands, enhancing the quality of reproduction when compared to traditional time domain methods. The method has been developed by starting from assumptions about the spatial resolution of human perception, and designing the reproduction system to follow these assumptions using multichannel microphone signals that have first-order spherical harmonics as their directional patterns.

The name of the system, "directional audio coding," includes the word "coding," although the technique is not primarily a method to compress the data rate. The motivation of the name comes from the processing principle of DirAC. The directional properties of the sound field are analyzed, and then used instantaneously in the synthesis of sound. The property values are thus encoded into a parametric stream with human time–frequency resolution, and later decoded into the reproduced sound field. This is done primarily to enhance the quality of spatial sound reproduction, and only in some applications is the system designed to compress the data rate.

This chapter will describe the basic principles, and it will discuss how the system has been implemented for different applications, and how it can be optimized for critical spatial sound scenarios. Finally, Matlab code is published that shows the implementational details in a typical application.

5.1 Representing Spatial Sound with First-Order B-Format Signals

The first-order B-format microphone signal is often the starting point for DirAC processing, as it allows easy estimation of the directional parameters needed in DirAC, namely the direction of arrival (DOA) of sound and diffuseness ψ of the sound field in the time–frequency domain. DirAC is not, though, exclusively restricted to the use

Parametric Time–Frequency Domain Spatial Audio, First Edition. Edited by Ville Pulkki,
Symeon Delikaris-Manias, and Archontis Politis.
© 2018 John Wiley & Sons Ltd. Published 2018 by John Wiley & Sons Ltd.
Companion Website: www.wiley.com/go/pulkki/parametrictime-frequency

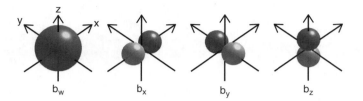

Figure 5.1 Directional patterns of the four coincident microphones of a B-format microphone. The lighter patterns indicate phase-inversed response. b_w has not been scaled by $1/\sqrt{2}$ for this illustration.

of a B-format microphone signal, as many other microphone arrays can be used for the same task. The first-order B-format signal contains a pressure microphone signal $b_w(n)$ and three pressure-gradient microphone signals $b_x(n)$, $b_y(n)$, and $b_z(n)$, where n denotes the discrete time index. The signals have ideally been captured in the same spatial position, and they are directionally orthogonal, as shown in Figure 5.1. It is a convention that the output signals of the pressure-gradient microphones are scaled in such a way that $b_w(n) = b_x(n)/\sqrt{2}$ for a plane wave arriving from the direction of the positive x-axis in free field, and correspondingly for $b_y(n)$ and $b_z(n)$ with their positive axes. Ideally, a pressure microphone has omnidirectional directional characteristics, and pressure-gradient microphones have dipole characteristics, independent of frequency.

Let us first formulate the physical parameters of sound pressure, $p(n)$, and sound particle velocity, $\mathbf{u}(n)$, from the B-format signals. The sound pressure can be computed as

$$p(n) = (1/s)b_w(n), \tag{5.1}$$

where s represents the sensitivity of the microphone in conversion from instantaneous pressure to voltage. In practice, s does not affect the outcome of the analysis of DOA and ψ, and thus it is often conveniently assumed that $s = 1$. By assuming a plane wave, the particle velocity can be estimated as

$$\mathbf{u}(n) = -\frac{1}{s\rho_0 c\sqrt{2}} \begin{bmatrix} b_x(n) \\ b_y(n) \\ b_z(n) \end{bmatrix}, \tag{5.2}$$

where ρ_0 is the mean density of air and c is the speed of sound.

In practice, in the context of DirAC, the time domain formulation is needed if the time–frequency transform is implemented with a time domain filter bank and thus real-valued narrow-band discrete time domain signals are to be analyzed. In most cases DirAC is implemented with transforms that utilize windowing or downsampling, where the signals to be analyzed are complex-valued samples and the pair (k, n) specifies the discrete frequency index k and discrete time index n associated with each sample. The use of complex numbers makes the equations slightly different from their counterparts in the time domain, and both versions are given in such cases. In other cases, the indices are dropped for simplicity.

From p and $\mathbf{u}$, the sound field intensity vector $\mathbf{i}$ (Fahy, 2002) can be obtained in the discrete time domain and the short-time frequency domain as

$$\mathbf{i}(n) = p(n)\mathbf{u}(n), \tag{5.3}$$

$$\mathbf{i}(k, n) = \mathrm{Re}[p^*(k, n)\,\mathbf{u}(k, n)]. \tag{5.4}$$

The operator * denotes the complex conjugate, and Re returns the real component of a complex number. The energy density e is computed as in (Stanzial *et al.*, 1996):

$$e = \frac{\rho_0}{2}||\mathbf{u}||^2 + \frac{|p|^2}{2\rho_0 c^2} \tag{5.5}$$

(the formulation is the same for the time and time–frequency domains); $||\cdot||$ denotes the vector norm operator, and $|\cdot|$ denotes the absolute value operator.

The intensity variable $\mathbf{i}$ expresses the direction and the magnitude of the flow of sound energy. The parameters in Equations (5.4) and (5.5) can be applied to estimate the most prominent DOA as follows:

$$\text{DOA} = \angle\,\text{E}[-\mathbf{i}], \tag{5.6}$$

where the operator $\angle$ gives the 3D angle of a vector, and E[ ] is the expectation operator, which can be implemented as a short-time temporal averaging, or with a suitable time window. In general, in DOA analysis the temporal constants and time window lengths are shorter than with ψ analysis. The diffuseness ψ is computed as

$$\psi = 1 - \frac{||\text{E}[\mathbf{i}]||}{c\text{E}[e]} \tag{5.7}$$

(Merimaa and Pulkki, 2005). Also, other conditions than diffuse sound fields, such as two plane waves arriving at the microphone from almost opposite directions with equal or almost equal amplitudes, can generate $\psi \approx 1$. In such a case, the DOA estimate does not necessarily express the organization of the sound scene, but simply the opposite of the net flow direction of the sound energy. Note that in this case the magnitude of the net flow of the energy is low compared to the amount of total energy in the field.

The constants s, ρ_0, and c do not contribute to the estimated DOA in Equation (5.6), and they also cancel out from the formulation of ψ in Equation (5.7) in cases in which the estimation of $p(n)$ and $\mathbf{u}(n)$ takes place, as in Equations (5.1) and (5.2). ψ can be directly computed from a B-format signal as

$$\psi(n) = 1 - \sqrt{2}\frac{\left|\left|\text{E}\left[b_w(n)\begin{pmatrix}b_x(n)\\b_y(n)\\b_z(n)\end{pmatrix}\right]\right|\right|}{\text{E}\left[b_w^2(n) + \frac{b_x^2(n)+b_y^2(n)+b_z^2(n)}{2}\right]}, \text{ or} \tag{5.8}$$

$$\psi(k,n) = 1 - \sqrt{2}\frac{\left|\left|\text{E}\left[\text{Re}\left(b_w^*(k,n)\begin{pmatrix}b_x(k,n)\\b_y(k,n)\\b_z(k,n)\end{pmatrix}\right)\right]\right|\right|}{\text{E}\left[|b_w(k,n)|^2 + \frac{|b_x(k,n)|^2+|b_y(k,n)|^2+|b_z(k,n)|^2}{2}\right]}. \tag{5.9}$$

Equations (5.6), (5.7), and (5.9) are valid only if the directional patterns of the B-format signals are relatively accurate and the sensitivity of the microphones is the same at the frequency band of interest. A reasonable accuracy in analysis is typically obtained in a certain spectral window, the width and the position of which depends on the setup of the microphone array and the characteristics of the microphones. The microphone array has to be designed specifically to obtain the quality needed for a given application.

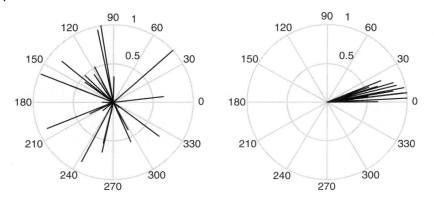

Figure 5.2 Temporally subsequent intensity vectors analyzed on one frequency channel from a real 2D microphone setup in highly diffuse [left] and non-diffuse [right] sound fields.

In some cases, the diffuseness cannot estimated with Equation (5.7) due to some constraints in the microphone array preventing unbiased estimation of the intensity vectors and energy density. Nevertheless, the intensity vector can still often be estimated with some assumptions about the sound field, even though the energy density is not available. Ahonen and Pulkki (2009) presented an alternative method to compute diffuseness, based on statistics of estimates of intensity (or direction) vectors – no knowledge of energy density is needed. When studying the vectors in a diffuse sound field in subsequent short-time windows, it can be seen that the vectors are pointing in random directions at different time instants, as shown in Figure 5.2. The right panel of the same figure shows the case with only a single plane wave, where the vectors of different time instants point in the same direction.

Based on this finding, the value of the diffuseness can be computed by dividing the length of the time-averaged intensity vectors by the averaged length of the same vectors (Ahonen and Pulkki, 2009; Del Galdo *et al.*, 2012):

$$\psi(n) = \sqrt{1 - \frac{||E[\mathbf{i}]||}{E[||\mathbf{i}||]}}. \tag{5.10}$$

This is termed the coefficient of variation method. In a diffuse field, the directional variation of the vectors makes their average tend to zero; however, the average length of the vectors tends to a finite value. In a field with a single plane wave the vectors have constant direction and length, and thus the length of the averaged vector has also a finite value, which equals the average of the lengths of the vectors. The diffuseness ψ is bound between 0 and 1 at each time instant, as $E[||\mathbf{i}||] \geq ||E[\mathbf{i}(n)]|| > 0$.

5.2 Some Notes on the Evolution of the Technique

The first author became interested in methods to capture and reproduce spatial sound over arbitrary loudspeaker layouts in late 1990s during his PhD studies. The first-order Ambisonics developed by Gerzon (1985) was perhaps the only well-defined technique targeting the reproduction of multi-microphone capture of a sound field over different loudspeaker systems while preserving the spatial characteristics. In the system, B-format signals are combined with time-invariant matrix operations in such a way that each loudspeaker receives a signal with a unique first-order directional pattern emphasizing the sound arriving from the direction in the sound field corresponding to the loudspeaker direction. Unfortunately, the audio quality obtained by Ambisonics was not found to be acceptable. The artifacts were pronounced with a high number of loudspeakers, and especially in listening conditions with low reverberation. The main causes of the relatively low audio quality in Ambisonics were found to be timbral artifacts such as "muffled sound" depending on listening position, referred to as "phasiness" in the Ambisonics literature, and studied further in Solvang (2008). Additionally, spatial artifacts manifested as generally vague perception of the directions of sources, and in some cases inside-head localization occurred despite the loudspeaker reproduction.

The main reason for these artifacts was identified to be the high coherence between loudspeaker signals. For example, the signal of a single plane wave will be applied to most loudspeakers, and hence the same signal arrives from multiple directions at the ear of the listener. Since the propagation delay and the acoustic response are different for each loudspeaker to ear canal path, the signals will be added in the ear canals with a phase relationship that depends on frequency. This causes a non-flat frequency response with deep dips at high frequencies, and consequently the perception of the sound is "muffled." This is also known as the "comb-filter effect." Also, this makes directional cues depend on frequency, causing spatial artifacts (Pulkki, 2002). A puzzling question was, then, "could we enhance the quality of Ambisonics reproduction?" While a branch of research targeted the utilization of higher-order B-format signals to achieve lower coherence by making the directional patterns narrower (Gerzon, 1973; Poletti, 2005; Moreau *et al.*, 2006; Zotter and Frank, 2012), methods to enhance the quality obtained with first-order B-format recordings were sought in this project.

The idea to utilize the limitations in the resolution of human directional hearing was considered in summer 2000. The auditory filters of the cochlea and the subsequent neural analysis of spatial cues are strong limiting factors in the perception of the spatial properties of sound. The mechanism of how a spatial auditory image is formed from spatially complex auditory input is not known in detail; an overview of the current knowledge can be found in Santala *et al.* (2015). However, it is known that only when all the spatial cues within a single critical band imply a coherent direction for a continuous human-localizable sound will the corresponding auditory event be well-localized in a single direction (Blauert, 1997). If the spatial cues within a single critical band suggest different directions, a spatially divergent and vague perception of the direction of the sound follows.

The first idea was to analyze the most prominent direction of arrival from B-format signals, and then to reproduce the omnidirectional signal in the analyzed direction with vector base amplitude panning (VBAP; Pulkki, 1997). The motivation was to reduce the coherence between the loudspeaker signals by applying the sound signal within each frequency band only to the most prominent direction. Unfortunately, the first tests in August 2000 failed because of the naïve method of DOA estimation. A solution for DOA analysis emerged later in discussions with Juha Merimaa, who suggested using the opposite direction of the sound intensity vector as an estimate of the DOA, and furthermore to use the diffuseness of the sound field to differentiate between plane waves and diffuse reverberation. This was immediately a much more plausible approach, and it was implemented first for the rendering of room impulse responses to loudspeaker channels, for application in convolution reverberators. In the first publication of the system (Pulkki *et al.*, 2004) at a conference, the formula for diffuseness parameter computation was incomplete, and the results of the processing were not very good. Fortunately, Prof. Angelo Farina noted the flaw, and pointed out the correct formulation (Stanzial *et al.*, 1996) in the conference. After this correction the system performed well with B-format room impulse responses, and the first journal articles were published (Merimaa and Pulkki, 2005), where the technique was called "spatial impulse response rendering" (SIRR).

In parallel with the present work with impulse responses, similar approaches were taken externally in the domain of upmixing and coding of audio, where time–frequency domain analysis of inter-channel parameters was used. The first applications were in the upmixing of stereo tracks to 5.1 surround by Avendano and Jot (2004), and for compression of multichannel audio content by Faller (2006). The publication of the methods and corresponding demonstrations motivated the implementation of a version of SIRR for continuous sound, which had actually been the original idea behind the experiments. Analyzing the parametric data directly from sound field recordings seemed challenging; however, it was known that the metadata in upmixing and compression was not very accurate, though it still produced high-quality results for human listeners. Such tolerance to inaccurate parametric metadata provided the motivation to pursue the development.

The first successful tests were completed in 2005 with the reproduction of continuous sound, and the system was named DirAC (Pulkki, 2007). It was successfully demonstrated at the 120th Audio Engineering Society (AES) Convention and the 28th AES Conference in 2006 for 2D and 3D sound reproduction, and for stereo upmixing. The development of DirAC was continued in collaboration with Fraunhofer IIS in 2007.

5.3 DirAC with Ideal B-Format Signals

The design of the SIRR and DirAC techniques is based on four assumptions about the interaction between sound field properties and the perceptual attributes that they produce (Merimaa and Pulkki, 2005). The assumptions are:

- Direction of arrival of sound transforms into interaural time difference (ITD), interaural level difference (ILD), and monaural localization cues.
- Diffuseness of sound transforms into interaural coherence cues.
- Timbre depends on the monaural (time-dependent) spectrum together with ITD, ILD, and interaural coherence of sound.
- The direction of arrival, diffuseness, and spectrum of sound measured in one position with the temporal and spectral resolution of human hearing determines the auditory spatial image the listener perceives.

According to this, we should then be able to reproduce spatial sound to human listeners with high fidelity, if the system reproduces the diffuseness in frequency bands, and for low-diffuseness cases also the direction of sound, correctly. In basic implementations, this is realized by dividing each frequency band of sound into two *streams* using the diffuseness parameter, as shown in Figure 5.3. The diffuse stream is reproduced as

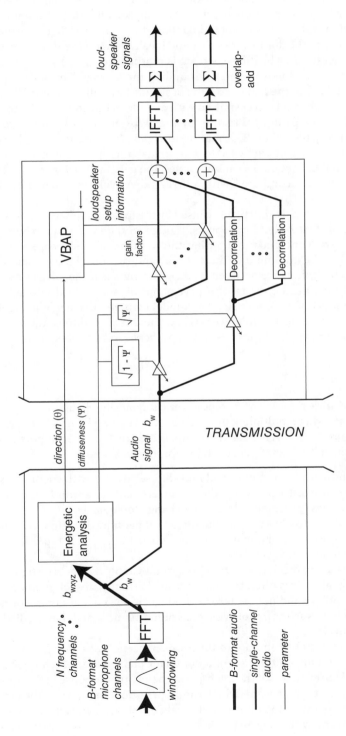

Figure 5.3 Flow diagram of DirAC with single-channel transmission.

evenly surrounding sound, and the non-diffuse (direct) stream is reproduced as point-like virtual sources in directions defined by the analyzed direction of sound.

Ideally, the non-diffuse stream should contain the sound arriving as plane waves at the microphone, and the diffuse stream should ideally contain the sound caused by surrounding reverberations. The diffuse stream is reproduced by applying the sound after decorrelation to all the loudspeakers surrounding the listener. This ideally produces a perfectly diffuse field to the listener. In the basic version, called "mono-DirAC," where only a single audio channel is transmitted with metadata, the non-diffuse stream is produced by steering the panning direction by the DOA analyzed from the corresponding time–frequency tile of the B-format input signal. In "virtual microphone DirAC," the amplitude panning gains are used to gate virtual microphone signals; which will be described in detail in Section 5.6. In both cases, the gain factors for frequency-specific panning directions are computed with VBAP, individually for each time–frequency position.

Both streams thus avoid the artifacts caused by the too-high coherence between loudspeaker signals in first-order Ambisonics, though in different ways. The non-diffuse stream reduces the coherence by limiting the number of loudspeakers producing a coherent signal to the minimum of three for 3D and two for 2D cases, and the diffuse stream reduces the coherence between loudspeaker signals by decorrelation.

The method thus can be thought to have a certain sound field where it performs the best, or a model of a sound field. The model consists of a single plane wave with power P_{dir} and diffuse field with power P_{diff}. If the direct-to-diffuse ratio Γ defined as $\Gamma = P_{\text{dir}}/P_{\text{diff}}$ is known, the diffuseness of sound field can be expressed as

$$\psi = \frac{1}{1 + \frac{P_{\text{dir}}}{P_{\text{diff}}}} = \frac{1}{1 + \Gamma}. \tag{5.11}$$

The analyzed direction and diffuseness values typically change rapidly with time, especially in cases where multiple sources alternate. As such changes directly affect how the sound is reproduced over the audio setup, there are a number of potential sources of the distortion, or "artifacts," perceived in reproduced sound:

- Division of the sound into diffuse and non-diffuse streams with coefficients $\sqrt{\psi}$ and $\sqrt{1 - \psi}$ may cause distortion when the coefficients change abruptly in time. To avoid this, the diffuseness parameter should not change too quickly.
- In the non-diffuse stream the rapid fluctuation of panning direction, and consequent fast changes in gain factors, may also cause distortion. This is typically avoided by temporally averaging the gain factors.
- In the diffuse stream the decorrelation process inevitably changes the temporal structure of the signal. This is often the main cause of artifacts in such methods. To avoid this, such sounds should not be decorrelated where the changes in temporal pattern are clearly audible, such as impulsive signals or harmonic signals having distinct peaks in the time envelope.
- The division into diffuse and non-diffuse streams is not ideal with such simple multiplication by ψ. Only in the case of a dominant plane wave, which typically occurs when a sound onset arrives at the microphone before the response of the room, is all of the sound routed to the non-diffuse stream. Correspondingly, when the direct sound is not present anymore, the diffuse sound field dominates, and all sound is routed to the diffuse stream, without crosstalk in the streams. When both direct sound and the

diffuse field are present at the same time with similar levels in a frequency band, the signal of the plane wave is partly routed to the diffuse stream, and the signal created by diffuse field sound is routed to the non-diffuse stream. In practice, the leakage of non-diffuse sound into the diffuse stream often causes more artifacts than the opposite case. When non-diffuse sound is partly decorrelated, artifacts emerge, such as the sources being perceived a little further away in reproduction than in reality. Also, in some cases an "added room effect" is heard, where, for example, a voice is perceived to be spectrally filtered by a room-like response, although there is no such response in the original sound. This kind of effect can also happen in the case when multiple spectrally overlapping plane waves arrive from different directions at the microphone, causing leakage of plane wave signals to the diffuse stream.

- The microphone patterns are not accurate and the sensitivities of the microphones are not matched. Such flaws typically cause the value of diffuseness to be overestimated, and DOA to be erroneous, which may produce the artifacts mentioned earlier in this list.

To minimize these artifacts, the development of the technique has concentrated on making the enhancements as prominent as possible, but keeping the non-linear artifacts under the threshold of audibility.

The following chapters discuss the overall principles in DirAC systems, and implementation details are kept to a minimum. To give a thorough walkthrough of the processing in detail, the Matlab code of first-order virtual microphone DirAC with single-resolution short-time Fourier transform (STFT) is provided in Section 5.9. In practice, the details of DirAC processing are relatively similar between different implementations, and the code shows a feasible set of implementational details and choices, which can then be further modified for different applications.

5.4 Analysis of Directional Parameters with Real Microphone Setups

An ideal B-format microphone does not exist, unfortunately. However, the B-format signals can be relatively easily estimated from basically any array of a few omnidirectional or directional microphones that are at a distance of 1–10 cm from each other. There are a number of tradeoffs, and the array has to be optimized specifically for different applications to produce directional metadata with sufficient accuracy. This section reviews the results obtained in optimizing such arrays.

5.4.1 DOA Analysis with Open 2D Microphone Arrays

A simple and inexpensive solution for capturing B-format signals for DirAC with reproduction over a horizontal loudspeaker setup is a square array of four omnidirectional miniature microphones, as shown in Figure 5.4 (Kallinger *et al.*, 2009; Ahonen, 2010). The velocity signals are derived as directional pressure gradients from the microphone signals as

$$b_x(n,k) = \sqrt{2} \cdot A(k) \cdot [p_1(n,k) - p_2(n,k)], \qquad (5.12)$$
$$b_y(n,k) = \sqrt{2} \cdot A(k) \cdot [p_3(n,k) - p_4(n,k)],$$

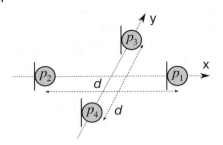

Figure 5.4 Square array of four omnidirectional microphones to derive horizontal B-format signals. Opposing microphones are spaced at a distance d apart, with typical values of d between 1 cm and 10 cm.

where $p_1(n, k)$, $p_2(n, k)$, $p_3(n, k)$, and $p_4(n, k)$ denote the STFT bins of the microphone signals, and the equalization constant

$$A(k) = -j\frac{cM}{2\pi k d f_s} \tag{5.13}$$

compensates for the frequency-dependent attenuation of the desired signals. Here, j is the imaginary unit, M is the number of frequency bins in the STFT, d is the distance between opposing microphones, and f_s is the sampling rate.

The velocity signals b_x and b_y, derived as directional pressure gradients in Equation (5.12), suffer from inadequate sampling in space, depending on the inter-microphone distance. This results in spatial aliasing and a consequent deformation of the directional patterns at high frequencies where the inter-microphone distance is smaller than the half-wavelength of the arriving sound. The spatial aliasing frequency, which gives a theoretical upper frequency limit, is defined as

$$f_{sa} = \frac{c}{2d}. \tag{5.14}$$

On the other hand, an undesired side effect of the equalization factor $A(f)$ in Equation (5.12) is that it also amplifies the self-noise of the microphones with -6 dB per octave frequency characteristics. This results in a noise component that is audible as low-frequency rumbling in velocity signals, and it also causes errors in directional analysis. With larger spacing d, the lower the level of the noise; unfortunately, the spatial aliasing frequency is also lower then. Such microphone arrays thus exhibit a spectral window wherein the derived signals have the desired fidelity in signal-to-noise ratio and in directional patterns.

Additionally, b_w has to be estimated, for example by taking an average over the signals of all four microphones. However, due to the spatially separated microphones the signal summation in the averaging process results in the comb-filter effect at frequencies above the spatial aliasing frequency. It has been found that better quality is obtained when the omnidirectional signal b_w is captured with only one microphone of the square array, which is then used to form the non-diffuse and diffuse streams. The microphone can be either predetermined or chosen adaptively by using the microphone that corresponds to the estimated DOA (Ahonen, 2010).

The performance of such microphone arrays in the analysis of direction performed in DirAC has been evaluated using square arrays with inter-microphone distances of 1 cm, 2 cm, and 4 cm (Ahonen *et al.*, 2012a). Since the microphone noise causes temporal variation in the analyzed DOA and ψ values, a temporal integration setting realistic in DirAC is used here. The values are integrated using a time constant determined as ten times the period time of the center frequency of each frequency channel.

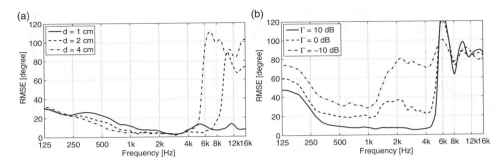

Figure 5.5 (a) RMS error in estimation of direction using square arrays of four omnidirectional microphones, when the distances between opposing microphones are 1 cm (solid line), 2 cm (dashed line), and 4 cm (dash-dotted line). (b) The effect of diffuse sound on directional analysis with a square array with 4 cm inter-microphone distances. Simulations were conducted with different energy ratios Γ between a plane wave and the diffuse sound field. In (a) and (b) the results for root mean square errors are averaged over various analysis directions. Source: Ahonen 2012a. Reproduced with permission of the Audio Engineering Society.

Figure 5.5 shows the results for the direction analysis with the root mean square errors (RMSE) averaged over four directions of plane waves. When using square arrays with inter-microphone distances of 2 cm and 4 cm, the direction is estimated significantly incorrectly starting at frequencies of about 8 kHz and 5 kHz, respectively, which match the theoretical frequency limits for spatial aliasing. Also, a smaller inter-microphone distance results in a less accurate direction estimation at low frequencies because of the poor SNR of the velocity signals.

The effect of the presence of diffuse background noise with three Γ values is shown in Figure 5.5(b). It can be seen that for $\Gamma = 10$ dB the result is relatively similar to the plane-wave field in Figure 5.5(a). At frequencies below 1 kHz and with cases 0 dB and -10 dB, the RMSE increases about 10° and 20°, respectively. Between 1.5 kHz and 4 kHz, the increase of the RMSE is notably higher. Above the aliasing frequency the noise naturally has no influence on the RMSE.

5.4.2 DOA Analysis with 2D Arrays with a Rigid Baffle

As stated, the use of the horizontal B-format signals created using a square microphone array leads to bandwidth-limited directional estimation. To achieve reliable results for a broader frequency band, a rigid object, such as a cylinder, has been proposed to be inserted between the omnidirectional microphones (Ahonen *et al.*, 2012a). Such a microphone array is illustrated in Figure 5.6; it is called a cylinder array.

The inclusion of the cylinder inside the array provides prominent level differences between the microphone signals at high frequencies due to acoustic shadowing. The method utilizes these differences to estimate the direction of arrival of a sound wave, assuming the dominance of a single plane wave. The power spectra of the microphone signals are subtracted from one another, approximating the instantaneous x- and y-axial sound intensities as

$$\tilde{i}_x(n, k) = |p_1(n, k)|^2 - |p_2(n, k)|^2 \text{ and} \tag{5.15}$$

$$\tilde{i}_y(n, k) = |p_3(n, k)|^2 - |p_4(n, k)|^2;$$

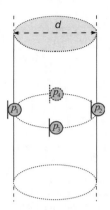

Figure 5.6 Cylinder microphone array. A rigid cylinder of diameter *d* casting an acoustic shadow and causing scattering is positioned between four omnidirectional microphones.

the sound intensity components $\tilde{i}_x$ and $\tilde{i}_y$ are employed to estimate the azimuth DOA. Consequently, the spatial aliasing issues in the analysis of DOA and ψ related to the phases of the microphones are solved with this method.

The estimation of direction based on the energy gradients is, however, incorrect without measurable inter-microphone level differences, and such shadowing effects are not present below a certain frequency, depending on the size of the baffle. Thus, instead of energy gradients, the pressure gradients used in the basic implementation of DirAC are employed at low frequencies, and such a combination is called the pressure–energy gradient (PEG) method. The frequency limit for using either the pressure or energy gradient needs to be equal to or smaller than the spatial aliasing frequency, and sufficient inter-microphone level differences must exist above the frequency limit. In practice, the above-mentioned requirements have been found to be satisfied in practice when the length of the circumference of the cylinder defines the frequency limit as

$$f_{\text{lim}} = \frac{c}{\pi d}. \tag{5.16}$$

For instance, an 8 cm diameter results in 1.3 kHz as the frequency limit, which is smaller than the corresponding spatial aliasing frequency of 2.1 kHz in Equation (5.14).

The results for the direction analysis with the PEG method are shown in Figure 5.7, simulated in the same way as the corresponding analysis in the previous section. The RMSE results reveal that the method provides reliable direction estimation at a wider frequency range than with an open microphone array, as shown in Figure 5.5(a). The use of the energy gradients results in significantly more accurate direction estimation at high frequencies than with the pressure gradients. The inclusion of the rigid cylinder causes shadowing effects between microphones, and the travel path of the sound is also longer as it has to circumvent the cylinder. In other words, the cylinder increases the effective inter-microphone distance and spectral magnitude differences as a function of its diameter. Consequently, a more accurate direction estimation is provided at low frequencies since the effectively larger diameter results in better SNR.

The influence of ambient noise on the error in DOA is shown in Figure 5.7(b). When compared with the results from the square array shown in Figure 5.5(b), it can be seen that diffuse sound degrades the estimated direction very similarly with and without the inclusion of the cylinder in the frequency window where the estimation of DOA is most

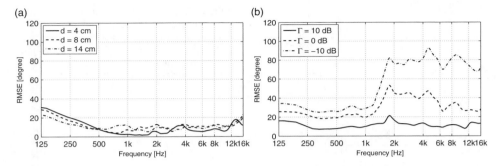

Figure 5.7 RMS error in DOA estimation with the pressure–energy gradient method. (a) Results for cylinder arrays with diameters 4 cm, 8 cm, and 14 cm. A white noise sample is reproduced alternatively at 0°, 15°, 30°, and 45° in an anechoic chamber. (b) The effect of diffuse sound on PEG directional analysis with a cylinder array of diameter 8 cm with $f_{lim} = 1.3$ kHz. The direction estimation is performed with different energy ratios Γ between a plane wave and the diffuse sound. The plane wave arrives at intervals of 5° from 0° to 45°. In (a) and (b), RMSEs are averaged over various directions. Source: Ahonen 2012a. Reproduced with permission of the Audio Engineering Society.

accurate. At frequencies below 250 Hz results with clearly lower RMSE are obtained, and at frequencies above the spatial aliasing frequency the accuracy of estimation is improved with higher values of Γ.

5.4.3 DOA Analysis in Underdetermined Cases

The number of microphones in the array does not often allow the capture of a full 3D B-format signal. If a 3D field is captured with a 2D horizontal B-format signal (b_w, b_x, b_y), the elevation angle of the DOA is not captured. Similarly, if only a 1D B-format signal (b_w, b_x) is available, only the angle between the plane perpendicular to the x-axis and the DOA is revealed. However, with some assumptions about the spatial arrangement of sources, useful analysis results can still be achieved. An important example of such arrays is an application where the housing of a visual display includes two microphones. Such an array provides only 1D B-format signals, and further assumptions are required to estimate the sound direction. In this case it is often assumed that all sources are in front of the display with the same elevation as the display, and with this restriction the analyzed angle then reveals unequivocally the DOA of direct sound in the frontal half-plane.

DOA can be estimated if it is assumed that the sound field has only a single plane wave present. In Figure 5.8, the plane wave arrives with a lateral angle α and the magnitude of the intensity vector $|\mathbf{i}|$. The sound intensity for the x-axis can thus be expressed as $i_x = |\mathbf{i}| \sin(\alpha)$. With a plane wave, $|\mathbf{i}|$ is equal to the power spectrum of the omnidirectional signal $|b_w|^2$, and i_x corresponds to multiplication of the pressure by the x-axis velocity component. Thus, the instantaneous lateral angle $\alpha(k, n)$ can be obtained in the time–frequency domain (Ahonen, 2010) as

$$\alpha(k, n) = \arcsin \left[\mathrm{Re} \left(\frac{b_w^*(k, n) b_x(k, n)}{|b_w(k, n)|^2} \right) \right]. \tag{5.17}$$

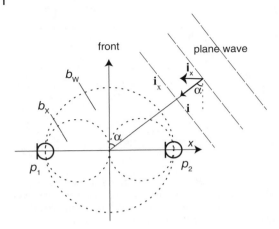

Figure 5.8 A one-dimensional microphone array is used to derive the omnidirectional signal b_w and dipole signal b_x whose directional patterns are shown with dashed lines. A plane wave arrives with lateral angle α. Adapted from Ahonen (2010).

Based on a similar formulation, 3D DOA analysis can be performed with 2D arrays, where the angle between the horizontal plane and the DOA is obtained (Kuech *et al.*, 2008). It is thus ambiguous whether the sound arrives from above or below the plane. However, this can still be useful since in many recording cases the sources are either in the horizontal plane or elevated, for example in typical concert hall or church recordings.

5.4.4 DOA Analysis: Further Methods

A wide variety of methods exists to estimate DOA for narrow-band signals from a microphone array (Benesty *et al.*, 2008; Brandstein and Ward, 2013), and basically any of them is appropriate to be used with DirAC. The requirement for the accuracy of the spatial analysis in DirAC-based sound reproduction is not very demanding, as correct estimation is only needed when the SNR is good, that is, only if a single plane wave dominates. The intensity vector has been used in DirAC since it is computationally efficient, and provides an unbiased estimate of DOA in most cases, and since in a low-SNR case the analyzed DOA is random.

Other techniques have indeed also been used in the context of DirAC. In the case of open arrays, the directional analysis cannot be conducted using the intensity vector above the spatial aliasing frequency, and alternative solutions have been found. However, in many cases the dominant signal is known to have a prominent temporal envelope, such as with speech, where strong level modulations in frequency bands are found. In Kratschmer *et al.* (2012), DOA is estimated by computing cross-correlation as a function of time delay between half-octave-bandpassed pressure-microphone signals. The delay that results in the highest correlation is assumed to correspond to the time delay of the arrival of the wavefront at the microphones, which can be used to compute the value of DOA in the left–right dimension. DOA can also be analyzed over the spatial aliasing frequency if shadowing effects are present in the microphone signals, such as with the cylindrical baffle in the center of the array as in Section 5.4.2, and with omnidirectional microphones with a large housing in an open array (Ahonen and Pulkki, 2011). In Politis *et al.* (2015a), the energetic DOA vector developed for cylinder arrays in Section 5.4.2

was generalized for symmetric 3D arrays of directional or baffled microphones with the focus on estimation of DOA above the aliasing frequency.

In Thiergart *et al.* (2011), more than two microphones are used in a linear array. This allows the measurement of the DOA of single or multiple incoming plane waves using the method called "estimation of signal parameters via rotational invariance techniques" (ESPRIT; Kailath, 1980). Compared to the traditional, intensity-based parameter estimators for linear arrays, the proposed alternatives provide higher estimation accuracy, but at the cost of a higher computational complexity. ESPRIT yields unbiased direction estimates, lower estimation variance for sound events arriving from the broadside, and a higher spatial aliasing frequency compared to the traditional methods. While the intensity-based estimators yield sufficient accuracy for many applications, ESPRIT approaches can improve high-quality applications of parametric spatial audio where computational complexity constraints are less demanding. A comparison of different analysis techniques was conducted by Thiergart (2015) in the context of parametric spatial audio.

5.4.5 Effect of Spatial Aliasing and Microphone Noise on the Analysis of Diffuseness

Any real microphone produces "inner noise" with or without a signal present, caused by Brownian motion of the membrane and by non-ideal electronic components. The effect of microphone inner noise can be characterized easily, if we assume that the inner noise has the same spectrum and level in all microphones, and that the noise could have been caused by a noise signal in the diffuse field with power P_{noise}.

We can then define $\Upsilon = P_{\text{dir}}/P_{\text{noise}}$ to be the signal-to-microphone-noise ratio, and extend Equation (5.11) with it. When both Γ and Υ are known, the diffuseness can then be expressed as

$$\psi = \frac{1}{1 + \dfrac{P_{\text{dir}}}{P_{\text{diff}} + P_{\text{noise}}}} = \frac{1}{1 + \dfrac{1}{1/\Gamma + 1/\Upsilon}}. \tag{5.18}$$

The effect of Υ on ψ is shown in Figure 5.9, where it can be seen that microphone noise biases ψ towards unity, that is, the analyzed diffuseness will be higher than the diffuseness of the sound field if the microphone has prominent inner noise. Note that typically

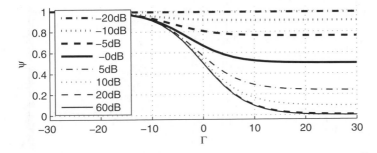

Figure 5.9 ψ as function of the direct-to-diffuse ratio Γ on abscissa, with different direct-to-microphone-noise ratio Υ values as denoted in the legend.

pressure and velocity microphones have different background noise levels, and Υ cannot easily be directly estimated. Nevertheless, as a rule of thumb it can be stated that a higher level of microphone noise causes ψ to be overestimated. The effect of the low-frequency noise typically present in pressure gradient microphones introduces error in the estimation of ψ at lower frequencies, as analyzed by Del Galdo *et al.* (2012), which in practice means that in DirAC low frequencies are routed to the diffuse stream with a higher level than ideally they should.

In open arrays, the directional patterns of **u** signals depart markedly from the first-order dipole at frequencies over the spatial aliasing frequency, and the signals no longer present the velocity components of the sound field. If ψ is computed with Equation (5.7), erroneous values are obtained. In many cases the computed value of diffuseness is biased towards unity. For example, Vilkamo (2008) measured that above the spatial aliasing limit the diffuseness analyzed with a Soundfield ST350 microphone varied linearly between values of about 0.5 and 1.0, depending on the actual diffuseness of the field, whereas at lower frequencies unbiased estimations were obtained.

Above the spatial aliasing frequencies, and in such cases where the pressure signal cannot be estimated accurately, the coefficient of variation method for diffuseness analysis, Equation (5.10), can still be used to compute ψ if a vector can be formed with direction corresponding to the DOA and length corresponding to the instantanous amplitude. The intensity vector can be estimated by shadowing effects, as shown earlier; alternatively, a general direction vector can be estimated by setting its direction to match the DOA analyzed by any method, and the length of the vector to match the level of sound in the corresponding time–frequency position. This approach has been successfully used when estimating the intensity vector by the shadowing effect of a baffle or microphone directional patterns (Politis and Pulkki, 2011; Ahonen *et al.*, 2012a; Politis *et al.*, 2015b), or with methods based on cross-correlation-based estimation of DOA from open microphone arrays (Kratschmer *et al.*, 2012).

In addition, in cases when underdetermined measurements of the 3D velocity field are available, the coefficient of variation method provides much better correspondence at all frequencies with the actual diffuseness of the sound field than energetic analysis does (Del Galdo *et al.*, 2012). This is demonstrated for simulated recordings with a simulated 1D recording of a 3D field. The results obtained with the coefficient of variation method are shown in Figure 5.10(b) as a function of Γ. The notable benefit of the method

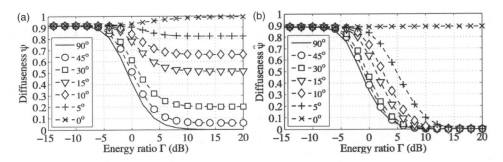

Figure 5.10 Diffuseness estimated with a two-microphone array as a function of the energy ratio Γ between a plane wave from a given azimuth angle and the diffuse field. (a) Results with energetic analysis, Equation (5.7). (b) Results with coefficient of variation analysis. Source: Ahonen 2009. Reproduced with permission of IEEE.

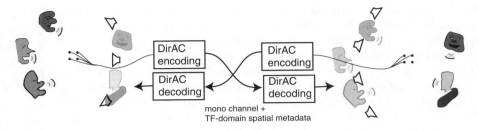

Figure 5.11 Setup for spatial teleconferencing with DirAC.

is clear when compared with the estimation results using the traditional method in Figure 5.10(a). The diffuseness estimation still depends on the arrival angle of the plane wave, but the estimation error is notably smaller than with energy-based analysis.

5.5 First-Order DirAC with Monophonic Audio Transmission

In the DirAC teleconferencing project, the application scenario was two geographically separated groups of people wanting to have a meeting: they place a microphone somewhere between a loudspeaker setup and themselves at each end, as shown in Figure 5.11. The DirAC metadata is analyzed from the microphone array, and sent with a single channel of audio to the other side. For this, variants of DirAC were developed for teleconferencing applications with a focus on minimizing the transmission data bandwidth and the complexity of the microphone array, while maximizing the plausibility of the spatial characteristics and the overall quality of the audio reproduction.

The implementations of DirAC developed for teleconferencing have been evaluated with different perceptual attributes. The first evaluations concerned how well the advantage of intelligibility given by spatial separation of sources in real conditions is replicated with DirAC. The second type of evaluation was to measure the overall audio quality obtained with DirAC when different microphone setups are used. In the third type of test, the effect of the data rate for metadata on the DirAC reproduction was investigated.

Ahonen and Pulkki (2010) studied the speech intelligibility of two temporally overlapping and spatially separated talkers in DirAC reproduction, with an optimal XY microphone technique for two-channel stereophonic reproduction as a reference. DirAC was tested with monophonic sound transmission from both 1D and 2D open arrays. It was a remarkable result that the speech intelligibility did not differ notably between the results from 1D and 2D arrays, and additionally the DirAC reproduction with one-channel audio transmission provided a level of intelligibility equal to the traditional two-channel coincident microphone techniques.

The overall audio quality of the DirAC teleconferencing application was studied by Ahonen (2010) in comparison with ideal coincident microphone techniques with both cardioid and dipole patterns. The recording setup was arranged orthogonally to be ideal for an XY dipole with microphone directions of $\pm 45°$ of azimuth, that is, the sources were positioned either in the nulls of the dipole patterns or directly in front. This resulted in directionally correct reproduction for XY recording. The DirAC versions were tested with open 1D and 2D arrays, and with 1D directional analysis using two cardioids. It was found that the audio quality obtained with these alternatives was as good as with the XY cardioid technique, but in off-sweet-spot listening, a bit surprisingly, the DirAC techniques provided noticeably better results than conventional methods, even though only one channel of audio was transmitted in DirAC compared to the two channels transmitted in the conventional methods.

The effect of the bit rate of the metadata for DirAC on the perceived quality was studied by Hirvonen *et al.* (2009) in the context of audiovisual teleconferencing. Informal tests with one-way reproduction, as well as usability testing where an actual teleconference was arranged, were utilized for this purpose. The frequency resolution, the number of bits allocated for each parameter, and also the update rate of DOA and ψ were varied. A still acceptable setting was obtained when using 3 direction bits and 2 diffuseness bits, 29 equivalent rectangular bandwidth (ERB) bands (Glasberg and Moore, 1990), and a 50 ms update period, equaling a metadata rate of 2.9 kbit s^{-1}. With these values the spatial quality was still high and suitable for teleconferencing, while meticulous listening could in some cases reveal perceptual differences with the metadata at a higher bit rate.

5.6 First-Order DirAC with Multichannel Audio Transmission

While testing the quality of reproduction of music and reverberant sound with teleconference versions of DirAC, it was found that the performance of the system varied depending on the spatial conditions in the recording phase. The performance dropped in some quality attributes in cases where any of following conditions were present: very strong early reflections in a time window of 0–30 ms, multiple continuous equally strong spectrally overlapping signals from spatially well-separated directions, or cases with strong reverberation in general.

In the teleconference version of DirAC, synthesis of the loudspeaker signals is performed by distributing the omnidirectional microphone signal to the loudspeakers according to the analyzed direction and diffuseness in the frequency bands. The fundamental idea in the method presented here is to utilize the non-linear synthesis methods on first-order directional virtual microphone signals instead of omnidirectional signals. In the implementation, a virtual microphone signal is created for each loudspeaker channel, with alignment towards the loudspeaker in question. Non-linear processing steered by DOA and ψ are then applied to these virtual microphone signals. The method is called "virtual microphone DirAC." The motivation is that when the directional selectivity of virtual microphone signals is also utilized in processing, the quality in problematic cases will be enhanced.

5.6.1 Stream-Based Virtual Microphone Rendering

An overview of DirAC virtual microphone synthesis, where the virtual microphone signals are divided into non-diffuse and diffuse streams, is shown in Figure 5.12. The first operation is the computation of the virtual microphone signals from band-passed B-format signals. The non-diffuse and diffuse sound streams are then created by multiplying each virtual microphone signal with coefficients $g_{\mathrm{nd}}\sqrt{1-\psi}$ and $g_{\mathrm{d}}\sqrt{\psi/N}$, respectively, where N is the number of loudspeakers, and g_{nd} and g_{d} are the gain factors to compensate for the energy loss caused by the virtual microphones. The gain factors will be defined in Equations (5.20) and (5.21).

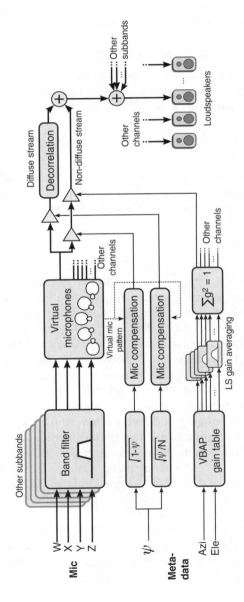

Figure 5.12 Flow graph of virtual microphone DirAC. Source: Vilkamo 2009. Reproduced with permission of the Audio Engineering Society.

A side effect of using virtual directional microphones in DirAC synthesis is their impact on the balance between the non-diffuse and the diffuse sound. A cardioid microphone, for example, has 4.8 dB attenuation for the diffuse sound, but 0 dB for a plane wave from the front (Schulein, 1976). A gain control scheme addressing this issue is required. In the derivation of the sceme it is assumed that in each time instant and in each frequency band, the sound field consists of one arriving plane wave superimposed with a sound field with a diffuseness value of unity. It is further assumed that the plane wave arrives exactly from a direction that corresponds to the direction of a loudspeaker, and that the gain of the corresponding virtual directional microphone is unity in this direction. With these assumptions, the energy of the synthesized non-diffuse sound with the virtual directional microphones, $e_{nd/virtual}$, can be expressed as

$$e_{nd/virtual} = [(1 - \psi) + \psi/Q]e_{nd/omni}, \tag{5.19}$$

where $e_{nd/omni}$ is the target energy (when the energy balance is correct with omnidirectional synthesis) and Q is the directivity factor of the microphone. The loss of energy of the non-diffuse sound signal is compensated by a gain factor

$$g_{nd} = \frac{1}{\sqrt{1 + \psi(1/Q - 1)}}. \tag{5.20}$$

Similarly, the energy of the synthesized diffuse sound with the virtual directional microphones is attenuated by a factor of $1/Q$ due to the directional microphone, and therefore the diffuse sound is compensated by a constant gain factor

$$g_d = \sqrt{Q}. \tag{5.21}$$

The non-diffuse sound stream consists of virtual microphone signals, which have been multiplied with the factor $g_{nd}\sqrt{1 - \psi}$. Amplitude panning, by definition, applies a monophonic input signal to several channels with different amplitudes. On the contrary, here the input signal is already a multichannel signal, being N virtual microphone signals computed for N loudspeakers. In the approach utilized, the gain factors computed with VBAP are used to gate the virtual microphone signals (Pulkki, 2007). This gating limits the number of loudspeakers to which the virtual microphone signals are applied, and thus the coherence problems typically present in first-order coincident microphone techniques are avoided. An at least theoretic drawback of such processing is that some of the energy of the signal may potentially be lost in processing. For example, in multisource scenarios DOA may point to a direction from which the sound is not actually coming, and the virtual microphones pointing to the directions of the sources are consequently gated off. This would result in reproduction of sound with a lower level than it should be ideally. This is the problem of not taking into account the energies and dependencies of the input signals, and what the energies and dependencies are that we want to achieve. This is better addressed with adaptive mixing utilized in covariance-domain rendering, which is discussed in Section 5.6.4.

The diffuse sound stream consists of virtual microphone signals that have been multiplied with the factor $g_d\sqrt{\psi/N}$, and thus the level of non-diffuse sound has been attenuated in the stream. The content of the diffuse stream is mostly the reverberant part of the sound, and also the sound coming from different directions and overlapping in time and frequency. In the approach used for synthesis of diffuse sound, the virtual microphone signals are applied to the loudspeakers after decorrelation. The decorrelation is performed to reduce the inter-channel coherence still present in the virtual microphone signals. Again, the process does not take into account how non-coherent the virtual microphone channels already are, which is better taken into account in covariance-domain rendering – see Section 5.6.4.

A version of virtual microphone DirAC is shown as Matlab code in Section 5.9, where the implementational details are shown.

5.6.2 Evaluation of Virtual Microphone DirAC

Listening tests measuring the audio quality obtained with virtual microphone DirAC with stream-based decoding have been conducted, both in an anechoic chamber and in a listening room, to assess the perceived spatial sound reproduction quality (Vilkamo *et al.*, 2009). A large number of loudspeakers was used with simulated acoustics to create different reference acoustic environments, called reference scenarios, in an anechoic room or a listening room. These scenarios were then recorded using a simulated ideal or a real B-format microphone and reproduced using smaller numbers of loudspeakers. This allows instantaneous switching between the reference scenario and the reproduction in a controlled acoustic environment.

For the tests, 11 different reference scenarios with different combinations of distributed sources and simulated rooms were constructed. The mean opinion scores for the test in an anechoic chamber for the tested reproduction methods are shown in Figure 5.13. The verbal score for *DirAC 16 (ideal mic)* was "excellent," and the scores for *DirAC 16 (real mic)* and *DirAC 5.0 (real mic)* were "good." Multiple comparison tests showed significantly different means between all the technologies, except for *DirAC 16 (real mic)* compared with *DirAC 5.0 (real mic)* and *Lowpass* compared with *Mono*. According to the results, the details of which are shown in Vilkamo *et al.* (2009), DirAC performed differently in reproducing different spaces, with the best performance in the reverberant spaces and in the free field. It is concluded that a factor that affects the quality of DirAC reproduction is the prominence of early reflections. The reproduction

Figure 5.13 Mean opinion scores for different reproduction methods in the anechoic chamber with 95% confidence intervals. Source: Vilkamo 2009. Reproduced with permission of the Audio Engineering Society.

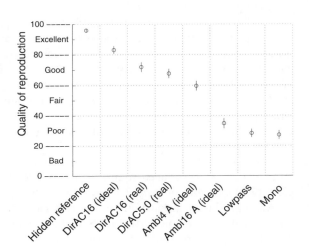

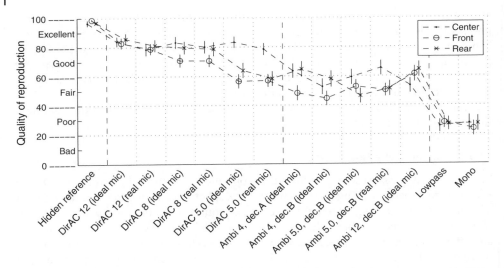

Figure 5.14 Mean opinion scores for different reproduction methods in the listening room with 95% confidence intervals. The legend denotes the listening position. Source: Vilkamo 2009. Reproduced with permission of the Audio Engineering Society.

quality also depended on the source signal. Small defects were noticed most easily with the snare drum sound. The enhancement compared to Ambisonics reproduction is clear in the test results.

A similar test was conducted in a listening room with a horizontal-only reproduction system, but in this case three listening positions were used. The mean opinion scores for the different test techniques conducted in the listening room are shown in Figure 5.14. The scores in the center listening position were similar to the scores in the anechoic chamber, which suggests that the room effect in the listening room has neither a major degrading nor enhancing effect on DirAC reproduction. The results show that DirAC performed equally well with the different loudspeaker setups in the center listening position. The reproduction quality was still preserved in the other positions with a 12-loudspeaker setup, but degraded to fair or good with a 5-loudspeaker setup. A likely explanation for this is that the low number of loudspeakers caused spatial concentration of the sound energy, and thus individual loudspeakers close to the listener may have dominated the spatial sound perception. Also in this case, the audio quality produced by DirAC depended on the sound material – the details are shown in Vilkamo *et al.* (2009). The trends were similar in the listening room to the anechoic chamber, although the differences between the scores with different sound materials were smaller. It seems that the room response of the listening room made the slight artifacts in DirAC reproduction less audible.

Virtual microphone DirAC has also been utilized in audiometry for speech intelligibility testing in reproduced sound scenes, needed to test how hearing-impaired people cope in everyday noisy situations. The method proposed in Koski *et al.* (2013) uses DirAC-reproduced background sound scenes augmented by target speech sources over a multichannel loudspeaker setup. Subjective listening tests were performed to validate the proposed method: speech reception thresholds (SRT) in noise were measured in a reference sound scene and in a room where the reference was reproduced by a loudspeaker setup. The listening tests showed that for normal-hearing test subjects the method provides nearly identical speech intelligibility compared to the real-life reference when using a nine-loudspeaker reproduction setup in anechoic conditions (<0.3 dB error in SRT).

5.6.3 Discussion of Virtual Microphone DirAC

Virtual microphone processing has several positive impacts. When the signals are already non-coherent to some degree, the non-linear processing needs to be less aggressive to obtain the desired non-coherence between loudspeaker signals when compared with mono-DirAC rendering. This also makes it easier to select temporal constants so that the artifacts are inaudible. Furthermore, the spatial distribution of the diffuse stream follows the distribution in recording conditions, as well as first-order patterns allow. Also, the conditions of amplitude panning gains in the non-diffuse stream range between "one gate open" and "all gates open," depending on the variation of DOA. In the case when all gates are open, the reproduction implements virtual microphone decoding; when few gates are open, the system in practice concentrates sound energy in certain directions. This makes the method more robust to analysis errors and artifacts.

According to the informal experience of the developers, using virtual directional microphones instead of only the omnidirectional microphone produces more authentic reproduction of spaciousness, source localization, and sound color. This appears contradictory to the assumption that the reproduction of the direction and diffuseness is sufficient for reproducing spatial sound. Focusing on this mismatch, two fundamental issues were identified. First, the B-format recording is performed in a single position, while human ears are located approximately 17 cm apart. Coherent wavefronts arriving from different directions (e.g., a direct sound and a reflection) cause different-magnitude responses in each ear, meaning that there is more spectral information available for a human listener than for a single-position omnidirectional measurement. Secondly, the temporal integration of the parameters has to be conducted with relatively long time windows to avoid audible distortions in the sound, which may cause the system to be too sluggish, and hence perceivable problems emerge when only a single audio channel is used in rendering, in cases with multiple concurrent sources. The virtual directional microphones enhance the separation of the signals directly from sources and the signals via reflections, and are also beneficial in diffuse sound synthesis due to the lower initial inter-channel coherence.

5.6.4 Optimized DirAC Synthesis

The covariance-domain rendering method described in Section 1.3.3 can be used in DirAC to minimize the level of decorrelated sound in the output and to minimize the energy losses due to interaction between panning gains and virtual microphones. This is obtained by estimating how much the linearly decoded virtual microphone signals meet the requirements of inter-channel coherences and channel energies in the frequency bands, and by non-linearly enhancing the output only when the requirements are not met. The method analyzes the covariance matrix of the input, and inserts decorrelated sound to the output only by such an amount that the targeted non-coherence is obtained. Since the covariance-domain rendering method is explained in detail earlier in this book, this section describes only how to compute the variables needed for it in the DirAC context.

A block diagram of the combined process of performing DirAC rendering using the covariance-rendering method is shown in Figure 5.15. DirAC analysis is applied to the input signal covariance matrix $\mathbf{C_b}$ to derive DOA and ψ, which in turn control the target covariance matrix $\mathbf{C_y}$, as described below.

The target covariance matrix $\mathbf{C_y}$ corresponding to the analyzed DOA and ψ can be built by formulating the non-diffuse and diffuse target covariance matrices $\mathbf{C_y^{ND}}$ and $\mathbf{C_y^D}$ separately, and adding them together. This is possible since the non-diffuse and diffuse components are non-coherent with respect to each other according to the applied sound field model.

The matrix $\mathbf{C_y^{ND}}$ is built with the following steps. Let us assume a column vector $\mathbf{v}(DOA)$ that contains the loudspeaker gain factors computed with VBAP (Pulkki, 1997) corresponding to the analyzed DOA. $\mathbf{v}(DOA)$ has a maximum of two non-zero values in the case of a horizontal loudspeaker setup, and three in the case of a 3D loudspeaker setup. The panning covariance matrix is formulated as $\mathbf{C_v}(DOA) = \mathbf{v}(DOA)\mathbf{v}^T(DOA)$, and

$$\mathbf{C_y^{ND}} = (1 - \psi)R\mathbf{C_v}(DOA), \tag{5.22}$$

where R is an estimate of the total sound energy of the time–frequency tile.

The matrix $\mathbf{C_y^D}$ is built by defining a diagonal $N_y \times N_y$ energy distributor matrix $\mathbf{D}$ with non-negative entries d_i on the diagonal, where i is the row index, and $\sum_{i=1}^{N_y} d_i = 1$:

$$\mathbf{C_y^D} = \psi R\mathbf{D}. \tag{5.23}$$

Matrix $\mathbf{C_y^D}$ is diagonal, which means that the diffuse part of the target covariance matrix is defined as non-coherent between all channel pairs. The distributor matrix $\mathbf{D}$ can be adjusted to fit the loudspeaker layout by setting the values d_i for each loudspeaker to a value that is inversely proportional to the loudspeaker density in the corresponding direction. For a uniformly distributed loudspeaker layout, all values $d_i = \frac{1}{N_y}$. Finally, the target covariance matrix for the time–frequency tile is

$$\mathbf{C_y} = \mathbf{C_y^{ND}} + \mathbf{C_y^D}. \tag{5.24}$$

The best choice for a prototype matrix $\mathbf{Q}$ defining the prototype signal $\hat{\mathbf{y}} = \mathbf{Qx}$, which is equivalent to the virtual microphone concept, depends on the microphone and the loudspeaker configurations. If these configurations are static, $\mathbf{Q}$ can be defined to be time invariant. For B-format rendering let us define $\mathbf{Q_B}$ for a loudspeaker layout with azimuths θ_i and elevations ϕ_i, where $1 \leq i \leq N_y$ is the channel index. A reasonable prototype matrix is a virtual directional microphone matrix with look directions towards the directions of the loudspeakers:

$$\mathbf{Q_B} = \begin{bmatrix} \mathbf{q}_1^T \\ \mathbf{q}_2^T \\ \vdots \\ \mathbf{q}_{N_y}^T \end{bmatrix}, \tag{5.25}$$

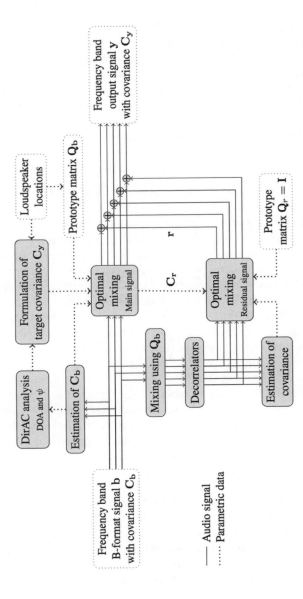

Figure 5.15 DirAC processing from a B-format signal using the covariance-rendering method. The main optimal mixing block uses the available signal components in the input channels to produce the output. The residual signal $\mathbf{r}$ is generated to compensate when there are not enough independent signal components available otherwise. Source: Vilkamo 2013. Reproduced with permission of the Audio Engineering Society.

where

$$q_i = \begin{bmatrix} \frac{2-\kappa}{2} \\ \frac{\kappa}{2\sqrt{2}} \cos(\theta_i) \cos(\phi_i) \\ \frac{\kappa}{2\sqrt{2}} \sin(\theta_i) \cos(\phi_i) \\ \frac{\kappa}{2\sqrt{2}} \sin(\phi_i) \end{bmatrix}, \tag{5.26}$$

where κ is a constant between 0 and 2 defining the directivity pattern of the virtual directional microphone, and θ_i is the azimuth and ϕ_i the elevation of the ith loudspeakers. A good choice is $\kappa = 1.5$, since it corresponds to a hypercardioid that maximizes the energy ratio of the look direction with respect to the other directions. The prototype signal in B-format rendering is then $\hat{y}_B = Q_B b$.

A listening test was performed in an anechoic chamber to evaluate the perceptual impact of the optimized level of decorrelated energy (Vilkamo and Pulkki, 2013). The results showed that significant improvement was provided by the covariance-rendering method when there are several microphone signals available, and the input signal is either surrounding applause or reverberant speech, that is, a signal that is more affected by the decorrelators that are critical in DirAC processing.

5.6.5 DirAC-Based Reproduction of Spaced-Array Recordings

As already noted, the major source of artifacts is the effect of the decorrelators, which are needed to scramble the phase spectrum of the signals to obtain sufficiently non-coherent loudspeaker signals. In surround recording techniques, the traditional audio engineering method to obtain such incoherence is to use a recording technique where a number of microphones are positioned with considerable distances between them. Direct playback of such spaced-microphone recordings does not suffer from coherence problems at mid and high frequencies, since the signals are non-coherent enough. However, Pulkki (2002) showed that combined time differences from all playback channels manifest themselves as localization blurring. In addition, the microphone signals are coherent at low frequencies, which causes timbral and spatial artifacts (Pulkki, 2007).

Time–frequency domain spatial analysis of a sound field can also be performed with such spaced arrays of microphones, and the DirAC synthesis can be modified for such cases (Politis *et al.*, 2015b). The processing then utilizes amplitude panning to reproduce the non-diffuse sound to reduce localization blurring, and performs decorrelation at the lowest frequency bands to avoid the high coherence problems there. This is expected to (i) improve the problematic cases of DirAC, by using the natural incoherence of the array recordings when possible and hence minimizing decorrelation; (ii) enhance the localization aspects of non-diffuse playback by the parametric approach of DirAC; and (iii) eliminate the coloration issues of non-diffuse playback by appropriate decorrelation at low frequencies.

The DOA estimation in DirAC, which is typically based on the direction of the intensity vector for each time–frequency block, cannot directly be conducted with such spaced arrays. Due to the large spacing of the microphones, spaced arrangements such as in Figure 5.16 allow estimation of the intensity only at very low frequencies. In Laitinen *et al.* (2011a) a coincident array was used for analysis of the directional parameters, and the microphone signals from the spaced array were used only for the diffuse

Figure 5.16 A surround microphone array used in DirAC processing. d_C, d_{LR}, and d_S define the dimensions of the setup, θ_1 defines the orientation of the L and R microphones, and θ_2 defines the orientation of the surround microphones. θ is the direction of incidence of the incoming wave. Source: Politis 2015b. Reproduced with permission of the Audio Engineering Society.

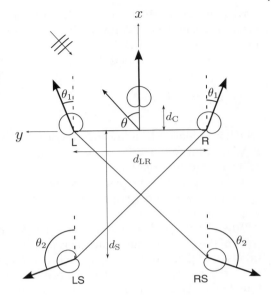

stream at high frequencies. With such an approach no decorrelation needs to be performed there, which improves the quality. A generalized version of the DOA estimation used for a square array with baffle in Section 5.4.2 can be used with spaced arrays. In the formulation, the following assumptions are made: A general left–right symmetric array is assumed, which is always true in practice. The orientations of the microphones are parameterized according to the angles θ_1 for the stereophonic pair and the angles θ_2 for the surround pair, as shown in Figure 5.16. Positive first-order patterns are assumed, such as cardioids or subcardioids, with their pattern given by $D(\theta) = \beta + (1 - \beta)\cos\theta$, with $1 \geq \beta \geq 0.5$. All operations are defined in a time–frequency transformed domain, such as an STFT, for the time frame index n and frequency index k. For a plane wave of amplitude $a(k, n)$ incident from an angle θ, the signals captured by the array microphones are

$$\mathbf{s}(k, n) = \begin{bmatrix} s_L(k, n) \\ s_R(k, n) \\ s_C(k, n) \\ s_{LS}(k, n) \\ s_{RS}(k, n) \end{bmatrix} = a(k, n) \begin{bmatrix} e^{-j\phi_L}D(\theta - \theta_1) \\ e^{-j\phi_C}D(\theta) \\ e^{-j\phi_R}D(\theta + \theta_1) \\ e^{-j\phi_{LS}}D(\theta - \theta_2) \\ e^{-j\phi_{RS}}D(\theta + \theta_2) \end{bmatrix}, \tag{5.27}$$

with ϕ_L, ϕ_C, ϕ_R, ϕ_{LS}, and ϕ_{RS} being the respective phase differences due to the position of each microphone and DOA of the plane wave. We further define the vector of magnitude spectra for the microphone signals as

$$\mathbf{m}(k, n) = \begin{bmatrix} |s_L(k, n)| \\ |s_R(k, n)| \\ |s_C(k, n)| \\ |s_{LS}(k, n)| \\ |s_{RS}(k, n)| \end{bmatrix} = |a(k, n)| \begin{bmatrix} D(\theta - \theta_1) \\ D(\theta) \\ D(\theta + \theta_1) \\ D(\theta - \theta_2) \\ D(\theta + \theta_2) \end{bmatrix}. \tag{5.28}$$

Assuming that the time frame length is significantly larger than the temporal dimensions of the array, $\mathbf{m}$ depends only on the magnitude of the plane wave and the gains of each microphone due to their directivity.

A DOA vector is defined,

$$\mathbf{i}_{\text{doa}} = |a(k, n)| \begin{bmatrix} \cos\theta \\ \sin\theta \end{bmatrix}, \tag{5.29}$$

which is proportional to the magnitude of the plane-wave signal and points to the DOA θ. When diffuse sound is also present, the direction of the DOA vector is going to fluctuate around the true DOA with a variance that depends on the power of the diffuse sound compared to the plane-wave power (Politis *et al.*, 2015a). This implies that the $\mathbf{i}_{doa}$ vectors can be directly used as substitutes for $\mathbf{i}$ in Equation (5.10) to estimate the value of diffuseness.

When a surround array is used, the DOA vector can be estimated by using opposite pairs of the front and surround microphones (e.g., the L–RS/R–LS pairs). The formulation is based on the magnitude spectra differences of these opposing pairs, which after some trigonometric manipulations result in dipole-like patterns scaled by the magnitude of the plane-wave signal, similar to Equation (5.29). More specifically,

$$\mathbf{i}_{\text{doa}}(k, n) = \mathbf{G}_{\text{msa}}\, \mathbf{m}(k, n), \tag{5.30}$$

where $\mathbf{G}_{\text{msa}}$ is the matrix that subtracts and normalizes the spectra to result in the form of Equation (5.29). With respect to the array parameters, they are given by

$$\mathbf{G}_{\text{msa}} = \begin{bmatrix} g_1 & 0 & g_1 & -g_1 & -g_1 \\ -g_2 & 0 & g_2 & -g_2 & g_2 \end{bmatrix}, \tag{5.31}$$

with

$$g_1 = \frac{1}{4(\beta - 1)\sin\left(\frac{\theta_1 + \theta_2}{2}\right)\sin\left(\frac{\theta_1 - \theta_2}{2}\right)}, \text{ and}$$

$$g_2 = \frac{1}{4(\beta - 1)\sin\left(\frac{\theta_1 + \theta_2}{2}\right)\cos\left(\frac{\theta_1 - \theta_2}{2}\right)}. \tag{5.32}$$

It is obvious from Equation (5.32) that the estimator requires the microphones to be directional and fails for omnidirectional ones with $\beta = 1$.

At low frequencies, where the spacing between the microphones is significantly smaller than the wavelength, the microphones can be assumed to be coincident. It is then possible to obtain B-format signals from the first-order microphones and perform an intensity-based estimation as normally done in DirAC.

According to the listening tests in Politis *et al.* (2015b), the method enhances the reproduction compared to both unprocessed spaced-array recordings and the standard B-format DirAC method. The processing scheme achieves improved localization at reproduction by combining the microphone directivities with amplitude panning, and retains pleasant enveloping qualities of spaced recordings by decorrelating the output signals only at the frequency range that is required.

5.7 DirAC Synthesis for Headphones and for Hearing Aids

5.7.1 Reproduction of B-Format Signals

DirAC is also a powerful tool for reproducing recorded sound scenes with headphones, where the movements of the listener can be taken into account effectively and efficiently (Laitinen and Pulkki, 2009). The DirAC streams are rendered to the listener's headphones using virtual loudspeakers simulated by a head-related transfer function (HRTF). The movements of the listener are updated instantaneously to the system, which noticeably increases the quality of reproduction.

Binaural reproduction is considered separately for the non-diffuse and diffuse streams. A straightforward way to synthesize the non-diffuse stream used in the implementation shown in Laitinen and Pulkki (2009) is to define a 3D setup of virtual loudspeakers, and to filter the DirAC output channels with corresponding static HRTF filters. The diffuse sound is reproduced analogously to loudspeaker-based DirAC reproduction. The decorrelated loudspeaker signals are created and also applied to the virtual loudspeakers evenly in the surrounding 3D setup. Laitinen and Pulkki (2009) found that an adequate number of virtual loudspeakers for diffuse sound was 12–20 in three dimensions, which agrees with a study on the spatial impression of the diffuse sound field (Hiyama *et al.*, 2002).

The target of head tracking is to keep the spatial properties of the reproduced auditory events constant with respect to the outer world, that is, the virtual source directions should not move according to the movements of the listener. There are two possibilities for including head tracking information in DirAC reproduction. Either the directions of the virtual loudspeakers are updated with the information, or the direction of origin is updated inside the DirAC processing with the information. The first alternative has not been found appealing, since it would require fast updating of the filters, which have relatively long impulse responses. Such updating is prone to artifacts, especially when the filters should be updated in a few milliseconds. The second alternative only requires updating the virtual microphone coefficients, and algebraic manipulations of metadata, and it was utilized by Laitinen and Pulkki (2009). The selected method was found to be free of artifacts even when updating the processing in a very fast manner, which provided good externalization and stable directional perception of sources.

In practice, the DOA parameter and the directions of the virtual microphones are updated to compensate for the changes in the orientation of the head of the listener. The virtual loudspeakers thus move with the head of the listener, but the panning directions and virtual microphone directions are kept constant within the external coordinate system. Relatively low latency between head movements and parameter updates is needed to obtain well-externalized perception of the sound scene. In Laitinen and Pulkki (2009), the updating is performed at a rate of 50 Hz. If the listener rotates her head rapidly, the temporal averaging of the gain factors of the virtual loudspeakers performed in the synthesis of the non-diffuse sound can cause sluggishness in the perceived direction, as there would be a discrepancy between the movement of the head and the perceived direction of the sound. This is avoided by updating the parameters faster when the head is rotated rapidly. In practice, such faster updates can be achieved by adjusting the temporal constants to be lower during fast head movements. When the rotation is slow enough, the default temporal smoothing is applied. This way the sound field is updated fast enough to respond to the movement of the head, but no artifacts are perceived due to the faster smoothing. More details on the effect of the influence of temporal and spatial resolution of head tracking in DirAC can be found in Laitinen *et al.* (2012).

The fidelity of binaural DirAC reproduction has been measured subjectively by Laitinen and Pulkki (2009). Four sound samples recorded in real acoustical environments with real B-format microphones were reproduced in the test. Six different techniques were tested that did or did not utilize head tracking. The results implied that, using binaural reproduction of DirAC, a realistic reproduction of spatial sound can be achieved, as both overall quality and spatial impression were rated with the highest scores in the case when head tracking was involved, with verbal anchors "excellent" and "truly believable," respectively. An important result is also that DirAC reproduction without head tracking was preferred to traditional coincident two-channel reproduction played directly over headphones without HRTF processing, both with and without head tracking.

The binaural reproduction method for DirAC was published in 2009. In 2013, a real-time implementation of it was combined with head-mounted reproduction of spherical video images, both with head tracking. Both spherical video and B-format audio are then recorded from the same position in a venue. Such a combination of 3D sound and surrounding video gives a truly immersive reproduction of reality, and the subject gets an immersive perception of "being there." This happens even though the HRTFs used in the demonstration are not individualized; instead, non-individual HRTFs measured from a real subject are used. When the subject sees the environment, and at the same time hears the sounds through the non-individual HRTFs with head tracking, the front–back confusions typically found with headphone listening are not present.

This section has not considered the covariance-domain implementation of DirAC headphone reproduction. The theory of the implementation is presented for the first time in this book in connection with higher-order DirAC theory in Chapter 6, as it can be efficiently used with any order of B-format input.

5.7.2 DirAC in Hearing Aids

In principle, the most typical hearing aids consist of a pair of devices attached to each ear. The device typically carries 1–3 microphones, electronics, and a headphone. In advanced solutions the devices are also able to communicate digitally with each other with a wireless link. The signals gathered by the microphones are reproduced after some more or less complex equalization, beam forming, and dynamic range control computations (Dillon, 2001). DirAC has also been suggested to be used in such scenarios (Ahonen *et al.*, 2012b). The microphone setup is shown in Figure 5.17, and is essentially

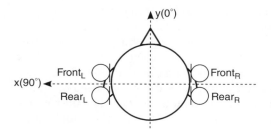

Figure 5.17 Four-channel microphone setup in binaural hearing aid.

a rectangular setup with an open array between the front and back microphones, and a rigid baffle between the left and right arrays. The intensity vectors can again be estimated differently for low and high frequencies, similarly to in context of cylindrical microphones in Section 5.4.2. However, in this case different frequency limits have to be used for the front and side directions, since the inter-microphone distance is very different there. The analysis results for DOA with such a microphone setup are shown in Ahonen *et al.* (2012b), and accuracy comparable to the results with a cylinder array are obtained at frequencies below about 4 kHz. Also, ψ here can be computed with the coefficient of variation method. The spatial parameters can then be used in beam forming, and signal enhancement applications in general.

The synthesis for hearing aids can be simplified from the DirAC rendering of B-format signals for headphones. As the microphone signals already have relevant spatial cues, they need not be amplitude panned or decorrelated. However, the signal at each side is divided into diffuse and non-diffuse streams, and their balance can be altered. Typically, the diffuse stream is attenuated to mitigate the negative effects of reverberation on speech intelligibility. Furthermore, the non-diffuse stream can be altered based on the DOA metadata: typically, all other directions except the frontal direction are attenuated.

The fact that the devices communicate over a digital link with a limited data rate makes it complicated to perform broad-band beam forming with such methods that require all the microphone signals to be present at the same location. With duplex transmission of uncompressed 16-bit PCM audio with a 16 kHz sample rate this would mean a 512 kbit s^{-1} rate, which may be too high for such applications. A technical advantage of using DirAC with hearing aids is that the computation of the intensity vector requires the transmission of frequency-smoothed spectral values with an update rate of about 100 Hz, which enables beam forming at all frequency bands. Ahonen *et al.* (2012b) estimate the rate of duplex transmission to be 104 kbit s^{-1}, which is a considerably lower value than with PCM transmission.

5.8   Optimizing the Time–Frequency Resolution of DirAC for Critical Signals

The first operation in DirAC is to transform the time-domain input audio signals into the time–frequency presentation. Optimally, the temporal resolution of the processing should be at least as good as the temporal resolution of human hearing. The hearing system can be interpreted to include a sliding temporal integrator for smoothing the ear canal signal (Moore, 1982). Thus, two closely spaced impulses can be perceived as one impulse if the gap between them is sufficiently small. According to Moore (1982), this gap is about 2–3 ms.

A B-format recording of a surrounding audience applausing is one of the most critical signals for DirAC. For example, if two claps from different directions arriving at slightly different times are analyzed within a single time window, the diffuseness will be analyzed high, as the diffuseness equation, Equation (5.7), detects sound coming from multiple directions. Adjusting the temporal window to be shorter makes such coincidences less frequent. It is also assumed that the window should not be shorter than the human temporal resolution, as shorter time windows will add noise to the DOA and ψ analysis. In this work, time windows as short as 2 ms were used.

The temporal resolution is increased by using smaller temporal windows for STFT analysis. However, for example with a 1 ms window, the frequency resolution of the transform is only 1 kHz, which is clearly of a larger bandwidth than the bandwidth of the critical bands of hearing at low frequencies. The solution applied by Laitinen *et al.* (2011b) is to divide the input signals prior to STFT into two or more frequency regions and to apply a separate STFT for each of them, as shown in Figure 5.18. At low frequencies, a long window is used to get proper frequency resolution; at high frequencies, a short window is used to get proper temporal resolution. The technique of using multiple STFTs with different window sizes is called "multiresolution STFT" here.

The design principle for using multiresolution STFT in DirAC is straightforward. As shown in Figure 5.18, the input signals are divided into a small number of frequency regions and separate DirAC processing with different temporal constants is run within each region. Note that a properly designed filter bank would also produce the desired accuracy in time and frequency.

In the implementation, the DirAC analysis is the same as in the conventional STFT implementation. However, as the window size is different in the different frequency regions, the update rate of the metadata is different. Furthermore, the DirAC analysis of a region needs to be performed only in the frequency range of that particular region. The synthesis parts of the DirAC blocks are also the same as in the traditional STFT version. The resulting loudspeaker signals are band-pass filtered to remove possible distortion outside the considered spectral region. Finally, the loudspeaker signals from the regions are simply summed to obtain the signals applied to the loudspeakers.

The subjective quality of the multiresolution-STFT-based DirAC reproduction of two 5.0 surround applause samples of different characteristics was evaluated in listening tests reported by Laitinen *et al.* (2011b). In the formal tests, the difference between the original samples and DirAC using optimal multiresolution STFT settings was graded as "Perceptible, but not annoying." In addition, the perceived quality was measured with different temporal resolutions. It was found that as the temporal resolution in the DirAC processing was made coarser, the quality of reproduction decreased.

This modification results in a significant audio quality improvement compared to a DirAC implementation using a frequency-independent window length of 10 ms. However, in the process it was found that sometimes two transients arrive from different directions at the microphone within the 2 ms time window, and the window will be analyzed to have high diffuseness, and consequently the signals are routed to the diffuse stream. Due to the decorrelation processing applied during DirAC synthesis, these transients are smeared in time and the claps in applause are not perceived to be as sharp as in the original signal. This can be prevented by processing transients in the diffuse stream separately. This modification provides an additional quality improvement in the reproduction (Laitinen *et al.*, 2011b).

5.9 Example Implementation

This chapter presents a Matlab implementation of stream-based virtual microphone DirAC. The implementation is based on an STFT with a window length of about 20 ms and a 50% hop size. Many of the variables are temporally smoothed with time constants that depend on frequency. This is to avoid the non-linear artifacts discussed earlier. The

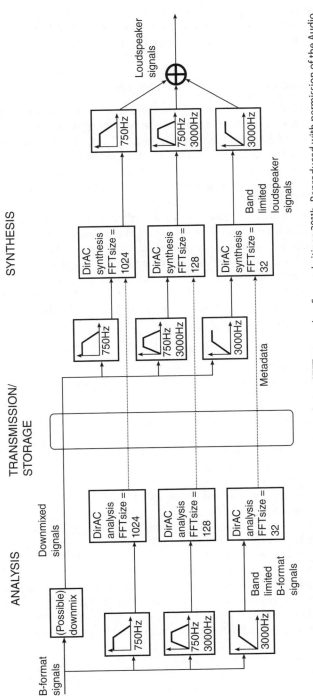

Figure 5.18 Block diagram of DirAC processing with multiresolution STFT processing. Source: Laitinen 2011b. Reproduced with permission of the Audio Engineering Society.

implementation does not show the maximum audio quality or data compression rate obtainable with DirAC; instead, it shows a decent implementation which already gives a notable improvement in audio quality when compared to simple matrixing solutions in multichannel reproduction. The reader is encouraged to test the code with different input files, and to change the parameters for educational purposes.

5.9.1 Executing DirAC and Plotting Parameter History

Listing 5.1 is the "main" script, which calls the initialization function, runs the programs, and stores the outputs. A synthetic sound scene with beeping sounds in moving directions and a background diffuse sound is simulated and stored in a B-format audio signal. The result is also made audible for headphones by a simple simulation of binaural listening as described in Section 5.9.4. For comparison, linear decoding of the same B-format signal is computed in a similar fashion to first-order Ambisonics techniques.

Note that in the 3D decoding a typical layout with horizontal and elevated loudspeakers is used, which implies that none of the loudspeakers is in the lower hemisphere. If the layout were defined as such to DirAC, the loudspeaker triangulation in VBAP would fail to compose any triangles below the horizontal plane. This would potentially cause some loss of energy, as DOA may also get values in the lower hemisphere. To avoid the loss of sound energy, the energy has to be routed to the existing loudspeakers. This is implemented simply by defining the loudspeaker layout with a loudspeaker directly below the listener. The signal that DirAC applies to that virtual loudspeaker is then divided equally between all loudspeakers.

Listing 5.1: DirAC Main Script

```
1
2   %% The main DirAC script
3   %% Archontis Politis and Ville Pulkki 2016
4
5   clear; close all; fs=48000;
6   siglen=12*fs; % length of signal
7
8   % Sawtooth signals with repeated exp-decaying temporal envelope
9   sig1=(mod([1:siglen]',200)/200-0.5) .* ...
        (10.^((mod([siglen:-1:1]',fs/5)/(fs/10)))-1)/10;
10  sig2=(mod([1:siglen]',321)/321-0.5) .* ...
        (10.^((mod([siglen:-1:1]',fs/2)/(fs/4)))-1)/10;
11  % Simulate B-format signals for the sources
12  azi1=[1:siglen]'/siglen*3*360; % changing source azimuth for sig1
13  ele1=[1:siglen]'*0; % constant elevation for sig1
14  azi2=round([1:siglen]'/siglen)*180-90; % azi for sig2
15  ele2=[1:siglen/2 siglen/2:-1:1]'/siglen*180; % changing elev ...
        for sig2
16  bw=(sig1+sig2)/sqrt(2);
17  bx=sig1.*cos(azi1/180*pi).*cos(ele1/180)+sig2.* ...
        cos(azi2/180*pi).*cos(ele2/180*pi);
```

```
18  by=sig1.*sin(azi1/180*pi).*cos(ele1/180)+sig2.* ...
        sin(azi2/180*pi).*cos(ele2/180*pi);
19  bz=sig1.*sin(ele1/180*pi)+sig2.*sin(ele2/180*pi);
20
21  % Add fading in diffuse low-passed noise about evenly in 3D
22  [b,a]=butter(1,[500/fs/2]);
23  for azi=0:10:1430 % four azi rotations in 10deg steps, random ...
        elevation
24      ele=asin(rand*2-1)/pi*180;
25      noise=filter(b,a,5*(rand(siglen,1)-0.5)).* ...
            (10.^((([1:siglen]'/siglen)-1)*2));
26      bw=bw+noise/sqrt(2);
27      bx=bx+noise*cos(azi/180*pi)*cos(ele/180*pi);
28      by=by+noise*sin(azi/180*pi)*cos(ele/180*pi);
29      bz=bz+noise*sin(ele/180*pi);
30  end
31  %% Alternatively, B-format audio load with bw scaled down by ...
        sqrt(2).
32  %% [xx,fs]=audioread('B_format_file.wav');
33  %% bw=xx(:,1);bx=xx(:,2);by=xx(:,3);bz=xx(:,4);
34
35  %% 2D PROCESSING %%%%
36  % Compose B-format and discard Z-component for 2D processing
37  bfsig_2D = [bw bx by bz*0]; bfsig_2D=bfsig_2D/max(max(abs ...
        (bfsig_2D)))/3;
38  bfsig_2D(end-500:end,:) = bfsig_2D(end-500:end,:) .* ...
        (linspace(1,0,501)'*[1 1 1 1]);
39  % Define the directions of loudspeakers
40  ls_dirs_2D = [-30 0 30 -110 110 -150 150]'; % 7.1 surround
41  %Initialize DirAC parameters
42  DirAC_struct = DirAC_init_stft(ls_dirs_2D, fs);
43  % Linear decoding + write loudspeaker signals to disk
44  LINsig_2D = bfsig_2D*DirAC_struct.decodingMtx/sqrt(7);
45  audiowrite(['Output2D-LIN.wav'],LINsig_2D,fs);
46
47  % DirAC processing + write loudspeaker signals to disk
48  [LSsig_2D, DIRsig_2D, DIFFsig_2D, DirAC_struct] = ...
        DirAC_run_stft(bfsig_2D, DirAC_struct);
49  audiowrite(['Output2D-DirAC.wav'],LSsig_2D,fs);
50  % Plot spatial metadata
51  figure; imagesc(DirAC_struct.parhistory(:,:,1)'); colorbar; ...
        title('Azimuth / 2D case'); xlabel('Time frame'); ...
        ylabel('Freq bin');set(gca,'YDir','normal');
52  figure; imagesc(DirAC_struct.parhistory(:,:,2)'); colorbar; ...
        title('Elevation / 2D case'); xlabel('Time frame'); ...
        ylabel('Freq bin');set(gca,'YDir','normal');
53  figure; imagesc(DirAC_struct.parhistory(:,:,4)'); colorbar; ...
        title('Diffuseness / 2D case'); xlabel('Time frame'); ...
        ylabel('Freq bin');set(gca,'YDir','normal');
```

```
54
55  % Simple simulation of LS listening with headphones
56  HPsig_LIN2D=Headphone_ITDILD_simulation(LINsig_2D, DirAC_struct);
57  HPsig_DirAC2D=Headphone_ITDILD_simulation(LSsig_2D, DirAC_struct);
58
59  %Listen to resulting headphone audio
60  %soundsc(bw,fs); disp('Mono reference'); pause(siglen/fs);
61  %soundsc(HPsig_DirAC2D,fs); disp('DirAC 2D'); pause(siglen/fs);
62  %soundsc(HPsig_LIN2D,fs); disp('Linear decoding 2D'); ...
        pause(siglen/fs)
63
64  %% 3D PROCESSING %%%%%%
65  ls_dirs_3D = [-30 0; 0 0; 30 0; -110 0; 110 0; -150 0;  150 0; ...
        -45 45; 45 45; -135 45; 135 45; 0 -90;];
66  % 7.1.4 surround + virtual channel below.
67
68  %B-format signal
69  bfsig_3D = [bw bx by bz];bfsig_3D=bfsig_3D/max(max(abs ...
        (bfsig_3D)))/3;
70  bfsig_3D(end-500:end,:) = bfsig_3D(end-500:end,:) .* ...
        (linspace(1,0,501)'*[1 1 1 1]);
71  DirAC_struct.parhistory = [];
72  DirAC_struct = DirAC_init_stft(ls_dirs_3D, fs);
73
74  % Linear decoding + store audio
75  LINsig_3D = bfsig_3D*DirAC_struct.decodingMtx/sqrt(12);
76  % Adding virtual channel to other loudspeakers
77  LINsig_3D=LINsig_3D(:,1:end-1) +  ...
        LINsig_3D(:,end)*ones(1,DirAC_struct.nOutChan-1)/ ...
        (DirAC_struct.nOutChan-1);
78  audiowrite('Output3D-LIN.wav',LINsig_3D,fs);
79
80  % DirAC processing + store audio
81  DirAC_struct.parhistory = [];
82  [LSsig_3D, DIRsig_3D, DIFFsig_3D, DirAC_struct] = ...
        DirAC_run_stft(bfsig_3D, DirAC_struct);
83  % Apply virtual loudspeaker to all other loudspeakers.
84  LSsig_3D=LSsig_3D(:,1:end-1) + ...
        LSsig_3D(:,end)*ones(1,DirAC_struct.nOutChan-1)/ ...
        (DirAC_struct.nOutChan-1);
85  audiowrite(['Output3D-DirAC.wav'],LSsig_3D,fs)
86
87  % Plot directional parameters
88  figure; imagesc(DirAC_struct.parhistory(:,:,1)'); colorbar; ...
        title('Azimuth / 3D case'); xlabel('Time frame'); ...
        ylabel('Freq bin');set(gca,'YDir','normal');
89  figure; imagesc(DirAC_struct.parhistory(:,:,2)'); colorbar; ...
        title('Elevation / 3D case'); xlabel('Time frame'); ...
        ylabel('Freq bin');set(gca,'YDir','normal');
```

```
90  figure; imagesc(DirAC_struct.parhistory(:,:,4)'); colorbar; ...
        title('Diffuseness / 3D case'); xlabel('Time frame'); ...
        ylabel('Freq bin');set(gca,'YDir','normal');
91
92  %% Create headphone signals with simple ILD-ITD modeling
93  HPsig_LIN3D=Headphone_ITDILD_simulation(LINsig_3D, DirAC_struct);
94  HPsig_DirAC3D=Headphone_ITDILD_simulation(LSsig_3D, DirAC_struct);
95
96  %Listen the output
97  %soundsc(bw,fs); disp('Mono reference'); pause(siglen/fs);
98  %soundsc(HPsig_DirAC3D,fs); disp('DirAC 3D'); pause(siglen/fs)
99  %soundsc(HPsig_LIN3D,fs); disp('Linear decoding 3D')
```

5.9.2 DirAC Initialization

Different parameters and settings for DirAC processing are set up in the initialization functions shown in Listing 5.2. When processing different recordings, there are two parameters that can be modified to obtain the best audio quality in the output. The parameter `DirAC_struct.diffsmooth_cycles` defines the time constant for temporal integration of the intensity vector and energy in the computation of the diffuseness parameter. The parameter is defined as depending on frequency as a number of cycles at each frequency band. The parameter `DirAC_struct.diffsmooth_limf` sets a frequency limit above which the time constant does not change with frequency, but the value at the limit is used. The parameters `DirAC_struct.gainsmooth_cycles` and `DirAC_struct.gainsmooth_limf` affect the time constant in the temporal smoothing of loudspeaker gain factors in the non-diffuse stream in a corresponding manner. In practice, if any artifacts such as roughness, musical noise, or "bubbling" are heard, they can typically be removed by adjusting these parameters. Note that in this example a very conservative value of 200 is used for `DirAC_struct.gainsmooth_cycles`, which should produce artifact-free rendering in most cases, even with the ideal B-format signal produced by the code. The B-format signal provided by the code is in practice a very critical signal, having sharp onsets on different sides of the microphone and diffuse background noise. If the reader applies the code with real B-format recordings with spatially less critical content, a substantially lower value for `DirAC_struct.gainsmooth_cycles` might be optimal; typically, values between 40 and 90 are used.

The need to adjust the parameters depending on the spatial conditions in recording is a suboptimal situation, naturally. However, this code is implemented as a simple illustrative demonstration with a single-resolution STFT without optimization of the level of decorrelated energy in the output. An implementation more tolerant to different recording conditions can be realized using a multiresolution STFT or filter banks, and by using covariance-domain rendering.

An example method to derive decorrelation filters is also included in the initialization of DirAC. The method is based on summing delayed band-passed impulses, and complicated heuristics are applied to avoid phase cancellations between frequency bands. The method has been developed by trial and error, and it can be further modified for different purposes.

Listing 5.2: DirAC Initialization

```
1   function DirAC_struct = DirAC_init_stft(ls_dirs, fs)
2   % Return different processing parameters for DirAC processing
3   % Archontis Politis and Ville Pulkki 2016
4
5   if fs == 44100 | fs == 48000
6       DirAC_struct.fs = fs; % Sample rate
7   else
8       disp('Sample rate has to be 44.1 or 48 kHz');
9       return
10  end
11
12  DirAC_struct.ls_dirs = ls_dirs;
13  % 2D/3D test
14  if min(size(ls_dirs))==1 || all(ls_dirs(:,2)==0)
15      DirAC_struct.dimension = 2;
16  else
17      DirAC_struct.dimension = 3;
18  end
19  nOutChan = length(ls_dirs);
20  DirAC_struct.nOutChan = nOutChan;
21
22  % compute VBAP gain table
23  DirAC_struct.VBAPtable = getGainTable(ls_dirs);
24  % compute virtual-microphone/ambisonic static decoding matrix
25  dirCoeff = (sqrt(3)-1)/2; % supercardioid virtual microphones
26  DirAC_struct.decodingMtx = computeVMICdecMtx(ls_dirs, dirCoeff);
27  % load/design decorrelating filters
28  [DirAC_struct.decorFilt, DirAC_struct.decorDelay] = ...
        computeDecorrelators(nOutChan, fs);
29  % winsize for STFT, with 50% overlap
30  DirAC_struct.winsize = 1024; % about 20ms
31  % smoothing parameters
32  DirAC_struct.dirsmooth_cycles = 20;
33  DirAC_struct.dirsmooth_limf = 3000;
34  DirAC_struct.diffsmooth_cycles = 50;
35  DirAC_struct.diffsmooth_limf = 10000;
36  DirAC_struct.gainsmooth_cycles = 200;
37  DirAC_struct.gainsmooth_limf = 1500;
38  % compute recursive smoothing coefficients for the given above ...
        values
39  freq = (0:DirAC_struct.winsize/2)'*fs/DirAC_struct.winsize;
40  period = 1./freq;
41  period(1) = period(2); % omit infinity value for DC
42
43  % diffuseness smoothing time constant in sec
44  tau_diff = period*DirAC_struct.diffsmooth_cycles;
45  % diffuseness smoothing recursive coefficient
46  alpha_diff = exp(-DirAC_struct.winsize./(2*tau_diff*fs));
```

```
47  % limit recursive coefficient
48  alpha_diff(freq>DirAC_struct.diffsmooth_limf) = ...
        min(alpha_diff(freq≤DirAC_struct.diffsmooth_limf));
49  DirAC_struct.alpha_diff = alpha_diff;
50
51  % direction smoothing time constant in sec
52  tau_dir = period*DirAC_struct.dirsmooth_cycles;
53  % diffuseness smoothing recursive coefficient
54  alpha_dir = exp(-DirAC_struct.winsize./(2*tau_dir*fs));
55  % limit recursive coefficient
56  alpha_dir(freq>DirAC_struct.dirsmooth_limf) = ...
        min(alpha_dir(freq≤DirAC_struct.dirsmooth_limf));
57  DirAC_struct.alpha_dir = alpha_dir;
58
59  % gain smoothing time constant in sec
60  tau_gain = period*DirAC_struct.gainsmooth_cycles;
61  % gain smoothing recursive coefficient
62  alpha_gain = exp(-DirAC_struct.winsize./(2*tau_gain*fs));
63  % limit recursive coefficient
64  alpha_gain(freq>DirAC_struct.gainsmooth_limf) = ...
        min(alpha_gain(freq≤DirAC_struct.gainsmooth_limf));
65  DirAC_struct.alpha_gain = alpha_gain * ones(1,nOutChan);
66
67  % Inverse directivity factor of vmics
68  DirAC_struct.invQ = dirCoeff^2 + (1/3)*(1-dirCoeff)^2;
69  Q = 1./DirAC_struct.invQ; % directivity factor of vmics
70  % correction factor for energy of diffuse sound
71  DirAC_struct.diffCorrection = sqrt(Q)*ones(1,nOutChan);
72
73  % Diffuse energy proportion to each loudspeaker.
74  DirAC_struct.lsDiffCoeff = sqrt(1/nOutChan)*ones(1,nOutChan);
75  DirAC_struct.parhistory = [];
76  end
77
78
79  %%%%%%%%%%%%%%%%%%%%%%%%%%%%%%%%%%%%%%%%%%%%%%%%%%%%%
80  function VMICdecMtx = computeVMICdecMtx(ls_dirs, alpha)
81  % virtual microphone type d(theta) = alpha + (1-alpha)*cos(theta)
82  % reshape ls_dirs, if 2D vector
83  if min(size(ls_dirs))==1
84      if isrow(ls_dirs), ls_dirs = ls_dirs'; end
85      ls_dirs(:,2) = zeros(size(ls_dirs));
86  end
87  % get the unit vectors of each vmic direction
88  Nvmic = size(ls_dirs, 1);
89  u_vmic = zeros(Nvmic, 3);
90  [u_vmic(:,1), u_vmic(:,2), u_vmic(:,3)] = ...
        sph2cart(ls_dirs(:,1)*pi/180, ls_dirs(:,2)*pi/180, ...
        ones(Nvmic, 1));
91  % divide dipoles with /sqrt(2) due to B-format convention
```

```
92      VMICdecMtx = [alpha*ones(Nvmic, 1) 1/sqrt(2)*(1-alpha)*u_vmic]';
93  end
94
95
96  %%%%%%%%%%%%%%%%%%%%%%%%%%%%%%%%%%%%%%%%%%%%%%%%%%%%%%%%%
97  function [decorFilt, decorDelay] = ...
            computeDecorrelators(nOutChan, fs)
98  % calls function that designs FIR filters for decorrelation
99  %decorFilt = compute_delay_decorrelation_response(fs,nOutChan);
100 decorDelay = 1500;
101
102 %This script creates decorrelation filters that have randomly
103 % delayed impulses at different frequency bands.
104 order = 3000;        %order of the bandpass filters
105 len = 1024*8;        %length of the decorrelation filters
106 maxdel = 80;         %maximum delay of the filters
107 mindel = 3;          %minimum delay of the filters
108 minmaxlocal = 30;    %above 1500Hz the value for delay upper limit
109 maxminlocal = 20;    %below 1500Hz the value for delay upper limit
110 mincycles = 10;      %mininum amount of delay in cycles
111 maxcycles = 40;      %maximum amount of delay in cycles
112
113 %compute the values in samples
114 maxdelN = round(maxdel/1000*fs);
115 mindelN = round(mindel/1000*fs);
116 minmaxlocalN = round(minmaxlocal/1000*fs);
117 maxminlocalN = round(maxminlocal/1000*fs);
118 if maxdelN > len-(order+1)
119     maxdelN = len-(order+1);
120 end
121 if minmaxlocalN > maxdelN
122     minmaxlocalN = maxdelN;
123 end
124 % Compute frequency band
125 [fpart, npart] = makepart_constcut(200, 2, nOutChan);
126 cutoff_f = fpart;
127 cutoff = cutoff_f/fs*2;
128 cycleN = fs./cutoff_f;
129 %compute the bandpass filters
130 h = zeros(order+1,npart,nOutChan);
131 for j = 1:nOutChan
132     h(:,1,j) = fir1(order, cutoff(1,j),'low');
133     for i = 2:npart
134         h(:,i,j) = fir1(order, [cutoff(i-1,j) cutoff(i,j)], ...
                    'bandpass');
135     end
136 end
137 % Compute the maximum and minimum delays
138 curveon = ones(npart,1);
```

```
139  mindellocalN = zeros(npart,1);
140  maxdellocalN = zeros(npart,1);
141  for i = 1:npart
142      maxdellocalN(i) = round(maxcycles*(1/cutoff_f(i))*fs);
143      mindellocalN(i) = round(mincycles*(1/cutoff_f(i))*fs);
144      if maxdellocalN(i) > maxdelN
145          maxdellocalN(i) = maxdelN;
146      end
147      if maxdellocalN(i) < minmaxlocalN
148          maxdellocalN(i) = minmaxlocalN;
149          curveon(i) = 0;
150      end
151      if mindellocalN(i) < mindelN
152          mindellocalN(i) = mindelN;
153      end
154      if mindellocalN(i) > maxminlocalN
155          mindellocalN(i) = maxminlocalN;
156      end
157  end
158  %convert to samples
159  maxdellocal = maxdellocalN/fs*1000;
160  mindellocal = mindellocalN/fs*1000;
161  delvariation = maxdellocal - mindellocal;
162  cycleT = cycleN/fs*1000;
163  %randomize the delays of the first band
164  decorFilt = zeros(len,nOutChan);
165  delayinit = ...
         (maxdelN-mindellocalN(1))*rand(1,nOutChan)+mindellocalN(1);
166  delay(1,:) = round(delayinit);
167  % Compute the frequency-dependent delay curve for each ...
         loudspeaker channel.
168  % A heuristic approach is used to form the curve, which limits
169  % how the delay varies between adjacent frequency channels.
170  for m = 1:nOutChan
171      for i = 2:npart
172          cycles = 0.5*i*i+1;
173          if curveon(i) == 0
174              delchange = ...
                     cycleN(i-1,m)*(round(rand(1,1)*cycles*2-cycles));
175          else
176              delchange = cycleN(i-1,m)* ...
                     (round(rand(1,1)*cycles*2-1.3*cycles));
177          end
178          delay(i,m) = delay(i-1,m) + delchange;
179          if delay(i,m) < mindellocalN(i)
180              k = 0;
181              while delay(i,m) < mindellocalN(i)
182                  delay(i,m) = delay(i,m) + cycleN(i-1,m);
183                  k = k+1;
```

```
184                end
185                if curveon(i) == 0
186                    delay(i,m) = ...
                         delay(i,m) + round(k/2)*cycleN(i-1,m);
187                end
188                while delay(i,m) > maxdellocalN(i)
189                    delay(i,m) = delay(i,m) - cycleN(i-1,m);
190                end
191            elseif delay(i,m) > maxdellocalN(i)
192                k = 0;
193                while delay(i,m) > maxdellocalN(i)
194                    delay(i,m) = delay(i,m) - cycleN(i-1,m);
195                    k = k+1;
196                end
197                if curveon(i) == 0
198                    delay(i,m) = ...
                         delay(i,m) - round(k/2)*cycleN(i-1,m);
199                end
200                while delay(i,m) < mindellocalN(i)
201                    delay(i,m) = delay(i,m) + cycleN(i-1,m);
202                end
203            end
204            delay(i,m) = round(delay(i,m));
205        end
206
207        % Summing up the response from band-pass impulse responses
208        hdelayed = zeros(len,npart);
209        for i = 1:npart
210            hdelayed(delay(i,m)+1:delay(i,m)+order+1,i) = h(:,i,m);
211        end
212        for i = 1:npart
213            decorFilt(:,m) = decorFilt(:,m) + hdelayed(:,i);
214        end
215 end
216 end
217
218 %%%%%%%%%%%%%%%%%%%%%%%%%%%%%%%%%%%%%%%%%%%%%%%%%%%%%%%%%%%%%%%%%%%%%%%%%
219 function [fpart npart] = makepart_constcut(first_band, ...
        bandsize, channels)
220 %%%% Compute auditory frequency bands
221 erb_bands = zeros(100,1); erb_bands(1) = first_band; ...
        lastband = 100; i = 2;
222 freq_upper_band = erb_bands(1);
223 while freq_upper_band < 20000
224     erb = 24.7 + 0.108 * freq_upper_band;   % Compute the width ...
            of the band.
225     % Compute the new upper limit of the band.
226     freq_upper_band = freq_upper_band + bandsize*erb;
227     erb_bands(i) = freq_upper_band;
```

```
228      i = i + 1;
229  end
230  lastband = min([lastband i-1]);
231  erb_bands = round(erb_bands);
232  erb_bands = erb_bands(1:lastband);
233  erb_bands(lastband) = 22000;
234  fpart = erb_bands*ones(1,channels);
235  npart = size(fpart,1);
236  end
```

5.9.3 DirAC Runtime

The runtime part of DirAC is based on a relatively simple STFT loop. The B-format signal is processed by subsequent time windows. The DOA parameter is computed without temporal integration, to enable it to respond as quickly as possible to temporal changes, for example in multi-source scenarios. The diffuseness parameter ψ and the computed gain factors are temporally smoothed using leaky integration using a frequency-dependent coefficient `alpha` computed from corresponding parameters. The parameters are then used to form the frequency-domain filters to be applied to the virtual microphone signals. To avoid time-aliasing artifacts in the processing, the filters are windowed in the time domain before applying them to the signals.

In the flow diagram of DirAC shown in Figure 5.3, the diffuse stream is summed with the non-diffuse stream after decorrelation and before the frequency to time domain transformation. This is a computationally efficient method, and it should be used particularly in real-time applications of DirAC. However, the implementation of decorrelation filters with temporal responses that exceed the time window lengths is not straightforward, and the streams cannot be monitored individually if a single audio file is produced. An implementationally simpler but computationally less efficient method is to generate the time-domain loudspeaker signals for the non-diffuse and diffuse streams separately, as shown in Figure 5.19. The loudspeaker signals for the diffuse stream are then

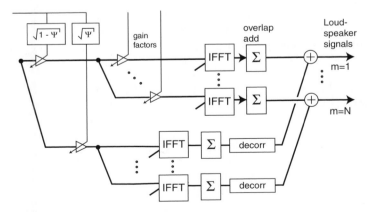

Figure 5.19 Implementation of decorrelation in the example code. The loudspeaker signals corresponding to the non-diffuse and diffuse streams are produced individually, and the diffuse stream is decorrelated after frequency to time domain transformation.

decorrelated in the time domain, which is easy to implement with simple FIR filtering. In addition, the loudspeaker signals for the streams can be monitored individually, and also their balance can be adjusted, which can be a useful audio effect. The Matlab implementation shown in Listing 5.3 follows this principle, and the output signals DIRsig and DIFFsig can be directly listened to as a technical demonstration of the functioning of DirAC.

Listing 5.3: DirAC Runtime

```
1   function [LSsig, DIRsig, DIFFsig, DirAC_struct] = ...
        DirAC_run_stft(insig, DirAC_struct)
2   %% Run-time processing of 2D or 3D virtual-microphone STFT DirAC
3   %% for loudspeaker output
4   %% Archontis Politis and Ville Pulkki  2016
5
6   lInsig = size(insig,1); % signal length
7   nInChan = size(insig,2); % normally 4 for B-format
8   nOutChan = DirAC_struct.nOutChan;
9
10  % STFT frame count and initialization
11  winsize = DirAC_struct.winsize;
12  hopsize = winsize/2;
13  fftsize = 2*winsize; % double the window size to suppress ...
        aliasing
14  Nhop = ceil(lInsig/hopsize) + 2;
15  insig = [zeros(hopsize,nInChan); insig; zeros(Nhop*hopsize - ...
        lInsig - hopsize,nInChan)]; % zero padding at start and end
16  % arrays for non-diffuse (direct) and diffuse sound output
17  dirOutsig = zeros(size(insig,1)+fftsize, nOutChan);
18  diffOutsig = zeros(size(insig,1)+fftsize, nOutChan);
19  % hanning window for analysis synthesis
20  window = hanning(winsize);
21  % zero pad both window and input frame to 2*winsize to
22  % suppress temporal aliasing from adaptive filters
23  window = [window; zeros(winsize,1)];
24  window = window*ones(1,nInChan);
25  % DirAC analysis initialization
26  DirAC_struct.Intensity_smooth = 0; % initial values for ...
        recursive smoothing
27  DirAC_struct.Intensity_short_smooth = 0; % initial values for ...
        recursive smoothing
28  DirAC_struct.energy_smooth = 0; % initial values for recursive ...
        smoothing
29  DirAC_struct.gains_smooth = 0;
30
31  % STFT runtime loop
32  for idx = 0:hopsize:(Nhop-2)*hopsize
33      % zero pad both window and input frame to 2*winsize for ...
            aliasing suppression
```

```
34      inFramesig = [insig(idx+(1:winsize),:); ...
            zeros(winsize,nInChan)];
35      inFramesig = inFramesig .* window;
36      % spectral processing
37      inFramespec = fft(inFramesig);
38      inFramespec = inFramespec(1:fftsize/2+1,:);
39      % save only positive frequency bins
40      % Analysis and filter estimation
41      % Estimate directional parameters from signal
42      %    using only non-interpolated spectrum
43      [pars,DirAC_struct] = ...
            computeDirectionalParameters(inFramespec(1:2:end,:), ...
            DirAC_struct);
44      pos=size(DirAC_struct.parhistory,1)+1;
45      DirAC_struct.parhistory(pos,:,:)=[pars];
46      % Non-diffuse (direct) and diffuse sound filters
47      directFilterspec = updateDirectFilters(pars, DirAC_struct);
48      diffuseFilterspec = updateDiffuseFilters(pars, DirAC_struct);
49      % Interpolate filters to fftsize
50      directFilterspec = interpolateFilterSpec(directFilterspec);
51      diffuseFilterspec = interpolateFilterSpec(diffuseFilterspec);
52      %%% Synthesis of non-diffuse/diffuse streams
53      % apply non-parametric decoding first (virtual microphones)
54      linOutFramespec = inFramespec*DirAC_struct.decodingMtx;
55      % adapt the linear decoding to the direct and diffuse streams
56      dirOutFramespec = directFilterspec .* linOutFramespec;
57      diffOutFramespec = diffuseFilterspec .* linOutFramespec;
58      % overlap-add
59      dirOutFramesig = real(ifft([dirOutFramespec; ...
            conj(dirOutFramespec(end-1:-1:2,:))]));
60      dirOutsig(idx+(1:fftsize),:) = ...
            dirOutsig(idx+(1:fftsize),:) + dirOutFramesig;
61      diffOutFramesig = real(ifft([diffOutFramespec; ...
            conj(diffOutFramespec(end-1:-1:2,:))]));
62      diffOutsig(idx+(1:fftsize),:) = ...
            diffOutsig(idx+(1:fftsize),:) + diffOutFramesig;
63  end
64  % remove delay caused by the intepolation of gains and circular ...
        shift
65  dirOutsig = dirOutsig(hopsize+1:end,:);
66  diffOutsig = diffOutsig(hopsize+1:end,:);
67  % apply decorrelation to diffuse stream and remove decorrelation
68  % delay if needed
69  if ~isempty(DirAC_struct.decorDelay) || DirAC_struct.decorDelay~=0
70      tempsig = [diffOutsig; zeros(DirAC_struct.decorDelay, ...
            nOutChan)];
71      tempsig = fftfilt(DirAC_struct.decorFilt, tempsig);
72      diffOutsig = tempsig(DirAC_struct.decorDelay+1:end,:);
73  else
```

```
74      diffOutsig = fftfilt(DirAC_struct.decorFilt, diffOutsig);
75   end
76   % remove delay due to windowing and truncate output to original ...
        length
77   DIRsig = dirOutsig(hopsize+(1:lInsig),:);
78   DIFFsig = diffOutsig(hopsize+(1:lInsig),:);
79   LSsig = DIRsig + DIFFsig;
80   end
81
82   %%%%%%%%%%%%%%%%%%%%%%%%%%%%%%%%%%%%%%%%%%%%%%%%%%%%%%%%%%%%%%%%%%
83   function [pars, DirAC_struct] = ...
        computeDirectionalParameters(insigSpec, DirAC_struct)
84
85      %%% B-format analysis
86      w = insigSpec(:,1); % omni
87      X = insigSpec(:,2:4)/sqrt(2);
88      % dipoles /cancel B-format dipole convention
89      Intensity = real(conj(w)*ones(1,3) .* X);
90      % spatially reversed normalized active intensity
91      energy = (abs(w).^2 + sum(abs(X).^2,2))/2;
92      % normalized energy density
93      % direction-of-arrival parameters
94      alpha_dir = DirAC_struct.alpha_dir;
95      Alpha_dir = alpha_dir*ones(1,3);
96      Intensity_short_smooth = ...
            Alpha_dir.*DirAC_struct.Intensity_short_smooth + ...
            (1-Alpha_dir).*Intensity;
97      azi = atan2(Intensity_short_smooth(:,2), ...
            Intensity_short_smooth(:,1))*180/pi;
98      elev = atan2(Intensity_short_smooth(:,3), ...
            sqrt(sum(Intensity_short_smooth(:,1:2).^2,2)))*180/pi;
99      % diffuseness parameter
100     alpha_diff = DirAC_struct.alpha_diff;
101     Alpha_diff = alpha_diff*ones(1,3);
102     Intensity_smooth = ...
            Alpha_diff.*DirAC_struct.Intensity_smooth + ...
            (1-Alpha_diff).*Intensity;
103     Intensity_smooth_norm = sqrt(sum(Intensity_smooth.^2,2));
104     energy_smooth = alpha_diff.*DirAC_struct.energy_smooth + ...
            (1-alpha_diff).*energy;
105     diffuseness = 1 - Intensity_smooth_norm./(energy_smooth + ...
            eps);
106     diffuseness(diffuseness<eps) = eps;
107     diffuseness(diffuseness>1-eps) = 1-eps;
108     % store parameters
109     pars = [azi elev energy diffuseness];
110     % update values for recursive smoothing
111     DirAC_struct.Intensity_short_smooth = Intensity_short_smooth;
112     DirAC_struct.Intensity_smooth = Intensity_smooth;
```

```
113        DirAC_struct.energy_smooth = energy_smooth;
114   end
115
116   %%%%%%%%%%%%%%%%%%%%%%%%%%%%%%%%%%%%%%%%%%%%%%%%%%%%%%%%%%%%
117   function [directFilterspec, DirAC_struct] = ...
           updateDirectFilters(pars, DirAC_struct)
118
119        nOutChan = DirAC_struct.nOutChan;
120        azi = pars(:,1);
121        elev = pars(:,2);
122        energy = pars(:,3);
123        diff = pars(:,4);
124        ndiff_sqrt = sqrt(1-diff); % diffuse sound suppresion filter
125        ndiff_energy = energy.*(1-diff); % non-diffuse energy amount
126        % Amplitude panning gain filters
127        Alpha = DirAC_struct.alpha_gain;
128        if DirAC_struct.dimension == 3
129            % look-up the corresponding VBAP gains from the table
130            aziIndex = round(mod(azi+180,360)/2);
131            elevIndex = round((elev+90)/5);
132            idx3D = elevIndex*181+aziIndex+1;
133            gains = DirAC_struct.VBAPtable(idx3D,:);
134        else
135            % look-up the corresponding VBAP gains from the table
136            idx2D = round(mod(azi+180,360))+1;
137            gains = DirAC_struct.VBAPtable(idx2D,:);
138        end
139        % recursive smoothing of gains (energy-weighted)
140        gains_smooth = Alpha.*DirAC_struct.gains_smooth + ...
               (1-Alpha).*(ndiff_energy * ones(1,nOutChan)).*gains;
141        % store smoothed gains for next update (before ...
               re-normalization)
142        DirAC_struct.gains_smooth = gains_smooth;
143        % re-normalization of smoothed gains to unity power
144        gains_smooth = gains_smooth .* ...
               (sqrt(1./(sum(gains_smooth.^2,2)+eps))*ones(1,nOutChan));
145        % Combine separation filters with panning filters, including
146        % approximate correction for the effect of virtual
147        % microphones to the direct sound
148        dirCorrection = (1./sqrt(1 + ...
               diff*(DirAC_struct.invQ-1)))*ones(1,nOutChan);
149        directFilterspec = gains_smooth .* ...
               (ndiff_sqrt*ones(1,nOutChan)) .* dirCorrection;
150   end
151
152   %%%%%%%%%%%%%%%%%%%%%%%%%%%%%%%%%%%%%%%%%%%%%%%%%%%%%%%%%%%%
153   function [diffuseFilterspec, DirAC_struct] = ...
           updateDiffuseFilters(pars, DirAC_struct)
154
```

```
155        diff = pars(:,4);
156        % Combine separation filters with approximate correction
157        % for the effect of virtual microphones to the diffuse sound
158        % energy, and energy weights per loudspeaker
159        diffuseFilterspec = sqrt(diff) * ...
                (DirAC_struct.diffCorrection.*DirAC_struct.lsDiffCoeff);
160   end
161
162   %%%%%%%%%%%%%%%%%%%%%%%%%%%%%%%%%%%%%%%%%%%%%%%%%%%%%%%%%%%%%%%
163   function intFilterspec = interpolateFilterSpec(filterspec)
164
165        nChan = size(filterspec,2);
166        hopsize = size(filterspec,1)-1;
167        winsize = hopsize*2;
168        % IFFT to time domain
169        filterimp = ifft([filterspec; ...
                conj(filterspec(end-1:-1:2,:))]);
170        % circular shift
171        filterimp = [filterimp(hopsize+1:end, :); ...
                filterimp(1:hopsize, :)];
172        % zero-pad to 2*winsize
173        filterimp = [filterimp; zeros(winsize, nChan)];
174        intFilterspec = fft(filterimp); % back to FFT
175        % save only positive frequency bins
176        intFilterspec = intFilterspec(1:winsize+1, :);
177   end
```

5.9.4 Simplistic Binaural Synthesis of Loudspeaker Listening

The function shown in Listing 5.4 simulates headphone listening of a multichannel loud-speaker setup. The signal path from a loudspeaker to each ear is modeled simply by the interaural time difference of arrival, and by the shadowing of the head. The shadowing is assumed to follow a cardioid pattern. In practice, this gives a realistic ITD cue, and ILD cues assuming frequency independence. No monaural cues are modeled. The spectral impairments caused by the linear techniques are shown clearly, since the different delay times from each loudspeaker to the ears cause comb-filter effects in the ear canal signals. On the other hand, the timbral effects due to decorrelation are also more audible in headphone listening with completely dry acoustics than with loudspeaker listening with loudspeakers, where the room response masks more or less the decorrelation artifacts.

Listing 5.4: Binaural Synthesis

```
1   function [HPsig] = Headphone_ITDILD_simulation(LSsig, DirAC_struct)
2   %% simulation of headphone listening with assumption of ...
            cardioid shadowing and ITD
3   ls_dirs=DirAC_struct.ls_dirs;
4
```

```
5   HPsig = zeros(size(LSsig,1),2);
6   LSsig = [LSsig; zeros(100,size(LSsig,2))];
7   initdelay = 0.00035*DirAC_struct.fs; % half of highest value ...
        for ITD
8   for i=1:size(LSsig,2) % delay each signal to get correct ITD, ...
        apply side-facing cardioid
9       delayL=round((sin(-ls_dirs(i,1)/180*pi)+1)*initdelay);
10      delayR=round((sin(ls_dirs(i,1)/180*pi)+1)*initdelay);
11      HPsig(:,1)=HPsig(:,1) + ...
            LSsig(1+delayL:end-100+delayL,i)*(1+sin(ls_dirs(i,1)))/2;
12      HPsig(:,2)=HPsig(:,2) + ...
            LSsig(1+delayR:end-100+delayR,i)*(1+sin(-ls_dirs(i,1)))/2;
13  end
```

5.10 Summary

Directional audio coding (DirAC) is a perceptually motivated time–frequency domain spatial sound reproduction method. It assumes that reproducing the parameters of the diffuseness of the sound field and the most prominent direction-of-arrival of sound provides high perceptual quality in reproduction of the spatial attributes of sound. This chapter has described how to use first-order DirAC in different applications with different theoretical and practical microphone setups for both loudspeaker and headphone playback. A clear conclusion is that DirAC indeed enhances the reproduction quality when compared to signal-independent decoding of the same first-order input signal with both loudspeakers and headphones.

The system performs well if the recorded spatial sound matches the implicit assumptions about the sound field in DirAC – that at each frequency band only a single source is dominant at one time, with a moderate level of reverberation. In cases where the recorded sound field strongly violates these assumptions, audible distortions, artifacts, may occur. Typical cases are, for example, surrounding applause, speech in the presence of broad-band noise from opposing directions, or strong early reflections within a single temporal analysis window. The artifacts are most often due to the temporal and spectral effects from decorrelation processing, which is needed to decrease the level of coherence between loudspeaker signals in the case of a sound field with high diffuseness. The reasons why decorrelation in some cases produces artifacts have been tracked and methods to mitigate the artifacts have been developed in the work presented. Practical microphone arrays already exhibit bias and noise effects in spatial metadata, which in some cases cause too high a level of decorrelated energy in the output. Some of the artifacts occurring in spatially complex sound fields can be avoided by optimizing the time–frequency resolution of the processing. In addition, covariance-domain rendering mitigates the artifacts by optimizing the level of decorrelated energy automatically. However, it seems that even in the best case the reproduced sound is still perceptually sightly different from the recording conditions with a spatially complex sound scene. It is our assumption that the quality can be enhanced further by utilization of higher-order microphones, as discussed in the next chapter.

References

Ahonen, J. (2010) Microphone configurations for teleconference application of directional audio coding and subjective evaluation. *Audio Engineering Society 40th International Conference: Spatial Audio: Sense the Sound of Space*. Audio Engineering Society.

Ahonen, J., Del Galdo, G., Kuech, F., and Pulkki, V. (2012a) Directional analysis with microphone array mounted on rigid cylinder for directional audio coding. *Journal of the Audio Engineering Society*, **60**(5), 311–324.

Ahonen, J. and Pulkki, V. (2009) Diffuseness estimation using temporal variation of intensity vectors. *IEEE Workshop on Applications of Signal Processing to Audio and Acoustics, WASPAA'09*, pp. 285–288. IEEE.

Ahonen, J. and Pulkki, V. (2010) Speech intelligibility in teleconference application of directional audio coding. *Audio Engineering Society 40th International Conference: Spatial Audio: Sense the Sound of Space*. Audio Engineering Society.

Ahonen, J. and Pulkki, V. (2011) Broadband direction estimation method utilizing combined pressure and energy gradients from optimized microphone array. *IEEE International Conference on Acoustics, Speech, and Signal Processing (ICASSP)*, pp. 97–100. IEEE.

Ahonen, J., Sivonen, V., and Pulkki, V. (2012b) Parametric spatial sound processing applied to bilateral hearing aids. *Audio Engineering Society 45th International Conference: Applications of Time–Frequency Processing in Audio*. Audio Engineering Society.

Avendano, C. and Jot, J.M. (2004) A frequency-domain approach to multichannel upmix. *Journal of the Audio Engineering Society*, **52**(7/8), 740–749.

Benesty, J., Chen, J., and Huang, Y. (2008) *Microphone Array Signal Processing*, vol. 1. Springer Science & Business Media, New York.

Blauert, J. (1997) *Spatial Hearing: The Psychophysics of Human Sound Localization* (rev. ed.), MIT Press, Cambridge, MA.

Brandstein, M. and Ward, D. (2013) *Microphone Arrays: Signal Processing Techniques and Applications*. Springer, Berlin.

Del Galdo, G., Taseska, M., Thiergart, O., Ahonen, J., and Pulkki, V. (2012) The diffuse sound field in energetic analysis. *Journal of the Acoustical Society of America*, **131**(3), 2141–2151.

Dillon, H. (2001) *Hearing Aids*, vol. 362, Boomerang Press, Sydney.

Fahy, F. (2002) *Sound Intensity*, CRC Press, Boca Raton, FL.

Faller, C. (2006) Multiple-loudspeaker playback of stereo signals. *Journal of the Audio Engineering Society*, **54**(11), 1051–1064.

Gerzon, M.A. (1973) Periphony: With-height sound reproduction. *Journal of the Audio Engineering Society*, **21**(1), 2–10.

Gerzon, M.A. (1985) Ambisonic in multichannel broadcasting and video. *Journal of the Audio Engineering Society*, **33**(11), 859–871.

Glasberg, B.R. and Moore, B.C.J. (1990) Derivation of auditory filter shapes from notched-noise data. *Hearing Research*, **47**(1–2), 103–38.

Hirvonen, T., Ahonen, J., and Pulkki, V. (2009) Perceptual compression methods for metadata in directional audio coding applied to audiovisual teleconference. *Audio Engineering Society Convention 126*. Audio Engineering Society.

Hiyama, K., Komiyama, S., and Hamasaki, K. (2002) The minimum number of loudspeakers and its arrangement for reproducing the spatial impression of diffuse sound field. *Audio Engineering Society 113th Convention*. Audio Engineering Society.

Kailath, T. (1980) *Linear Systems*, Prentice Hall Information and System Sciences series, Prentice-Hall, Englewood Cliffs, NJ.

Kallinger, M., Del Galdo, G., Kuech, F., Mahne, D., and Schultz-Amling, R. (2009) Spatial filtering using Directional Audio Coding parameters. *IEEE International Conference on Acoustics, Speech and Signal Processing*, pp. 217–220. IEEE Computer Society.

Koski, T., Sivonen, V., and Pulkki, V. (2013) Measuring speech intelligibility in noisy environments reproduced with parametric spatial audio. *Audio Engineering Society Convention 135*. Audio Engineering Society.

Kratschmer, M., Thiergart, O., and Pulkki, V. (2012) Envelope-based spatial parameter estimation in directional audio coding. *Audio Engineering Society Convention 133*. Audio Engineering Society.

Kuech, F., Kallinger, M., Schultz-Amling, R., Galdo, G.D., Ahonen, J., and Pulkki, V. (2008) Directional audio coding using planar microphone arrays. *IEEE Conference on Hands-Free Speech Communication and Microphone Arrays*, pp. 37–40. IEEE.

Laitinen, M.V., Kuech, F., and Pulkki, V. (2011a) Using spaced microphones with directional audio coding. *Audio Engineering Society Convention 130*. Audio Engineering Society.

Laitinen, M.V., Kuech, F., Disch, S., and Pulkki, V. (2011b) Reproducing applause-type signals with directional audio coding. *Journal of the Audio Engineering Society*, **59**(1/2), 29–43.

Laitinen, M.V., Pihlajamäki, T., Loesler, S., and Pulkki, V. (2012) Influence of resolution of head tracking in synthesis of binaural audio. *Audio Engineering Society Convention 132*. Audio Engineering Society.

Laitinen, M.V., and Pulkki, V. (2009) Binaural reproduction for Directional Audio Coding. *IEEE Workshop on Applications of Signal Processing to Audio and Acoustics (WASPAA)*, New Paltz, NY, USA.

Merimaa, J. and Pulkki, V. (2005) Spatial impulse response rendering 1: Analysis and synthesis. *Journal of the Audio Engineering Society*, **53**(12), 1115–1127.

Moore, B.C.J. (1982) *An Introduction to the Psychology of Hearing*, Academic Press, New York.

Moreau, S., Bertet, S., and Daniel, J. (2006) 3D sound field recording with higher order Ambisonics: Objective measurements and validation of spherical microphone. *120th Convention of the Audio Engineering Society*, Paris, France.

Poletti, M.A. (2005) Three-dimensional surround sound systems based on spherical harmonics. *Journal of the Audio Engineering Society*, **53**(11), 1004–1025.

Politis, A., Delikaris-Manias, S., and Pulkki, V. (2015a) Direction-of-arrival and diffuseness estimation above spatial aliasing for symmetrical directional microphone arrays. *IEEE International Conference on Acoustics, Speech and Signal Processing (ICASSP)*, pp. 6–10. IEEE.

Politis, A., Laitinen, M.V., Ahonen, J., and Pulkki, V. (2015b) Parametric spatial audio processing of spaced microphone array recordings for multichannel reproduction. *Journal of the Audio Engineering Society*, **63**(4), 216–227.

Politis, A. and Pulkki, V. (2011) Broadband analysis and synthesis for directional audio coding using A-format input signals. *Audio Engineering Society Convention 131*. Audio Engineering Society.

Pulkki, V. (1997) Virtual sound source positioning using vector base amplitude panning. *Journal of the Audio Engineering Society*, **45**(6), 456–466.

Pulkki, V. (2002) Microphone techniques and directional quality of sound reproduction. *112th Audio Engineering Society Convention*, Munich, Germany.

Pulkki, V. (2007) Spatial sound reproduction with Directional Audio Coding. *Journal of the Audio Engineering Society*, **55**(6), 503–516.

Pulkki, V., Merimaa, J., and Lokki, T. (2004) Multichannel reproduction of measured room responses. *International Congress on Acoustics*, pp. 1273–1276.

Santala, O. *et al.* (2015) Perception and auditory modeling of spatially complex sound scenarios. PhD thesis. Aalto University.

Schulein, R.B. (1976) Microphone considerations in feedback-prone environments. *Journal of the Audio Engineering Society*, **24**(6), 434–445.

Solvang, A. (2008) Spectral impairment of two-dimensional higher order Ambisonics. *Journal of the Audio Engineering Society*, **56**(4), 267–279.

Stanzial, D., Prodi, N., and Schiffrer, G. (1996) Reactive acoustic intensity for general fields and energy polarization. *Journal of the Acoustical Society of America*, **99**(4), 1868–1876.

Thiergart, O. (2015) Flexible multi-microphone acquisition and processing of spatial sound using parametric sound field representations. PhD thesis. Friedrich-Alexander-Universitat Erlangen-Nurnberg.

Thiergart, O., Kratschmer, M., Kallinger, M., and Del Galdo, G. (2011) Parameter estimation in directional audio coding using linear microphone arrays. *Audio Engineering Society Convention 130*. Audio Engineering Society.

Vilkamo, J. (2008) Spatial sound reproduction with frequency band processing of b-format audio signals. Master's thesis. Aalto University.

Vilkamo, J., Lokki, T., and Pulkki, V. (2009) Directional audio coding: Virtual microphone-based synthesis and subjective evaluation. *Journal of the Audio Engineering Society*, **57**(9), 709–724.

Vilkamo, J. and Pulkki, V. (2013) Minimization of decorrelator artifacts in directional audio coding by covariance domain rendering. *Journal of the Audio Engineering Society*, **61**(9), 637–646.

Zotter, F. and Frank, M. (2012) All-round ambisonic panning and decoding. *Journal of the Audio Engineering Society*, **60**(10), 807–820.

6

Higher-Order Directional Audio Coding

Archontis Politis and Ville Pulkki

Department of Signal Processing and Acoustics, Aalto University, Finland

6.1 Introduction

The first-order model of directional audio coding (FO-DirAC), detailed in Chapter 5, estimates a single direction of arrival (DoA) and a global diffuseness estimate that depend on the interactions of all the sound wave contributions arriving in a single observation window. If two source signals with similar spectra and incident from different directions are captured with nearly equal powers at the same time–frequency point, three scenarios can occur during FO-DirAC reproduction with perceivable effects. If the analysis of the B-format signals is computed through very short time averages, then the estimated DoA will fluctuate between the two true DoAs, with most estimates clustered around the true values. If no temporal smoothing is performed on the respective time–frequency gains applied during synthesis, then this directional distribution will be reproduced appropriately, with most power spatialized around the two DoAs, but with a high chance of perceived image instability and musical noise. If longer temporal smoothing is used on the spatialization gains, then most likely multiple loudspeakers will be activated between the two DoAs, resulting in a stable rendering but with an increased spatial blurring similar to linear or ambisonic decoding. On the other hand, if the DoA and diffuseness are computed using longer temporal windows or temporal averaging, then the analyzed DoA will be stable but on the mean of the two true DoAs, while the diffuseness will be overestimated. While an averaged DoA may not be a problem from a perceptual point of view, since it will be equivalent to a panned virtual source representing the two true ones, the high diffuseness will have a detrimental perceptual effect, due to more sound being assigned to the diffuse stream and decorrelated. Undesired decorrelation effects may include smearing of transients (Laitinen *et al.*, 2011), timbral coloration at high frequencies, and an added reverberation effect for multi-source but otherwise non-reverberant scenes (Laitinen and Pulkki, 2012). Some of these cases are depicted schematically in Figure 6.1.

Reproduction of such cases can benefit from techniques that estimate multiple narrowband directional components, such as the multi-source beamforming or blind source separation (BSS) approaches detailed in Chapters 7 and 9, and the references

Parametric Time–Frequency Domain Spatial Audio, First Edition. Edited by Ville Pulkki,
Symeon Delikaris-Manias, and Archontis Politis.
© 2018 John Wiley & Sons Ltd. Published 2018 by John Wiley & Sons Ltd.
Companion Website: www.wiley.com/go/pulkki/parametrictime-frequency

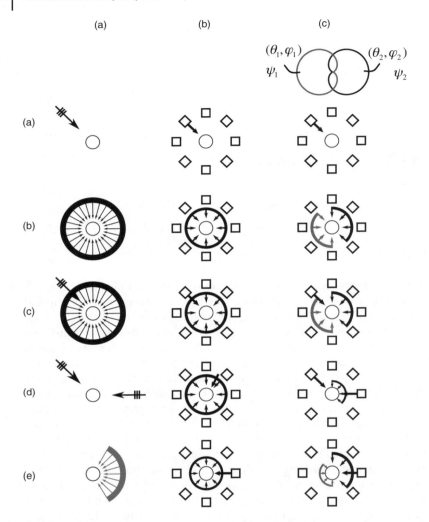

Figure 6.1 Some basic sound-field scenarios, and their interpretation and loudspeaker reproduction using FO-DirAC and HO-DirAC. The rows depict: (a) a single plane wave, (b) an isotropic diffuse field, (c) a single plane wave mixed with a diffuse field, (d) two plane waves, and (e) a diffusely distributed extended source. The columns depict (a) the original field, (b) FO-DirAC, and (c) HO-DirAC. The HO-DirAC analysis here is demonstrated with two angular sectors only.

in Chapter 4. However, retaining the energetic analysis model of DirAC is advantageous due to its perceptual effectiveness and computational simplicity. Such a method has been presented by Pulkki *et al.* (2013) and Politis *et al.* (2015), and is termed here "higher-order DirAC" (HO-DirAC). HO-DirAC corresponds to a generalization of FO-DirAC to higher orders of input, and hence it requires additional input channels to the minimum four of B-format. HO-DirAC can also be seen as an application of parametric spatial audio processing to higher-order Ambisonics (HOA).

The HO-DirAC approach is based on a spherical harmonic (SH) representation of the sound field, equivalent to what is used in non-parametric HOA processing. The terms

"SH signals" or "HOA signals" are used interchangeably in this context. HOA signals naturally encode the directional properties of the sound field with increased spatial resolution and allow estimation of additional parameters to describe the acoustic scene, compared to a first-order representation. In HO-DirAC, such a solution is formulated with the following interrelated objectives: (i) resolve the issues of FO-DirAC by appropriate use of HOA signals, (ii) retain the energetic analysis/synthesis scheme of DirAC due to its robustness and perceptual effectiveness, (iii) achieve (i) and (ii) while preserving the energy of the recording at reproduction equally well for all directions.

HO-DirAC uses the additional HOA signals to divide the sound scene into angular sectors within which the single DoA and diffuseness model parameters are estimated. The sectors and the energetic analysis inside them rely on beam patterns with appropriate directional properties, while their number depends on the available order. The method resolves the issues of first-order DirAC by reducing the effects of simultaneous sources or reflections incident from directions outside each sector. This is sketched in Figure 6.1 for a simple two-sector case. A further benefit is that the diffuse component has a directional distribution and therefore it covers the cases of non-uniform reverberation and sources with spatial extent. At reproduction, the synthesis of the multiple per-sector parameters is optimized using the optimal least-squares mixing technique proposed by Vilkamo *et al.* (2013); see also Section 5.6.4. The overall operational principle of HO-DirAC analysis and synthesis in sectors is shown in Figure 6.2.

Since the increased spatial resolution of HOA signals also improves the reconstruction of spatial auditory cues for non-parametric linear decoding, the perceptual advantage of parametric over non-parametric processing seems to get smaller at higher orders. However, considering the listening test results of Politis *et al.* (2015) and Section 6.5, there is a clear benefit in HO-DirAC processing up to fourth order under anechoic

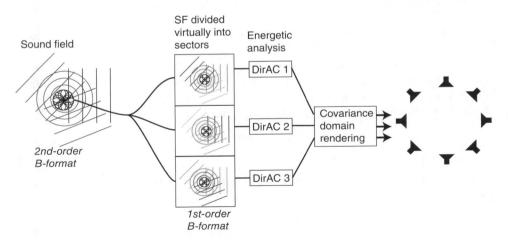

Figure 6.2 Principle of operation of HO-DirAC using second-order signals. The sound field is virtually divided into three directionally emphasized components, called sectors, and modified pressure and velocity signals are computed that would occur in the sectors. The typical energetic analysis of standard DirAC is performed on the modified pressure–velocity signals for the three sectors, and the parameters are combined and utilized during synthesis using the optimal mixing technique with covariance rendering.

playback conditions and headphone rendering. These orders are commonly the case of interest, due to the number of channels being still manageable for 3D audio delivery and transmission, and due to compact and practical recording arrays not delivering orders higher than that. As a drawback, HO-DirAC has a higher computational complexity than the highest-quality first-order DirAC variants, virtual microphone DirAC (VM-DirAC), or the optimal-mixing DirAC (OM-DirAC) presented in the previous chapter. The perceptual benefit depends on the availability of higher-order signals and on the acoustic scene itself. In general, VM-DirAC and OM-DirAC provide a good trade-off between complexity and performance, being effective for a majority of scenarios with the minimum number of four B-format channels for 3D playback. If, however, the application requires maximum quality for all scenarios and the material or recordings deliver HOA signals, then the HO-DirAC method should be preferred.

6.2 Sound Field Model

We assume that a real spatial sound scene is captured or a simulated one is encoded in a spherical harmonic representation. Such a representation is based on the formalism of a continuous band-limited amplitude distribution describing the incident sound field, and capturing all the contributions of the sound scene, such as multiple sources, reflections, and reverberation. The band limitation refers to the spatial, or angular, variability of the sound amplitude distribution, and a low-order representation approximates only coarsely sounds incident with high spatial concentration.

Let us assume that this continuous spatial distribution describing the sound field at time t is expressed by $a(t, \gamma)$, where γ is a unit vector pointing at azimuth ϕ and elevation θ. The vector of SH (or HOA) signals corresponds to the sound field coefficients $\mathbf{a}(t)$ given by the spherical harmonic transform of the sound field distribution,

$$\mathbf{a}(t) = SHT\{a(t, \gamma)\} = \int_\gamma a(t, \gamma)\mathbf{y}(\gamma)\, d\gamma, \tag{6.1}$$

where integration occurs over the unit sphere with $d\gamma = \cos\theta d\theta d\phi$. The vector $\mathbf{y}(\gamma)$ contains the spherical harmonic functions $Y_{nm}(\gamma)$ of order n and degree m. For a representation band-limited to order N there are $M = (N+1)^2$ signals and SHs in the vectors above. The ordering of the components in these vectors is:

$$[\mathbf{y}]_q = Y_{nm} \quad \text{with} \quad n = 0, 1, \ldots, N, \quad m = -n, \ldots, n,$$
$$[\mathbf{a}]_q = a_{nm}, \quad \text{and} \quad q = n^2 + n + m + 1. \tag{6.2}$$

To be compatible with HOA conventions, the real form of the SHs is used here, specified as

$$Y_{nm}(\theta, \phi) = \sqrt{(2n+1)\frac{(n-|m|)!}{(n+|m|)!}}P_{n|m|}(\sin\theta)y_m(\phi), \tag{6.3}$$

with

$$y_m(\phi) = \begin{cases} \sqrt{2}\sin|m|\phi & m < 0, \\ 1 & m = 0, \\ \sqrt{2}\cos m\phi & m > 0, \end{cases} \tag{6.4}$$

and P_{nm} the associated Legendre functions of degree n. The SHs are orthonormal with

$$\int_\gamma \mathbf{y}(\gamma)\mathbf{y}^T(\gamma)\,\mathrm{d}\gamma = 4\pi\mathbf{I}, \tag{6.5}$$

where $\mathbf{I}$ is the $M \times M$ identity matrix. Using this power normalization, the zeroth-order ambisonic signal $[\mathbf{a}]_1 = a_{00}$ is equivalent to an omnidirectional signal at the origin.

Beamforming in the spherical harmonic domain (SHD) reduces to a simple weight-and-sum of the SH signals with the SH coefficients of the beam pattern. A real-valued beam pattern, described by the directivity pattern $w(\gamma)$, has its SHD representation as the coefficient vector $\mathbf{w} = SHT\{w(\gamma)\}$. The coefficients in this case constitute the beamforming weights, and the beamformer's output is given by

$$y(t) = \mathbf{w}^T\mathbf{a}(t). \tag{6.6}$$

Note that in the Ambisonics literature, such beamforming on the HOA signals is termed a virtual microphone.

When a set of beamformers as in Equation (6.6) are used to distribute the sound to a playback system of L channels for sound reproduction, the $L \times M$ matrix of weights $\mathbf{D}$ is termed a "decoding matrix." It is derived according to the spatial properties of the reproduction system and is commonly frequency independent (a matrix of gains), even though frequency-dependent schemes are sometimes used (a matrix of filters). The signal vector $\mathbf{b} = [b_1, \dots, b_L]^T$ of the output channels is given by

$$\mathbf{b}(t) = \mathbf{D}\,\mathbf{a}(t). \tag{6.7}$$

Contrary to the linear processing of HOA signals in the ambisonic framework, where all operations can be performed with multiplications or convolutions and summations of the time-domain signals, parametric spatial sound processing is performed based on second-order statistics of the HOA signals in the time–frequency domain. Such transforms are, for example, the short-time Fourier transform (STFT) or a quadrature mirror filterbank (QMF) – see also Chapter 1. Assuming frequency and temporal indices k, l respectively, the statistics of interest are captured in the spatial covariance matrix (SCM) of the HOA signals,

$$\mathbf{C}_\mathbf{a}(k, l) = \mathbb{E}\{\mathbf{a}_{kl}\mathbf{a}_{kl}^H\}, \tag{6.8}$$

where $\mathbb{E}\{\cdot\}$ denotes statistical expectation. The statistics are updated constantly to capture spatial variations in the sound scene and are utilized for the extraction of the spatial parameters.

6.3   Energetic Analysis and Estimation of Parameters

The method consists of two main stages: the analysis stage, where energetic spatial sound field parameters are estimated in the SHD, and the synthesis stage, where these parameters for each time frame are used to adaptively mix the signals in such a way that the spatial characteristics of the original sound scene are reconstructed in a perceptual sense. In the following, the time–frequency indices l, k are dropped for clarity.

6.3.1 Analysis of Intensity and Diffuseness in the Spherical Harmonic Domain

Considering the amplitude density formulation of the sound scene presented above, the pressure and the acoustic particle velocity at the origin due to it are given by

$$p = \int_\gamma a(\gamma)\,d\gamma, \tag{6.9}$$

$$\mathbf{u} = -\frac{1}{Z_0}\int_\gamma a(\gamma)\gamma\,d\gamma = -\frac{1}{Z_0}\mathbf{v}. \tag{6.10}$$

The signal vector $\mathbf{v} = [v_x,\ v_y,\ v_z]^T$ corresponds to the opposite-sign unnormalized Cartesian components of the particle velocity, and $Z_0 = c\rho_0$ is the characteristic impedance of air. The pressure signal can be captured with an omnidirectional pattern, while the velocity, as is obvious from Equation (6.10), can be captured with three dipole patterns $x(\gamma)$, $y(\gamma)$, and $z(\gamma)$ corresponding to the components of γ as

$$\gamma = \begin{bmatrix} x(\gamma) \\ y(\gamma) \\ z(\gamma) \end{bmatrix} = \begin{bmatrix} \cos\theta\cos\varphi \\ \cos\theta\sin\varphi \\ \sin\theta \end{bmatrix}. \tag{6.11}$$

Let us denote a pressure–velocity signal vector as $\mathbf{a}_{pv} = [p,\ \mathbf{v}^T]^{T}.$[1] As was mentioned already, the pressure signal is equivalent to the zeroth-order SH signal $p = a_{00} = [\mathbf{a}]_1$. The dipole signals are also directly related to the first-order SH signals through reordering and scaling. More specifically, the pressure–velocity signal vector is related to the SH signals through the matrix

$$\mathbf{a}_{pv} = \begin{bmatrix} 1 & 0 & 0 & 0 \\ 0 & 0 & 0 & 1/\sqrt{3} \\ 0 & 1/\sqrt{3} & 0 & 0 \\ 0 & 0 & 1/\sqrt{3} & 0 \end{bmatrix}\mathbf{a}_1, \tag{6.12}$$

where $\mathbf{a}_1 = [a_{00},\ a_{1(-1)},\ a_{10},\ a_{11}]^T$ are the first-order SH signals.

From the pressure and velocity signals, the active intensity, sound field energy, and diffuseness are computed as follows:

$$\mathbf{i}_a = -\mathrm{Re}\{p\mathbf{v}^H\}, \tag{6.13}$$

$$E = \frac{1}{2}[|p|^2 + \mathbf{v}^H\mathbf{v}], \tag{6.14}$$

$$\psi = 1 - \frac{\|\mathbf{i}_a\|}{E} = 1 - \frac{2\|\mathrm{Re}\{p\mathbf{v}^H\}\|}{|p|^2 + \mathbf{v}^H\mathbf{v}}, \tag{6.15}$$

and the DoA from $(\theta_{\mathrm{DoA}}, \phi_{\mathrm{DoA}})$ is taken as the direction opposite to the intensity vector:

$$\gamma_{\mathrm{DoA}} = -\frac{\mathbf{i}_a}{\|\mathbf{i}_a\|} = \frac{\mathrm{Re}\{p\mathbf{v}^H\}}{\|\mathrm{Re}\{p\mathbf{v}^H\}\|}. \tag{6.16}$$

---

[1] This signal set is essentially the Ambisonics traditional first-order B-format, without the $1/\sqrt{2}$ scaling of the pressure signal.

6.3.2 Higher-Order Energetic Analysis

The above relations show that computation of the standard FO-DirAC parameters does not require signals of order higher than one. The HO-DirAC analysis exploits the fact that if HOA signals of order $N > 1$ are available, the sound field can be weighted directionally before the parameters are estimated. Such a weighting $w(\gamma)$ results in a new sound field $a_w(\gamma) = a(\gamma)w(\gamma)$ in which the DirAC parameters mostly capture contributions from sources that are not affected by it. In the HO-DirAC formulation this directional weighting is referred to as a "sector pattern," and the corresponding parameter estimation as "sector analysis." The shape of the sector patterns naturally depends on the application; in this implementation, only real axisymmetric sector patterns are considered.

The sector signals are formed by the beamforming relation Equation (6.6). Due to the axisymmetry of the patterns, their beamforming weights can be factorized into a direction-dependent and a pattern-dependent part as

$$[\mathbf{w}(\gamma_0)]_q = w_{nm}(\gamma_0) = c_n Y_{nm}(\gamma_0), \tag{6.17}$$

where γ_0 is the look direction of the sector pattern, and c_n are coefficients which define its shape. Some useful axisymmetric patterns suitable for the analysis presented here are:

$$c_n = \frac{N!N!}{(N+n+1)!(N-n)!} \qquad \text{cardioid,} \tag{6.18}$$

$$c_n = \frac{1}{(N+1)^2} \qquad \text{hypercardioid,} \tag{6.19}$$

$$c_n = \frac{P_n(\cos \kappa_N)}{\sum_{n=0}^{N}(2n+1)P_n(\cos \kappa_N)} \qquad \text{max-rE,} \tag{6.20}$$

with $\kappa_N = \cos(2.407/(N+1.51))$, as given by Zotter and Frank (2012).

Now, assuming that such a sector pattern is applied to the sound field distribution, the pressure signal is given by

$$p_w = \int_\gamma a(\gamma)w(\gamma)\,\mathrm{d}\gamma = \mathbf{w}^\mathsf{T}\mathbf{a}, \tag{6.21}$$

and the velocity signals by

$$\mathbf{u}_w = -\frac{1}{Z_0}\int_\gamma w(\gamma)\gamma\,a(\gamma)\,\mathrm{d}\gamma = -\frac{1}{Z_0}\mathbf{v}_w, \tag{6.22}$$

where the signal vector $\mathbf{v}_w$ corresponds to the signals captured with the directional patterns $w(\gamma)\gamma$.

In agreement with the estimation of the standard intensity and diffuseness, the last relation shows that the velocity components of the weighted sound field can be captured with beam patterns that are products of the sector pattern and the three orthogonal dipoles, as in

$$w(\gamma)\gamma = \begin{bmatrix} w^x(\gamma) \\ w^y(\gamma) \\ w^z(\gamma) \end{bmatrix} = \begin{bmatrix} w(\theta, \varphi)\cos\theta\cos\varphi \\ w(\theta, \varphi)\cos\theta\sin\varphi \\ w(\theta, \varphi)\sin\theta \end{bmatrix}. \tag{6.23}$$

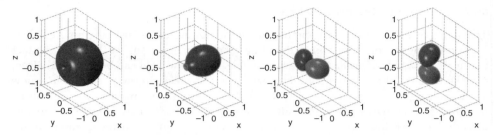

(a) First-order cardioid sector oriented at front, and the resulting velocity patterns.

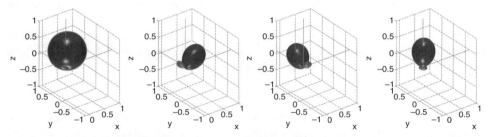

(b) Second-order cardioid sector oriented at (45,45), and the resulting velocity patterns.

Figure 6.3 Example HO-DirAC velocity patterns based on cardioid sectors. The dark and light tones depict positive and negative polarity respectively.

The beamforming weights of such weighted-velocity patterns are linearly related to the weights of the sector through

$$\mathbf{w}_N^i = \mathbf{A}_N^i \mathbf{w}_{N-1}, \quad \text{with } i = \{x, y, z\}. \tag{6.24}$$

The velocity patterns are of one order greater than the sector pattern, since they are products of it with the first-order dipoles. The matrices $\mathbf{A}^i$ are $(N+1)^2 \times N^2$ matrices that are deterministic and independent of the sector pattern. They can be pre-computed up to some maximum order of interest (Politis *et al.*, 2015). For more information on their structure, the reader is referred to Politis *et al.* (2015) and Politis and Pulkki (2016). An example of a sector pattern and the corresponding velocity patterns are shown in Figure 6.3.

Finally, after the weights for the sector and velocity patterns have been derived, the weighted pressure–velocity signal vector $\mathbf{a}_{pv,w}$ is given for order-N signals $\mathbf{a}_N$ as

$$\mathbf{a}_{pv,w} = \mathbf{W}_{pv,w}^{T} \mathbf{a}_N, \tag{6.25}$$

with $\mathbf{W}_{pv,w} = [\mathbf{w}, \mathbf{w}^x, \mathbf{w}^y, \mathbf{w}^z]$, and with the sector weights $\mathbf{w}$ zero-padded to a length of one order higher. With the generation of the patterns of Equation (6.23) and the capture of the sector and velocity signals p_w and $\mathbf{v}_w$, it is possible to estimate the energetic quantities of the previous section in a non-global manner but instead with a certain directional selectivity. More specifically, a local active intensity $\mathbf{i}_{a,w}$, energy density E_w, and local diffuseness ψ_w can be estimated with the exact same formulas of Equations (6.13)–(6.15) if the total pressure and velocity power and cross-spectra are replaced by their spatially filtered versions. Note that in a purely diffuse field the local sector diffuseness

$\psi_{w,df}$ is less than unity in this case, due to the directional energy concentration, and it is given by

$$\psi_{w,df} = \sum_{n=0}^{N} 2(n+1)c_n c_{n+1},$$ (6.26)

as shown by Zotter and Frank (2012). On the contrary, in the presence of a single plane, the local diffuseness is always zero. For a more thorough discussion on the properties of energetic analysis on a weighted sound field, the reader is referred to Politis and Pulkki (2016).

6.3.3   Sector Profiles

The higher-order intensity and diffuseness have the potential to resolve multiple sound events. Combined with directional scanning for DoA estimation, for example, it possesses some interesting properties not found in standard beamforming techniques. A comparison can be made with the steered-response power (SRP) approach, where a beamformer is oriented sequentially on a specified grid of directions, and its power output is used to form a directional map which is then used to estimate the most probable DoAs by peak finding. The resolution of the output will depend on the beam width and the side lobes of the beamformer, and for a low available order the power will be distributed in a wide range around the true estimates.

Steering the sector pattern on the same grid, and getting the intensity, diffuseness, and energy estimates, allows one to:

- weight the energy estimate with a diffuseness-based weight that is unity when the diffuseness is zero and lower when the diffuseness is higher, resulting in a sharpened power distribution for every grid point;
- redirect the sharpened power estimate of every grid point to a new grid point that is determined by the direction of the local intensity vector.

The second property is quite unique, and distorts the original scanning grid to a new one where most points are concentrated around the true DoAs, resulting in a very high-resolution directional map of the sound field. This operation resembles somewhat the time–frequency reassignment method (Auger and Flandrin, 1995) for sharpening time–frequency distributions, and we refer to it here as *directional reassignment*. Furthermore, the operation is a purely spatial one and does not depend on signal statistics as with signal-dependent beamformers such as minimum variance distortionless response (MVDR), or subspace DoA estimation methods such as MUSIC (Brandstein and Ward, 2013). It can be applied even on the time–frequency signals of a single observation window. Figure 6.4 demonstrates the reassignment effect more clearly. Three plane waves of unit amplitude are incident from different directions, in the presence of a weak diffuse field with -30 dB power. Figure 6.4(a) shows a very high-order ($N = 30$) representation of sound field distribution, and the power map is highly concentrated around the true DoAs. Assuming that we have only a relatively low-order representation of the same scene ($N = 5$), the power map is widely spread around the true DoAs due to the decreased spatial resolution, as seen in Figure 6.4(b). However, using the same low-order signals with the sector-based intensity, and by reassigning the grid power estimates, we get a very sharp directional distribution similar to the very high-order one – Figure 6.4(c).

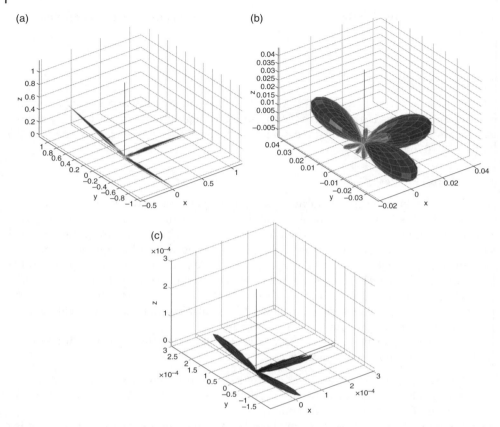

Figure 6.4 Directional distributions of three unit-amplitude plane waves incident from $(0, 0)$, $(-90, 0)$, and $(90, 30)$. (a) A 30th-order representation of the amplitude distribution. (b) A 5th-order representation. (c) The reassigned distribution using sector analysis on the 5th-order signals.

In the HO-DirAC context, it is necessary to determine a practical number and arrangement of sectors appropriate to the sound reproduction task. The following conditions should be met:

- The analysis performance is equal in all directions, meaning that similar axisymmetric sectors should be used, uniformly covering the sphere.
- The analyzed sector energies preserve the energy of the recording in any direction, meaning that the combined sectors are energy preserving,

$$\sum_{j=1}^{J} \beta \, w_j^2(\gamma) = 1 \; \forall \, \gamma, \tag{6.27}$$

where J is the number of uniformly oriented sectors, $\beta = Q_w / J$ is a normalization constant, and Q_w is the directivity factor of the sector pattern.
- The minimum number of sectors that fulfill the above conditions should be used, as additional sectors increase the computational load without additional benefits.

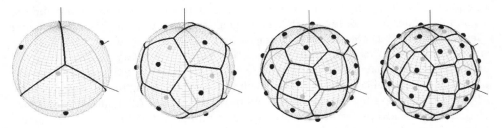

Figure 6.5 Sector profiles for analysis orders $N = 2 - 5$, based on t-designs, illustrating the sector centers and the areas of points closer to them. Source: Politis 2015. Reproduced with permission of IEEE.

Considering the sector pattern, any axisymmetric design of order $N - 1$ is suitable for order-N available HOA signals. Considering the arrangement and number, the above conditions are fulfilled by uniform arrangements capable of integrating exactly spherical polynomials of degree $2N - 2$, such as t-designs of $t = 2N - 2$ with the smaller number of points (Hardin and Sloane, 1996). For a proof of this condition and a more detailed discussion on suitable sector profiles, refer to Politis *et al.* (2015). Figure 6.5 shows such sector arrangements for different available orders of HOA signals.

6.4 Synthesis of Target Setup Signals

The analysis stage provides a set of parameters describing the sound field as a function of time and frequency. DirAC aims to utilize these parameters in a lossless sense, meaning that during synthesis using a number of surrounding loudspeakers, if the B-format or HOA signals are recorded again at the sweet spot and reanalyzed, they would produce the original sound field parameters. In detail, the non-diffuse portions of the sound energy are reproduced at their estimated directions. The diffuse portions of the sound energy are reproduced with all loudspeakers with mutual non-coherence. Although the sound field parameters are known, the sound signals corresponding to the non-diffuse or diffuse parameters are not independently available.

In first-order VM-DirAC, the synthesis is split into two parts: a non-parametric ambisonic decoding stage in which the B-format signals are distributed to the playback system, and a parametric synthesis stage in which the output channels are enhanced with sharper directional cues and appropriate decorrelated energy in diffuse conditions (Vilkamo *et al.*, 2009); see also Chapter 5. The synthesis stage forces spatialization on the output channels as dictated by the analyzed parameters, which would otherwise have been impossible to achieve just by linear decoding. Assuming that the DoA, diffuseness, and energy analyzed at a certain time–frequency point (k, l) are $\mathbf{p}_{kl} = [\gamma_{kl}, \psi_{kl}, E_{kl}]$, the first-order input B-format signals are $\mathbf{a}_{kl}^{(1)}$, and the L output signals are $\mathbf{b}_{kl}$, a simplified version of VM-DirAC synthesis can be formulated as

$$\mathbf{z}_{kl} = \mathbf{D}\mathbf{a}_{kl}^{(1)}, \tag{6.28}$$

$$\mathbf{d}_{kl} = \mathcal{D}\left[\sqrt{\psi_{kl}}\mathbf{z}_{kl}\right], \tag{6.29}$$

$$\mathbf{b}_{kl} = \sqrt{1 - \psi_{kl}}\mathbf{G}(\gamma_{kl})\mathbf{z}_{kl} + \mathbf{U}\mathbf{d}_{kl}. \tag{6.30}$$

In Equation (6.30), the first part corresponds to rendering of directional sound and the second to rendering of diffuse sound. The signals $\mathbf{z}$ of the initial ambisonic decoding specified in matrix $\mathbf{D}$ are used in both stages, and $\mathcal{D}[\mathbf{z}]$ is a decorrelated version of the same signals used to achieve diffuse reproduction. Furthermore, $\mathbf{G}(\boldsymbol{\gamma}_{kl}) = \mathrm{diag}[\mathbf{g}(\boldsymbol{\gamma}_{kl})]$ is an $L \times L$ diagonal matrix of the appropriate spatialization gains that spatially concentrates the directional component to the analyzed DoA according to the target system. These can be, for example, vector base amplitude panning (VBAP) gains (Pulkki, 1997) for loudspeaker setups, or head-related transfer functions (HRTFs) for headphones. Finally, the $L \times L$ matrix $\mathbf{U}$ is a diagonal diffuse energy distributor matrix with the property $\mathrm{tr}[\mathbf{U}] = 1$, $\mathrm{tr}[\cdot]$ being the matrix trace. In the simplest case, assigning equal energy to all loudspeakers, all entries of $\mathbf{U}$ are equal to $1/L$.

One issue for VM-DirAC is that it does not take into account what is already achieved by the ambisonic decoding in terms of directional spatialization and diffuse reproduction. Instead, time-variant panning and diffuse gains are applied blindly to the linearly decoded signals, with only an average correction for energy preservation (Vilkamo *et al.*, 2009). More recently, Vilkamo and Pulkki (2013) showed that higher quality of reproduction can be obtained in a perceptual sense by first building the output channel characteristics in the parametric domain, and then processing the non-diffuse and diffuse sounds in a single combined least-squares optimized mixing step, as used in OM-DirAC. In its general form, the output signals of OM-DirAC are given by

$$\mathbf{d}_{kl} = \mathcal{D}\big[\mathbf{Da}_{kl}^{(N)}\big], \tag{6.31}$$

$$\mathbf{b}_{kl} = \mathbf{A}_{kl}(\mathbf{D}, \mathbf{p}_{kl})\mathbf{a}_{kl}^{(N)} + \mathbf{B}_{kl}(\mathbf{D}, \mathbf{p}_{kl})\mathbf{d}_{kl}. \tag{6.32}$$

The two adaptive matrices $\mathbf{A}, \mathbf{B}$ perform the task of optimally mixing the input signals so that they meet the target spatial statistics of the output channels, as dictated by the analyzed parameters. The matrix $\mathbf{B}$, which mixes decorrelated signals, is optional; if mixing of input $\mathbf{a}$ through $\mathbf{A}$ cannot achieve the target spatial properties, only then is the minimum required amount of decorrelation injected.

The optimal mixing matrices $\mathbf{A}, \mathbf{B}$ are updated at each time–frequency block according to the solution of Vilkamo *et al.* (2013), and they depend on the SCM of the input signals, $\mathbf{C_a}$, as defined in Equation (6.8), the linear/ambisonic decoding matrix of choice $\mathbf{D}$, and the target SCM of the output signals,

$$\mathbf{C_b}(k, l) = \mathbb{E}\big\{\mathbf{b}_{kl}\mathbf{b}_{kl}^{\mathrm{H}}\big\}. \tag{6.33}$$

The input covariance matrix $\mathbf{C_a}$ is estimated and updated at each time–frequency point from the input signals, by block or recursive averaging. The target covariance matrix is completely determined by the analyzed sound field parameters, conveniently incorporating the multiple estimates coming from the HO-DirAC analysis. A basic block diagram of the processing flow is presented in Figure 6.6.

6.4.1 Loudspeaker Rendering

HO-DirAC has primarily been applied to loudspeaker rendering, and is suitable for large setups such as the NHK 22.2 layout specification (Hamasaki *et al.*, 2008). The perceptual performance of ambisonic decoding is limited by the available order of the signals, without necessarily taking advantage of the full capabilities of the layout (Solvang, 2008). FO-DirAC and HO-DirAC, however, due to the parametric analysis and synthesis,

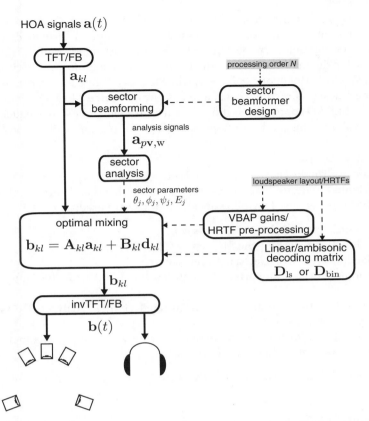

Figure 6.6 Basic block diagram of HO-DirAC processing, following the notation introduced in the text.

will perform an upmixing of directional and diffuse sounds to the full layout using the tools of panning and decorrelation appropriately. With regards to analysis order, HO-DirAC achieves a better spatial image stability and higher single-channel quality than FO-DirAC.

Definition of Non-Parametric Decoding

Apart from the target SCM, the optimal mixing requires a linear decoding matrix that is used to produce signals that serve as a reference in order to constrain the optimization and produce a unique solution. This linear decoding should already achieve reproduction of the sound scene as well as possible. An appropriate example is an ambisonic decoding matrix. Here we list three straightforward approaches to generating such ambisonic decoding matrices $\mathbf{D}_{\mathrm{ls}}$ for L loudspeakers and HOA signals of order N:

$$\text{Sampling:} \quad \mathbf{D}_{\mathrm{ls}} = \frac{1}{L}\mathbf{Y}_L^{\mathrm{T}}, \tag{6.34}$$

$$\text{Mode-matching:} \quad \mathbf{D}_{\mathrm{ls}} = \left(\mathbf{Y}_L^{\mathrm{T}}\mathbf{Y}_L + \lambda^2\mathbf{I}\right)^{-1}\mathbf{Y}_L^{\mathrm{T}}, \tag{6.35}$$

$$\text{ALLRAD:} \quad \mathbf{D}_{\mathrm{ls}} = \frac{1}{K}\mathbf{G}_{\mathrm{td}}\mathbf{Y}_{\mathrm{td}}^{\mathrm{T}}, \tag{6.36}$$

where $\mathbf{Y}_L = [\mathbf{y}(\boldsymbol{\gamma}_1), \ldots, \mathbf{y}(\boldsymbol{\gamma}_L)]$ is the $(N+1)^2 \times L$ matrix of SHs in the loudspeaker directions. In the mode-matching approach, the least-squares solution is usually constrained with a regularization value λ. In the ALLRAD method (Zotter and Frank, 2012), $\mathbf{Y}_{\text{td}} = [\mathbf{y}(\boldsymbol{\gamma}_1), \ldots, \mathbf{y}(\boldsymbol{\gamma}_K)]$ is the matrix of SHs in the K directions of a uniform spherical t-design (Hardin and Sloane, 1996), of $t \geq 2N + 1$, while the $\mathbf{G}_{\text{td}}$ is an $L \times K$ matrix of VBAP gains, with the t-design directions considered as virtual sources. In HO-DirAC implementations the ALLRAD approach has mostly been used, being robust to irregular loudspeaker layouts while having good energy-preserving properties for all directions.

Definition of Target Signal Statistics

Assuming J sectors are used in the analysis, there are J sets of parameters $\mathbf{p}_j = [\boldsymbol{\gamma}_j, \psi_j, E_j]$. Using these, we can build the desired target covariance matrix that will determine the optimal mixing of the HOA signals. The following assumptions are used:

- The energy of the diffuse part for the jth sector is $\psi_j E_j$.
- The energy of the directional part for the jth sector is $(1 - \psi_j)E_j$.
- The diffuse components are uncorrelated between sectors.
- The directional components are uncorrelated between them and with the diffuse components.
- The directional components are maximally concentrated at their analyzed DoAs using appropriate panning functions.
- The diffuse components are distributed non-coherently between output channels, with the directional power determined by the analysis.

Based on these assumptions, the covariance matrix of the directional component of a single sector is

$$\mathbf{C}_j^{\text{dir}} = \mathbf{g}_{\text{ls}}(\boldsymbol{\gamma}_j)\mathbf{g}_{\text{ls}}^{\text{H}}(\boldsymbol{\gamma}_j)(1 - \psi_j)E_j, \tag{6.37}$$

where $\mathbf{g}_{\text{ls}}$ is the $L \times 1$ vector of panning gains for the analyzed direction, for example VBAP gains. Similarly, the covariance matrix of the diffuse component of a single sector is

$$\mathbf{C}_j^{\text{diff}} = \psi_j E_j \mathbf{U}_j, \tag{6.38}$$

where $\mathbf{U}_j$ is a diagonal diffuse energy distributor matrix, as defined in Equation (6.30). Due to the uncorrelated property of all the components, constructing the overall covariance matrix of the output channels becomes a direct summation of the sector ones:

$$\mathbf{C}_{\mathbf{b}} = \sum_{j=1}^{J} \left(\mathbf{C}_j^{\text{dir}} + \mathbf{C}_j^{\text{diff}}\right)$$

$$= \sum_{j=1}^{J} E_j \left[\mathbf{g}_{\text{ls}}(\boldsymbol{\gamma}_j)\mathbf{g}_{\text{ls}}^{\text{H}}(\boldsymbol{\gamma}_j) \cdot (1 - \psi_j) + \mathbf{U}_j \cdot \psi_j\right]. \tag{6.39}$$

Regarding the definition of the diffuse distribution matrix $\mathbf{U}_j$, the sector energies should be distributed spatially according to the sector orientations. Such an example activates only the loudspeakers that are closer to the sector center compared to the rest of the sectors, with entries only at the respective channels and zeros otherwise. Another approach is to consider the power distribution determined by the non-parametric decoding $\mathbf{D}$ as a distribution matrix, as used for example by Politis *et al.* (2015).

6.4.2 Binaural Rendering

Binaural rendering with HO-DirAC is essentially the same as with loudspeaker rendering. It differs only in using HRTFs instead of panning functions, and in the definition of diffuse rendering that takes into account binaural coherence under diffuse conditions.

Definition of Non-Parametric Decoding

Linear decoding of HOA signals to binaural signals can be done using HRTFs. Since HRTFs are frequency dependent, so are the decoding matrices in this case. More specifically, the binaural signals $\mathbf{x}_{kl} = [x_L(k,l), x_R(k,l)]^T$ are given by

$$\mathbf{x}_{kl} = \mathbf{D}_{\text{bin}}(k)\mathbf{a}_{kl}^{(N)}, \tag{6.40}$$

with $\mathbf{D}_{\text{bin}}$ being the $2 \times (N+1)^2$ linear decoding matrix.

The linear decoding matrix can be constructed in a few different ways. A straightforward one considers the sound scene amplitude distribution $a(\gamma)$ and, for example, the left HRTF $h_L(\gamma)$, and models the left-ear signal as

$$
\begin{aligned}
x_L(k,l) &= \int_\gamma a(k,l,\gamma)h_L(k,\gamma)\,d\gamma \\
&= SHT\{a(k,l,\gamma)\} \cdot SHT\{h_L(k,\gamma)\} \\
&= \mathbf{h}_L^T(k)\,\mathbf{a}_{kl}^{(N)},
\end{aligned} \tag{6.41}
$$

where $\mathbf{h}_L$ are the coefficients of the HRTF expanded into spherical harmonics. Equation (6.41) states that the binaural signals are the result of the inner product between the ambisonic coefficients and the SH coefficients of the HRTFs. Hence, the decoding matrix in this case is simply $\mathbf{D}_{\text{bin}}(k) = [\mathbf{h}_L(k), \mathbf{h}_R(k)]^T$. Note that due to the commonly low-order representation of the sound field, the HRTF coefficients are truncated and their high-frequency variability is lost, with perceivable effects (Bernschütz *et al.*, 2014).

Construction of Target Setup Signal Statistics

In the case of loudspeaker rendering, it is assumed that the HOA material would be reproduced to more loudspeakers than HOA channels, with $L \geq (N+1)^2$. In the binaural case, the number of SH signals will always be higher than two (the number of ear signals). In this case, the computational complexity can be reduced if the optimal mixing stage is applied to the output of the linear decoding $\mathbf{D}_{\text{bin}}$ instead of the HOA signals, with equivalent results. Since $\mathbf{x}$ of Equation (6.40) serves as input to the optimal mixing stage, the 2×2 identity matrix $\mathbf{I}_2$ is used as a reference decoding matrix to constrain the mixing matrices. This complexity reduction is important in the binaural case since for high-quality headphone rendering, head tracking should be integrated, which imposes requirements for lightweight real-time operation of the method.

Similar to the loudspeaker case, the covariance matrix of the directional component of a single sector is defined as

$$\mathbf{C}_j^{\text{dir}} = \mathbf{g}_{\text{bin}}(\theta_j, \varphi_j)\mathbf{g}_{\text{bin}}^H(\theta_j, \varphi_j)(1 - \psi_j)E_j, \tag{6.42}$$

where $\mathbf{g}_{\text{bin}} = [h_L(\gamma), h_R(\gamma)]^T$ is the 2×1 vector of HRTFs for the analyzed direction and the specific frequency band. Regarding modeling the diffuse covariance matrix, instead of forcing a completely non-coherent distribution as in the loudspeaker case, it is beneficial to consider the binaural coherence under diffuse conditions. The diffuse-field binaural coherence c_{bin} is given by

$$c_{\text{bin}}(k) = \frac{\int_\gamma h_L(k,\gamma)h_R^*(k,\gamma)\,d\gamma}{\sqrt{\int_\gamma |h_L(k,\gamma)|^2\,d\gamma \int_\gamma |h_R(k,\gamma)|^2\,d\gamma}}, \tag{6.43}$$

or, by using the SHT version of the HRTFs, it can be conveniently computed in the SHD, as shown by Politis (2016), by

$$c_{\text{bin}}(k) = \frac{\mathbf{h}_R^H(k)\mathbf{h}_L(k)}{\sqrt{||\mathbf{h}_L(k)||^2||\mathbf{h}_R(k)||^2}}. \tag{6.44}$$

The diffuse binaural coherence has a negligible imaginary part, due to the inter-aural phase differences having an axisymmetric dipole-like response that vanishes across integration over all directions. Its real part is $c_{\text{bin}} \in [-1, 1]$, and it decays rapidly from unity at low frequencies towards zero above around 700 Hz. The binaural coherence can be computed from measured or modeled HRTFs, or from simpler models such as the one in Borß and Martin (2009), and it can be personalized or based on generic listener properties. An example of the diffuse binaural coherence curve from a measured set of HRTFs is presented in Figure 6.7.

Finally, the covariance matrix of the diffuse component is

$$\mathbf{C}_j^{\text{diff}}(k, l) = \psi_j(k, l)E_j(k, l)\mathbf{U}_{\text{bin}}(k), \tag{6.45}$$

where the diffuse distribution matrix in this case is

$$\mathbf{U}_{\text{bin}}(k) = \begin{bmatrix} \alpha & \sqrt{\alpha\beta}c_{\text{bin}}(k) \\ \sqrt{\alpha\beta}c_{\text{bin}}(k) & \beta \end{bmatrix}. \tag{6.46}$$

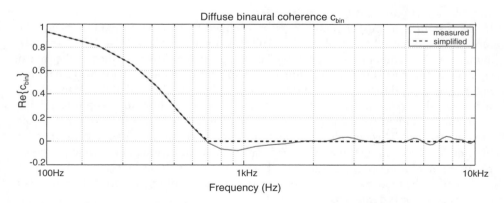

Figure 6.7 Binaural diffuse-field coherence example, based on the first author's measured HRTFs. The simplified curve shows an approximation used during parametric processing.

The factors α, β, with $\alpha + \beta = 1$, determine the portion of the diffuse energy distributed to the left and right ear. A straightforward way to determine them is to assume that the diffuse sound follows approximately the powers of the left and right signals from the linear decoding as

$$[\alpha, \beta] = \frac{\text{diag}[\mathbf{D}_{\text{bin}}\mathbf{C}_{\mathbf{a}}\mathbf{D}_{\text{bin}}^{\text{T}}]}{\text{tr}[\mathbf{D}_{\text{bin}}\mathbf{C}_{\mathbf{a}}\mathbf{D}_{\text{bin}}^{\text{T}}]}. \tag{6.47}$$

6.5   Subjective Evaluation

A listening experiment was organized to evaluate the perceptual benefits provided by HO-DirAC, using complex synthetic sound scenes encoded into fourth-order HOA signals. Two conditions were tested. The first was the effect of using HO-DirAC processing compared to the standard FO-DirAC processing. The second was the benefit, if any, of using a parametric technique over direct linear reproduction such as the ambisonic decoding of Section 6.4.1, even at orders as high as four.

The listening tests were conducted in an anechoic chamber using 28 loudspeakers on a sphere. The configuration resembled the recent 22.2 3D layout described in Hamasaki *et al.* (2008). The HOA signals were processed with the proposed method set to fourth-order processing at all frequency bands (P_full), and by the ALLRAD ambisonic decoding outlined in Section 6.4.1 (L_full), using all loudspeakers.

First-order B-format was processed with FO-DirAC (P_1st) and, additionally, with a first-order ambisonic (L_1st) decoding included as a low-quality anchor, using a quasi-regular subset of 12 loudspeakers. This subset was a compromise between the minimum of four loudspeakers, which is too sparse for consistent localization, and higher numbers, which are known to induce spectral coloration (Solvang, 2008). A more detailed description of the listening conditions, the generation of the different reproduction modes, the test setup, and the statistical analysis can be found in Politis *et al.* (2015).

The listening test results, summarized in Figure 6.8 indicate that:

- higher-order SH signals can be exploited to clearly improve the performance regarding reproduction accuracy in parametric rendering, such as in HO-DirAC;
- using ideal fourth-order SH signals, HO-DirAC achieves almost perceptually transparent results for all scenarios;
- for both low and high orders, parametric processing such as in DirAC is perceived to be closer to the reference than the linear decoding, using the same order of HOA signals.

6.6   Conclusions

In higher-order DirAC the sound field is captured into higher-order spherical-harmonic components and is virtually divided into several directionally emphasized sectors. In each sector, the partial pressure and velocity signals are computed from the spherical harmonic signals, and subsequently used for parametric reproduction of the sound field over loudspeakers or head-tracked headphones. The estimated spatial parameters are

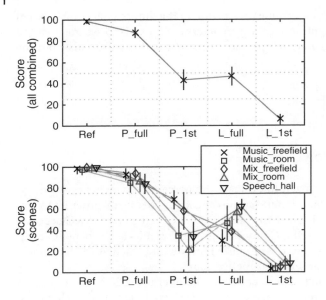

Figure 6.8 Means and 95% confidence intervals of the listening test with idealized reproduction assuming the first or fourth order of available spherical harmonics. P denotes parametric reproduction, and L denotes linear reproduction. Source: Politis 2015. Reproduced with permission of IEEE.

robust to spatially complex sound scenarios, thus mitigating unwanted processing artifacts that occur with first-order DirAC in such cases. Results from a subjective evaluation show that, in a variety of sound scenes, HO-DirAC processing outperforms linear decoding and improves on lower-order parametric processing. The perceptual effectiveness of the method, approximating transparent reproduction, implies the potential of such parametric processing as a sound field coding and compression method. The results suggest that perceptually transparent reproduction can be achieved with fourth-order signals for any sound scene, making encoding of the sound field at higher orders with increased bandwidth requirements unnecessary. Future work should further evaluate the compression capabilities of the method, and its applications to coding of virtual sound scenes and recording and reproduction of spatial room impulse responses.

References

Auger, F. and Flandrin, P. (1995) Improving the readability of time–frequency and time–scale representations by the reassignment method. *IEEE Transactions on Signal Processing*, **43**(5), 1068–1089.

Bernschütz, B., Giner, A.V., Pörschmann, C., and Arend, J. (2014) Binaural reproduction of plane waves with reduced modal order. *Acta Acustica united with Acustica*, **100**(5), 972–983.

Borß, C. and Martin, R. (2009) An improved parametric model for perception-based design of virtual acoustics. *35th International Conference of the AES: Audio for Games*, London, UK.

Brandstein, M. and Ward, D. (2013) *Microphone Arrays: Signal Processing Techniques and Applications*. Springer, New York.

Hamasaki, K., Nishiguchi, T., Okumura, R., Nakayama, Y., and Ando, A. (2008) A 22.2 multichannel sound system for ultra-high-definition TV (UHDTV). *SMPTE Motion Imaging Journal*, **117**(3), 40–49.

Hardin, R.H. and Sloane, N.J. (1996) McLaren's improved snub cube and other new spherical designs in three dimensions. *Discrete & Computational Geometry*, **15**(4), 429–441.

Laitinen, M.V., Kuech, F., Disch, S., and Pulkki, V. (2011) Reproducing applause-type signals with directional audio coding. *Journal of the Audio Engineering Society*, **59**(1/2), 29–43.

Laitinen, M.V. and Pulkki, V. (2012) Utilizing instantaneous direct-to-reverberant ratio in parametric spatial audio coding. *133rd Convention of the AES*. Audio Engineering Society.

Politis, A. (2016) Diffuse-field coherence of sensors with arbitrary directional responses. arXiv:1608.07713.

Politis, A. and Pulkki, V. (2016) Acoustic intensity, energy-density and diffuseness estimation in a directionally-constrained region. arXiv:1609.03409.

Politis, A., Vilkamo, J., and Pulkki, V. (2015) Sector-based parametric sound field reproduction in the spherical harmonic domain. *IEEE Journal of Selected Topics in Signal Processing*, **9**(5), 852–866.

Pulkki, V. (1997) Virtual sound source positioning using vector base amplitude panning. *Journal of the Audio Engineering Society*, **45**(6), 456–466.

Pulkki, V., Politis, A., Del Galdo, G., and Kuntz, A. (2013) Parametric spatial audio reproduction with higher-order B-format microphone input. *134th Convention of the Audio Engineering Society*, Rome, Italy.

Solvang, A. (2008) Spectral impairment of two-dimensional higher order Ambisonics. *Journal of the Audio Engineering Society*, **56**(4), 267–279.

Vilkamo, J., Bäckström, T., and Kuntz, A. (2013) Optimized covariance domain framework for time–frequency processing of spatial audio. *Journal of the Audio Engineering Society*, **61**(6), 403–411.

Vilkamo, J., Lokki, T., and Pulkki, V. (2009) Directional audio coding: Virtual microphone-based synthesis and subjective evaluation. *Journal of the Audio Engineering Society*, **57**(9), 709–724.

Vilkamo, J. and Pulkki, V. (2013) Minimization of decorrelator artifacts in directional audio coding by covariance domain rendering. *Journal of the Audio Engineering Society*, **61**(9), 637–646.

Zotter, F. and Frank, M. (2012) All-round ambisonic panning and decoding. *Journal of the Audio Engineering Society*, **60**(10), 807–820.

7

Multi-Channel Sound Acquisition Using a Multi-Wave Sound Field Model

Oliver Thiergart and Emanuël Habets

International Audio Laboratories Erlangen, a Joint Institution of the Friedrich-Alexander-Universität Erlangen-Nürnberg (FAU) and Fraunhofer IIS, Erlangen, Germany

7.1 Introduction

During the last decade, different devices with two or more microphones have emerged that enable multi-channel sound acquisition. Typical examples include mobile phones, digital cameras, and smart television screens, which can be used for a huge variety of audio applications including hands-free communication, spatial sound recording, and speech-based human–machine interaction. Key to the realization of these applications with different devices is flexible and efficient sound acquisition and processing. "Flexible" means that the sound can be captured with different microphone configurations while being able to generate one or more desired output signals at the reproduction side depending on the application. "Efficient" means that only a few audio signals, in comparison to the number of microphones used, need to be transmitted to the reproduction side while maintaining full flexibility.

This flexible and efficient sound acquisition and processing can be achieved using a parametric description of the spatial sound. This approach is used, for example, in directional audio coding (DirAC) as discussed in Chapter 5 (Pulkki, 2007) or high angular resolution plane-wave expansion (HARPEX; Berge and Barrett, 2010a,b), which represent two well-known parametric approaches to the analysis and reproduction of spatial sound. See Chapter 4 for a broader introduction to the techniques. In DirAC, it is assumed that for each time-frequency instance the sound field can be decomposed into a direct sound component and a diffuse sound component. In practice, model violations may occur when multiple sources are active simultaneously (Thiergart and Habets, 2012). To reduce such model violations, a higher-order extension was proposed by Politis *et al.* (2015) as discussed in the previous chapter, which performs a multi-directional energetic analysis. This approach requires spherical harmonics of higher orders as input signals, which can be obtained using a spherical microphone array. To reduce the model violations with more practical and almost arbitrary microphone setups, we introduced in Thiergart *et al.* (2014b) a multi-wave sound field model in

Parametric Time–Frequency Domain Spatial Audio, First Edition. Edited by Ville Pulkki, Symeon Delikaris-Manias, and Archontis Politis.
© 2018 John Wiley & Sons Ltd. Published 2018 by John Wiley & Sons Ltd.
Companion Website: www.wiley.com/go/pulkki/parametrictime-frequency

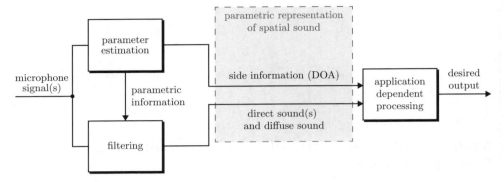

Figure 7.1 Block scheme of the parametric processing of spatial sound. Source: Thiergart 2015. Reproduced with permission of Oliver Thiergart.

which multiple direct sound components plus a diffuse sound component are assumed per time–frequency instance.

As illustrated in Figure 7.1, the parametric processing is performed in two successive steps. In the first step (parameter estimation and filtering), the sound field is analyzed in narrow frequency bands using multiple microphones to obtain the parametric representation of the spatial sound in terms of the direct sound and diffuse sound components and parametric side information, namely the direction of arrival (DOA) of the direct sounds. In the second step (application-dependent processing), one or more output signals are synthesized from the parametric representation depending on the application. As in DirAC, the described scheme allows for an efficient transmission of sound scenes to the reproduction side. In fact, instead of transmitting all microphone signals to the reproduction side for the application-dependent processing, only the compact parametric representation of the spatial sound is transmitted.

The direct sounds and diffuse sound in Figure 7.1 are extracted using optimal multi-channel filters. These filters can outperform the single-channel filters that are typically used in state-of-the-art (SOA) approaches such as DirAC. For example, multi-channel filters can extract the direct sounds while at the same time reducing the microphone noise and diffuse sound. Moreover, when extracting the diffuse sound using multi-channel filters, the leakage of undesired direct sounds into the diffuse sound estimate can be greatly reduced, as shown later. The multi-channel filters presented in this chapter are computed using instantaneous information on the underlying parametric sound field model, such as the instantaneous DOA or second-order statistics (SOS) of the direct sounds and diffuse sound. Incorporating this information enables the filters to adapt quickly to changes in the acoustic scene, which is of paramount importance for typical acoustic scenes where multiple sources are active at the same time in a reverberant environment.

The remainder of this chapter is organized as follows:

- Section 7.2 explains how the parametric description of the spatial sound can be employed to realize the desired flexibility with respect to different audio applications.
- Section 7.3 explains the parametric sound field model and introduces the corresponding microphone signals.

- Section 7.4 introduces the multi-channel filters for the extraction of the direct sounds and diffuse sound.
- Section 7.5 discusses the estimation of the parametric information, which is required to compute the filters.
- Section 7.6 shows an example application of the approaches presented, namely the application to spatial sound reproduction.
- Section 7.7 summarizes the chapter.

Each section also provides a brief overview of relevant SOA approaches. Note that throughout this work we focus on omnidirectional microphone configurations. However, in most cases it is straightforward to extend the discussed filters and estimators to directional microphone configurations.

7.2   Parametric Sound Acquisition and Processing

To derive the aforementioned filters and realize the different applications, we assume a rather dynamic scenario where multiple sound sources are active at the same time in a reverberant environment. The number of existing sound sources and their positions are unknown and sources may move, emerge, or disappear. To cope with such a scenario, we model the sound field in the time–frequency domain. We assume that the sound field $P(k, n, \mathbf{r})$ at time instant n, frequency band k, and position $\mathbf{r}$ is composed of L direct sound components and a diffuse sound component:

$$P(k, n, \mathbf{r}) = \sum_{l=1}^{L} P_{\mathrm{s},l}(k, n, \mathbf{r}) + P_{\mathrm{d}}(k, n, \mathbf{r}), \tag{7.1}$$

where the L direct sounds $P_{\mathrm{s},l}(k, n, \mathbf{r})$ model the direct sound of the existing sources, and the diffuse sound $P_{\mathrm{d}}(k, n, \mathbf{r})$ models the reverberant or ambient sound. As discussed in more detail in Section 7.3, we assume that each direct sound $P_{\mathrm{s},l}(k, n, \mathbf{r})$ can be represented as a single plane wave with DOA expressed by the unit-norm vector $\mathbf{n}_l(k, n)$ or azimuth angle $\varphi_l(k, n)$ and elevation angle $\vartheta_l(k, n)$. The DOA of the direct sounds represents a crucial parameter in this work. An example plane wave is depicted in Figure 7.2(a), while Figure 7.2(b) shows an example diffuse field. The resulting sound field $P(k, n, \mathbf{r})$ is shown in Figure 7.2(c). The sound field model in Equation (7.1) is similar to the one in DirAC (Pulkki, 2007) and HARPEX (Berge and Barrett, 2010a,b). However, DirAC only assumes a single plane wave ($L = 1$) plus diffuse sound, while HARPEX assumes $L = 2$ plane waves and no diffuse sound.

7.2.1   Problem Formulation

The ultimate goal throughout this work is to capture the L direct sounds $P_{\mathrm{s},1\ldots L}(k, n, \mathbf{r})$ and the diffuse sound $P_{\mathrm{d}}(k, n, \mathbf{r})$ at a reference position denoted by $\mathbf{r}_1$ with a specific, application-dependent response. This means that the target signal $Y(k, n)$ we wish to obtain is a weighted sum of the L direct sounds and the diffuse sound at $\mathbf{r}_1$:

$$Y(k, n) = \mathbf{d}_{\mathrm{s}}^{\mathrm{H}}(k, n)\mathbf{p}_{\mathrm{s}}(k, n, \mathbf{r}_1) + D_{\mathrm{d}}(k, n)P_{\mathrm{d}}(k, n, \mathbf{r}_1) \tag{7.2a}$$

$$= Y_{\mathrm{s}}(k, n) + Y_{\mathrm{d}}(k, n). \tag{7.2b}$$

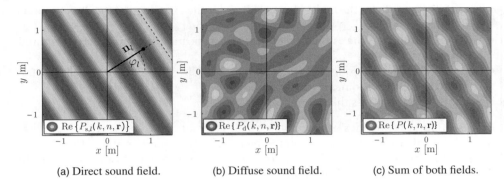

| (a) Direct sound field. | (b) Diffuse sound field. | (c) Sum of both fields. |

Figure 7.2 Example showing a direct sound field (plane wave), a diffuse sound field, and the sum of both fields, for a frequency of $f = 500\,\mathrm{Hz}$. Source: Thiergart 2015. Reproduced with permission of Oliver Thiergart.

The signal $Y_s(k, n)$ is the target direct signal which is the weighted sum of the L direct sounds at $\mathbf{r}_1$ contained in $\mathbf{p}_s(k, n, \mathbf{r}_1) = [P_{s,1}(k, n, \mathbf{r}_1), \ldots, P_{s,L}(k, n, \mathbf{r}_1)]^{\mathrm{T}}$. The potentially complex target responses for the direct sounds are contained in the vector $\mathbf{d}_s^{\mathrm{H}}(k, n) = [D_{s,1}(k, n), \ldots, D_{s,L}(k, n)]$. The signal $Y_d(k, n)$ is the target diffuse signal, with the target diffuse response given by $D_d(k, n)$.

The problem formulation in Equation (7.2) covers a huge variety of applications. The target direct responses $D_{s,l}(k, n)$ and target diffuse response $D_d(k, n)$ are the values of application-dependent response functions, for example $D_{s,l}(k, n) = d_s(k, \varphi_l)$ and $D_d(k, n) = d_d(k)$. In our applications, the direct sound response function $d_s(k, \varphi)$ depends on the time- and frequency-dependent DOA of the direct sound, for example, on the azimuth angle $\varphi(k, n)$. Both $d_s(k, \varphi)$ and the diffuse sound response function $d_d(k)$ are discussed below for different applications.

Speech Enhancement for Hands-Free Communication

In this application, we aim to extract direct sounds from all directions unaltered, while attenuating the diffuse sound that would reduce the speech intelligibility (Naylor and Gaubitch, 2010). For this purpose, we set the direct response function to 1 for all DOAs, that is, $d_s(k, \varphi) = 1$, while the diffuse response function $d_d(k)$ is set to 0. Alternatively, we can adjust the direct response function depending on the loudness of the direct sounds to achieve a spatial automatic gain control (AGC), as proposed by Braun *et al.* (2014).

Source Extraction

In source extraction we wish to extract the direct sounds from specific desired directions unaltered, while attenuating the direct sounds from other, interfering, directions. The diffuse sound is undesired. Therefore, we set the diffuse response function to 0, that is, $d_d(k) = 0$. The direct response function $d_s(k, \varphi)$ corresponds to a spatial window function that possesses a high gain for the desired directions and a low gain for the undesired interfering directions. For instance, if the desired sources are located around $60°$, we can use the window function depicted in Figure 7.3(a), which only extracts direct

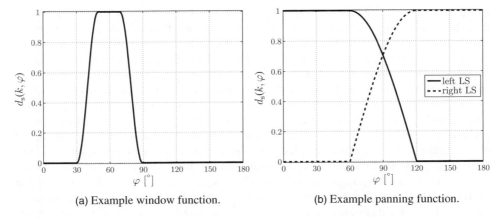

(a) Example window function. (b) Example panning function.

Figure 7.3 Example target response functions for the direct sound. The left response function can be used, for instance, in sound extraction applications. The right response functions can be used in spatial sound reproduction employing a stereo loudspeaker (LS) setup. Source: Thiergart 2014b. Reproduced with permission of IEEE.

sounds arriving close to 60°. The source extraction application is explained, for example, in Thiergart *et al.* (2014b).

Spatial Sound Reproduction

In spatial sound reproduction we aim to recreate the original spatial impression of the sound field that was present during recording. On the reproduction side we typically use multiple loudspeakers (for example, a stereo or 5.1 setup), and $Y(k, n)$ in Equation (7.2) is one of the loudspeaker signals. Note that aim not at recreating the original physical sound field, but at reproducing the sound such that it is *perceptually* identical to the original field in $\mathbf{r}_1$. Similarly to DirAC (Pulkki, 2007), this is achieved by reproducing the L direct sounds $P_{s,l}(k, n, \mathbf{r})$ from the original directions, indicated by the DOAs $\varphi_l(k, n)$, while the diffuse sound $P_d(k, n, \mathbf{r})$ is reproduced from all directions. To reproduce the direct sounds from the original directions, the direct sound response function $d_s(k, \varphi)$ for the particular loudspeaker corresponds to a so-called loudspeaker panning function. Example panning functions for the left and right loudspeaker of a stereo reproduction setup following the vector base amplitude panning (VBAP) scheme (Pulkki, 1997) are depicted in Figure 7.3(b). Alternatively to using loudspeaker panning functions, we can also use head-related transfer functions (HRTFs) to achieve spatial sound reproduction with headphones (Laitinen and Pulkki, 2009). The diffuse sound response function is set to a fixed value for all loudspeakers such that the diffuse sound is reproduced with the same power from all directions. This means that in this application the diffuse sound represents a desired component.

Acoustical Zooming

In acoustical zooming, we aim to mimic acoustically the visual zooming effect of a camera, such that the acoustical image and the visual image are aligned. For instance, when zooming in, the direct sound of the visible sources should be reproduced from the directions where the sources are visible in the video, while sources outside the visual image should be attenuated. Moreover, the diffuse sound should be reproduced from all directions but the signal-to-diffuse ratio (SDR) on the reproduction side should be increased to mimic the smaller opening angle of the camera. To achieve the correct reproduction of the direct sounds, we use the same approach as for the spatial sound reproduction explained above: the direct sound response function $d_s(k, \varphi)$ corresponds to a loudspeaker panning function. However, the loudspeaker panning functions are modified depending on the zoom factor of the camera to increase or decrease the width of the reproduced spatial image according to the opening angle of the camera. Moreover, we include an additional spatial window to attenuate direct sounds of sources that are not visible in the video. To achieve a plausible reproduction of the diffuse sound, we vary the diffuse response function $d_d(k)$ depending on the zoom factor. For small zoom factors, where the opening angle of the camera is large, we set $d_d(k)$ to 1, which means that we reproduce the diffuse sound with the original strength. When zooming in, we lower $d_d(k)$, that is, less diffuse sound is reproduced leading to a larger SDR at the reproduction side. Acoustical zooming is explained in Thiergart *et al.* (2014a).

7.2.2 Principal Estimation of the Target Signal

To estimate the target signal $Y(k, n)$ in Equation (7.2), we capture the sound $P(k, n, \mathbf{r})$ in Equation (7.1) with M microphones located at the positions $\mathbf{r}_{1...M}$. One microphone is located at the reference position $\mathbf{r}_1$ and is referred to as the reference microphone. In general, the target signal $Y(k, n)$ is estimated by applying linear multi-channel filters to the recorded microphone signals. We have two principal possibilities to estimate the target signal $Y(k, n)$ from the captured microphone signals, namely:

(i) direct estimation of the target signal $Y(k, n)$, and
(ii) separate estimation of the target direct signal $Y_s(k, n)$ and the target diffuse signal $Y_d(k, n)$, and then computing $Y(k, n)$ using Equation (7.2b).

Possibility (i) means that we jointly extract the direct sounds and diffuse sound with the desired target responses from the microphone signals. This means applying a single filter to the microphone signals to obtain the target signal $Y(k, n)$. Possibility (ii) means that we extract the direct and diffuse sounds separately with the desired target responses. This means that we apply two separate filters to the microphone signals to obtain $Y_s(k, n)$ and $Y_d(k, n)$ independently of each other.

Both possibilities have different advantages and disadvantages. In general, (i) is more accurate and the computational complexity is lower since only a single filter is computed. However, since the filter depends on the application (for example, the spatial window in source extraction, the loudspeaker setup in spatial sound reproduction, or the zooming factor in acoustical zooming), the filter must be computed and applied at the reproduction side. This means that we need to store or transmit all M microphone signals to the reproduction side, which requires high bandwidth. We can overcome this drawback when using (ii). With this approach, the two filters can be decomposed into two application-independent filters, which can be computed and applied at the recording side, and application-dependent target responses that can be applied at the reproduction side. This is illustrated in Figure 7.4. The filters on the recording side extract the direct sounds $P_{s,1...L}(k, n, \mathbf{r}_1)$ and the diffuse sound $P_d(k, n, \mathbf{r}_1)$, which are transmitted to the reproduction side and then combined depending on the application. Therefore, we

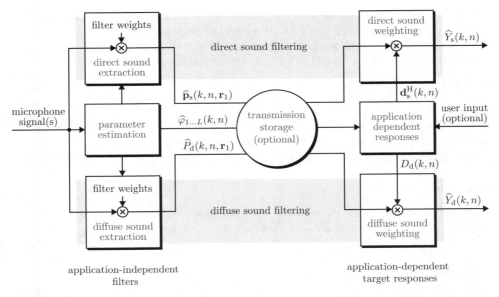

Figure 7.4 Principal estimation of the target signal $Y(k, n)$ using the indirect approach. Source: Thiergart 2015. Reproduced with permission of Oliver Thiergart.

need to transmit only $L + 1$ audio signals instead of M microphone signals,[1] and still maintain full flexibility with respect to the sound preproduction. The extracted direct sounds and diffuse sound together with the DOA information represent a compact parametric representation of the spatial sound that enables a fully flexible sound reproduction, independent of how the sound was recorded. Additionally, we have access to the separate target diffuse signal $Y_d(k, n)$. This is required in spatial sound reproduction applications, where we usually apply decorrelation filters to the diffuse sound to enable a realistic reproduction of the ambient sound (Pulkki, 2007).

7.3   Multi-Wave Sound Field and Signal Model

This section explains the sound field model in Equation (7.1) in more detail. As explained in Section 7.2, the L plane waves $P_{s,l}(k, n, \mathbf{r})$ represent the direct sound of multiple sound sources in a reverberant environment, and the diffuse sound $P_d(k, n, \mathbf{r})$ represents the reverberation. The number of plane waves L is usually smaller than the actual number of active sources, assuming that the source signals are sufficiently sparse in the time–frequency domain. This is typically the case for speech signals (Yilmaz and Rickard, 2004). The impact of the assumed number L is discussed in Section 7.4.1. The sound is captured with $M > L$ omnidirectional microphones positioned at $\mathbf{r}_{1...M}$. The microphone signals are written as

$$\mathbf{x}(k, n) = \sum_{l=1}^{L} \mathbf{x}_{s,l}(k, n) + \mathbf{x}_d(k, n) + \mathbf{x}_n(k, n), \tag{7.3}$$

---

[1] We also need to transmit the DOAs of the direct sound to compute the target responses.

where $\mathbf{x}_{s,l}(k,n)$ is the lth plane wave measured with the different microphones, $\mathbf{x}_d(k,n)$ is the measured diffuse field, and $\mathbf{x}_n(k,n)$ is a noise component.

Assuming that the three terms in Equation (7.3) are mutually uncorrelated, we can express the power spectral density (PSD) matrix of the microphone signals as

$$\mathbf{\Phi}_x(k,n) = E\{\mathbf{x}(k,n)\,\mathbf{x}^H(k,n)\} \tag{7.4a}$$

$$= \mathbf{\Phi}_s(k,n) + \mathbf{\Phi}_d(k,n) + \mathbf{\Phi}_n(k), \tag{7.4b}$$

where $\mathbf{\Phi}_s(k,n)$ is the direct sound PSD matrix, $\mathbf{\Phi}_d(k,n)$ is the diffuse sound PSD matrix, and $\mathbf{\Phi}_n(n)$ is the stationary noise PSD matrix. The different signal components (direct sound, diffuse sound, and noise) and corresponding PSD matrices are explained below.

7.3.1 Direct Sound Model

Without loss of generality, we consider the microphone located at $\mathbf{r}_1$ as the reference microphone. The direct sound $\mathbf{x}_{s,l}(k,n)$ in Equation (7.3) is written as

$$\mathbf{x}_{s,l}(k,n) = \mathbf{a}_l(k,n)P_{s,l}(k,n,\mathbf{r}_1), \tag{7.5}$$

where $\mathbf{a}_l(k,n) = [1, A_{2,l}(k,n), \ldots, A_{M,l}(k,n)]^T$ is the propagation vector of the lth plane wave with respect to the first microphone. The mth element of the lth propagation vector, $A_{m,l}(k,n)$, is the relative transfer function (RTF) between the first and mth microphones for the lth plane wave. Using the plane wave model for omnidirectional microphones, the RTF can be written as

$$A_{m,l}(k,n) = e^{J\kappa\,\mathbf{r}_{1m}^T\mathbf{n}_l(k,n)}, \tag{7.6}$$

where J is the imaginary unit, $\mathbf{r}_{m'm} = \mathbf{r}_m - \mathbf{r}_{m'}$, and κ is the wavenumber. The propagation vector depends on the DOA of the plane wave, which is expressed by the unit-norm vector $\mathbf{n}_l(k,n)$. For plane waves propagating in three-dimensional space, $\mathbf{n}_l(k,n)$ is defined as

$$\mathbf{n}_l(k,n) = \begin{bmatrix} \cos(\varphi_l)\cos(\vartheta_l) \\ \sin(\varphi_l)\cos(\vartheta_l) \\ \sin(\vartheta_l) \end{bmatrix}, \tag{7.7}$$

where $\varphi_l(k,n)$ denotes the azimuth and $\vartheta_l(k,n)$ is the elevation angle. Note that, especially in dynamic multi-source scenarios, the DOAs of the plane waves can vary rapidly across time and frequency.

Given the direct sound model above, we can express $\mathbf{\Phi}_s(k,n)$ in Equation (7.4b) as

$$\mathbf{\Phi}_s(k,n) = \mathbf{A}(k,n)\mathbf{\Psi}_s(k,n)\mathbf{A}^H(k,n), \tag{7.8}$$

where $\mathbf{A}(k,n) = [\mathbf{a}_1(k,n), \mathbf{a}_2(k,n), \ldots, \mathbf{a}_L(k,n)]$ is the propagation matrix. Moreover, the PSD matrix of the L plane waves is given by

$$\mathbf{\Psi}_s(k,n) = E\{\mathbf{p}_s(k,n,\mathbf{r}_1)\mathbf{p}_s^H(k,n,\mathbf{r}_1)\}, \tag{7.9}$$

where $\mathbf{p}_s(k,n,\mathbf{r}_1)$ contains the L plane wave signals at $\mathbf{r}_1$, as explained in Section 7.2. For mutually uncorrelated plane waves, $\mathbf{\Psi}_s(k,n)$ is a diagonal matrix where

$$\mathrm{diag}\{\mathbf{\Psi}_s(k,n)\} = \{\Psi_{s,1}(k,n), \ldots, \Psi_{s,L}(k,n)\} \tag{7.10}$$

are the powers of the L plane waves.

7.3.2 Diffuse Sound Model

As mentioned before, the diffuse sound $P_d(k, n, \mathbf{r}_1)$ models the reverberant sound that is present at the recording location. Reverberation is commonly modeled as a homogenous and isotropic, time-varying, diffuse field (Nélisse and Nicolas, 1997; Jacobsen and Roisin, 2000). Such a diffuse field can be modeled as the sum of $N \gg L$ mutually uncorrelated plane waves arriving with equal power and random phases from uniformly distributed directions (Nélisse and Nicolas, 1997; Jacobsen and Roisin, 2000). In this case, the mth microphone signal corresponding to the diffuse sound can be expressed as

$$X_{d,m}(k, n) = \sqrt{\frac{\Psi_d(k, n)}{N}} \sum_{i=1}^{N} e^{j\kappa \mathbf{r}_m^T \mathbf{m}_i(k,n) + j\theta_i(k,n)}, \tag{7.11}$$

where $\mathbf{m}_i(k, n)$ describes the DOA of the ith plane wave forming the diffuse sound, $\theta_i(k, n)$ is the phase of the wave in the origin of the coordinate system, and $\Psi_d(k, n)$ is the mean diffuse power. The phase terms $\theta_{1 \ldots N}(k, n)$ are mutually uncorrelated, uniformly distributed, random variables, that is, $\theta_{1 \ldots N}(k, n) \sim \mathcal{U}(0, 2\pi)$. A spherically isotropic diffuse field is obtained when the directions $\mathbf{m}_i$ are uniformly distributed on a sphere. The mean power of the diffuse field $\Psi_d(k, n)$ varies rapidly across time and frequency for a typical reverberant field. An example diffuse field computed with Equation (7.11) is depicted in Figure 7.2(b).

Given the model in Equation (7.11), the PSD matrix $\mathbf{\Phi}_d(k, n)$ of the diffuse sound in (7.4b) can be written as

$$\mathbf{\Phi}_d(k, n) = \Psi_d(k, n)\,\mathbf{\Gamma}_d(k). \tag{7.12}$$

The (m', m)th element of $\mathbf{\Gamma}_d(k)$, denoted by $\gamma_{d,m'm}(k)$, is the spatial coherence between microphones m' and m for a purely diffuse sound field, which is time invariant and known *a priori* (Elko, 2001). For instance, for a spherically isotropic diffuse field and omnidirectional microphones, we have (Nélisse and Nicolas, 1997)

$$\gamma_{d,m'm}(k) = \mathrm{sinc}(\kappa \, \|\mathbf{r}_{m'm}\|). \tag{7.13}$$

In the following, we define the vector

$$\mathbf{u}(k, n) \equiv \mathbf{x}_d(k, n) P_d^{-1}(k, n, \mathbf{r}_1), \tag{7.14}$$

which relates the diffuse sound captured by the different microphones to the diffuse field at the reference position. Clearly, as the diffuse field for each (k, n) is formed by a new realization of random plane waves in Equation (7.11), the elements of $\mathbf{u}(k, n)$ must be assumed non-deterministic and unobservable. As shown by Thiergart and Habets (2014), the expectation of $\mathbf{u}(k, n)$ is known, as it is given by

$$E\{\mathbf{u}(k, n)\} = \boldsymbol{\gamma}_{d,1}(k), \tag{7.15}$$

where $\boldsymbol{\gamma}_{d,1}(k)$ is the diffuse coherence vector, that is, the first column of $\mathbf{\Gamma}_d(k)$. The definitions in Equations (7.14) and (7.15) are helpful for the derivation of the filters in Section 7.4.2.

7.3.3 Noise Model

With the noise $\mathbf{x}_n(k, n)$ in Equation (7.3) we can model, for instance, a stationary background noise such as fan noise or the microphone self-noise. In contrast to the direct sound and diffuse sound, the noise PSD matrix $\mathbf{\Phi}_n(k)$ given in Equation (7.4) is assumed to be time invariant. This assumption allows us to estimate $\mathbf{\Phi}_n(k)$ directly from the microphone signals, for example during speech pauses, as discussed in Section 7.5.4. Unless otherwise stated, we make no further assumptions on the noise besides the stationary assumption. In practice, this assumption can be relaxed to slowly time-varying.

7.4 Direct and Diffuse Signal Estimation

This section explains the estimation of the target direct signal $Y_s(k, n)$ and target diffuse signal $Y_d(k, n)$ in Equation (7.2b) using multi-channel filters. We consider estimators for which closed-form solutions exist such that these can be computed efficiently for each time and frequency with current information on the DOA of the L direct sounds and SOS of the sound components.

7.4.1 Estimation of the Direct Signal $Y_s(k, n)$

State of the Art

Existing spatial filtering approaches, which can be used to estimate $Y_s(k, n)$, can be divided into *signal-independent* spatial filters and *signal-dependent* spatial filters (Van Veen and Buckley, 1988; Van Trees, 2002). Signal-independent filters extract desired sounds without taking time-varying SOS of the sound components into account. Typical examples are the delay-and-sum beamformer in Van Veen and Buckley (1988), the filter-and-sum beamformer in Doclo and Moonen (2003), differential microphone arrays (Elko *et al.*, 1998; Elko, 2000; Teutsch and Elko, 2001; Benesty *et al.*, 2008), or the superdirective beamformer (Cox *et al.*, 1986; Bitzer and Simmer, 2001). Computing these filters only requires the array steering vectors $\mathbf{a}_{1...L}(k, n)$ that specify the desired or undesired directions from which to capture or attenuate the sound. Unfortunately, the signal-independent filters cannot adapt to changing acoustic situations (that is, time-varying SOS of the sound components). Therefore, the superdirective beamformer, for example, performs well when the sound field is diffuse and the spatially white noise is low, but it performs poorly in noisy situations since its robustness against noise is low (Doclo and Moonen, 2007).

Signal-dependent spatial filters overcome this drawback by taking the time-varying SOS of the desired or undesired sound components into account. Examples are the minimum variance distortionless response (MVDR) beamformer (Capon, 1969) and the linearly constrained minimum variance (LCMV) filter (Van Trees, 2002), both of which minimize the residual power of the undesired components at the filter output subject to one or more linear constraints, respectively. The multi-channel Wiener filter represents a signal-dependent filter that does not satisfy a linear constraint but minimizes the mean squared error (MSE) between the true and estimated desired signal (Doclo and Moonen, 2002). This filter provides a stronger reduction of the undesired components compared to the linearly constrained filters, but introduces signal distortions. To achieve a trade-off between signal distortions and noise reduction, the parametric Wiener filter was proposed (Spriet *et al.*, 2004; Doclo and Moonen, 2005; Doclo *et al.*, 2005). Later, this filter was derived for the multi-wave case (Markovich-Golan *et al.*, 2012).

The optimal signal-dependent filter weights can be obtained directly or iteratively (Frost, 1972; Affes *et al.*, 1996; Herbordt and Kellermann, 2003; Gannot and Cohen, 2008). The latter filters are referred to as adaptive filters and typically achieve a strong reduction of undesired signal components. Unfortunately, the adaptive filters can suffer from a strong cancelation of the desired signal in practice. Therefore, many approaches have been proposed to improve the robustness of these filters (Cox *et al.*, 1987; Nordholm *et al.*, 1993; Hoshuyama *et al.*, 1999; Gannot *et al.*, 2001; Reuven *et al.*, 2008; Talmon *et al.*, 2009; Markovich *et al.*, 2009; Krueger *et al.*, 2011). Another drawback of these filters is the inability to adapt sufficiently quickly to the optimal solution in dynamic situations, for example, source movements, competing speakers that become active when the desired source is active, or changing power ratios between the noise and reverberant sound. In contrast, the signal-dependent filters, which can be computed directly, can adapt quickly to changes in the acoustic scene. However, this requires that the array steering vectors $a_{1...L}(k, n)$ and SOS of the desired and undesired components, which are required to compute the filters, are estimated with sufficient accuracy and temporal resolution from the microphone signals.

Informed Direct Sound Filtering

In the following we consider signal-dependent filters to estimate $Y_s(k, n)$. The filters are computed for each time and frequency with current information on the underlying parametric sound field model. This includes current information on the SOS of the desired and undesired signal components, but also current directional information on the L plane waves. This enables the filters to quickly adapt to changes in the acoustic scene. These filters are referred to as informed spatial filters (ISFs) in the following.

In general, an estimate of the desired signal $Y_s(k, n)$ is obtained by a linear combination of the microphone signals $\mathbf{x}(k, n)$:

$$\widehat{Y}_s(k, n) = \mathbf{w}_s^H(k, n)\mathbf{x}(k, n). \tag{7.16}$$

As shown later, the multi-channel filter $\mathbf{w}_s(k, n)$ can be decomposed as

$$\mathbf{w}_s(k, n) = \mathbf{H}_s(k, n)\mathbf{d}_s(k, n), \tag{7.17}$$

where $\mathbf{H}_s(k, n)$ is a filter matrix and $\mathbf{d}_s(k, n)$ are the target direct sound responses introduced in Section 7.1. Inserting the previous equation into Equation (7.16) yields

$$\widehat{Y}_s(k, n) = \mathbf{d}_s^H(k, n)\widehat{\mathbf{p}}_s(k, n, \mathbf{r}_1), \tag{7.18}$$

where

$$\widehat{\mathbf{p}}_s(k, n, \mathbf{r}_1) = \mathbf{H}_s^H(k, n)\mathbf{x}(k, n). \tag{7.19}$$

The filter matrix $\mathbf{H}_s(k, n)$ is application independent and can be applied to the microphone signals $\mathbf{x}(k, n)$ at the recording side. Thus, we need to transmit only the L estimated plane waves in $\widehat{\mathbf{p}}_s(k, n, \mathbf{r}_1)$ to the reproduction side, instead of the M microphone signals, and still maintain full flexibility with respect to the direct sound reproduction.[2] A corresponding block scheme which visualizes the recording and reproduction processing is depicted in Figure 7.4.

---

[2] Note that the DOAs of the L plane waves also need to be transmitted to the reproduction side to compute the target responses in $\mathbf{d}_s(k, n)$.

To obtain an accurate estimate of $Y_s(k,n)$ using Equation (7.16), a filter $\mathbf{w}_s(k,n)$ is required that can capture multiple source signals, namely the L plane waves, with the desired direct sound responses $\mathbf{d}_s(k,n)$. In Thiergart and Habets (2013), the filter weights $\mathbf{w}_s(k,n)$ in Equation (7.16) were derived as an informed LCMV filter, which minimizes the diffuse sound and stationary noise at the filter output. In Thiergart *et al.* (2013), the weights $\mathbf{w}_s(k,n)$ were derived as an informed minimum mean square error (MMSE) filter, which minimizes the mean square error between $Y_s(k,n)$ and $\hat{Y}_s(k,n)$. As explained later, both filters have specific advantages and disadvantages.

Parametric Multi-Channel Wiener Filter

In the following, we introduce an optimal multi-wave filter that represents a generalization of the informed LCMV filter and the informed MMSE filter. The proposed spatial filter is referred to as the informed parametric multi-wave multi-channel Wiener (PMMW) filter and is found by minimizing the stationary noise and diffuse sound at the filter output subject to a constraint that limits the signal distortion of the extracted direct sound. Expressed mathematically, the filter weights $\mathbf{w}_s(k,n)$ are computed as

$$\mathbf{w}_{\text{sPMMW}}(k,n) = \arg\min_{\mathbf{w}} \mathbf{w}^H \boldsymbol{\Phi}_u(k,n)\mathbf{w} \tag{7.20}$$

subject to

$$E\{|Y_{s,l}(k,n) - \hat{Y}_{s,l}(k,n)|^2\} \le \sigma_l^2 \quad \forall l, \tag{7.21}$$

where $\boldsymbol{\Phi}_u(k,n) = \boldsymbol{\Phi}_n(k) + \boldsymbol{\Phi}_d(k,n)$ is the undesired noise-plus-diffuse sound PSD matrix and $Y_{s,l}(k,n) = D_{s,l}(k,n)P_{s,l}(k,n,\mathbf{r}_1)$ is the desired filter output for the lth plane wave. Moreover, $\hat{Y}_{s,l}(k,n) = \mathbf{w}_s^H(k,n)\mathbf{x}_{s,l}(k,n)$ is the actual filter output for the lth plane wave, which potentially is distorted. The desired maximum distortion of the lth plane wave is specified with σ_l^2. A higher maximum distortion means that we can better attenuate the noise and diffuse sound in Equation (7.20). As shown by Thiergart *et al.* (2014b), a closed-form solution for $\mathbf{w}_{\text{sPMMW}}(k,n)$ is given by Equation (7.17), where $\mathbf{H}_s(k,n)$ is computed as

$$\mathbf{H}_{\text{sPMMW}}(k,n) = \boldsymbol{\Phi}_u^{-1}(k,n)\mathbf{A}\left[\boldsymbol{\Omega}_s(k,n)\boldsymbol{\Psi}_s^{-1}(k,n) + \mathbf{A}^H\boldsymbol{\Phi}_u^{-1}(k,n)\mathbf{A}\right]^{-1}. \tag{7.22}$$

Computing the filter requires the DOA of the L plane waves – to compute the propagation matrix $\mathbf{A}(k,n)$ with Equation (7.6) – as well as $\boldsymbol{\Psi}_s(k,n)$ and $\boldsymbol{\Phi}_u(k,n)$. The estimation of these quantities is discussed in Section 7.5. The real-valued, positive $L \times L$ diagonal matrix $\boldsymbol{\Omega}_s(k,n)$ contains time- and frequency-dependent control parameters,

$$\text{diag}\{\boldsymbol{\Omega}_s(k,n)\} = \{\omega_{s,1}(k,n), \dots, \omega_{s,L}(k,n)\}, \tag{7.23}$$

which allow us to control the trade-off between noise suppression and signal distortion for each plane wave. The trade-off is illustrated in Figure 7.5:

- For $\boldsymbol{\Omega}_s(k,n) = \mathbf{0}$, $\mathbf{H}_{\text{sPMMW}}(k,n)$ reduces to the informed LCMV filter, denoted by $\mathbf{H}_{\text{sLCMV}}(k,n)$, as proposed by Thiergart and Habets (2013), which extracts the L direct sounds without distortions ($\sigma_{1\dots L}^2 = 0$) when $\mathbf{A}(k,n)$ does not contain estimation errors. The informed LCMV filter provides a trade-off between white noise gain (WNG) and directivity index (DI) depending on the diffuse-to-noise ratio (DNR), that is, depending on which of the two undesired components (diffuse sound or noise) is more prominent.

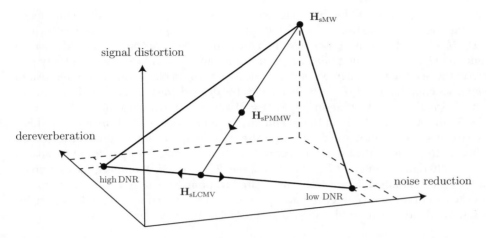

Figure 7.5 Relation of the different multi-channel filters for direct sound estimation in terms of noise reduction, diffuse sound reduction, and signal distortions. Source: Thiergart 2015. Reproduced with permission of Oliver Thiergart.

- For $\mathbf{\Omega}_s(k, n) = \mathbf{I}_L$, with $\mathbf{I}_L$ being the $L \times L$ identity matrix, $\mathbf{H}_{\text{sPMMW}}(k, n)$ reduces to the informed MMSE filter $\mathbf{H}_{\text{sMW}}(k, n)$ proposed by Thiergart *et al.* (2013). This filter provides a stronger suppression of the noise and diffuse sound compared to the informed LCMV but introduces undesired signal distortions ($\sigma_{1...L}^2 > 0$).
- For $0 < \omega_{s,l}(k, n) < 1$ we can achieve for each plane wave a trade-off between the LCMV filter and MMSE filter such that we obtain a strong attenuation of the noise while still ensuring a tolerable amount of undesired signal distortions. This is discussed in more detail shortly. For $\omega_{s,l}(k, n) > 1$, we can achieve an even stronger attenuation of the noise and diffuse sound than with the MMSE filter, at the expense of stronger signal distortions.

As an example of an ISF, Figure 7.6 (solid line) shows the magnitude of an arbitrary (for example, user-defined) desired response function $d_s(k, \varphi)$ as a function of the azimuth angle φ. This function means that we aim to capture a plane wave arriving

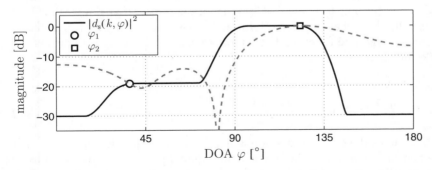

Figure 7.6 Solid line: Example desired response function $d_s(k, \varphi)$. Dashed line: Resulting directivity pattern of an example spatial filter that assumes $L = 2$ plane waves with DOAs φ_1 and φ_2. Source: Thiergart 2014b. Reproduced with permission of IEEE.

from $\varphi_1 = 37°$ (indicated by the circle) with a gain of $D_{s,1}(k, n) = -19\,\text{dB}$, while a second plane wave arriving from $\varphi_2 = 123°$ (indicated by the square) is captured with a gain of $D_{s,2}(k, n) = 0\,\text{dB}$. Both gains would then form the desired response vector $\mathbf{d}_s(k, \varphi)$ in Equation (7.17) – assuming both waves are simultaneously active. The directivity pattern of the resulting spatial filter, which would capture both plane waves with the desired responses contained in $\mathbf{d}_s(k, \varphi)$, is depicted by the dashed line in Figure 7.6. Here, we are considering the LCMV solution ($\mathbf{\Omega}_s(k, n) = \mathbf{0}$) and a uniform linear array (ULA) with $M = 5$ omnidirectional microphones and microphone spacing $r = 3\,\text{cm}$ at $f = 3.3\,\text{kHz}$. As we can see in the plot, the directivity pattern of the spatial filter exhibits the desired gains for the DOAs of the two plane waves. Note that the directivity pattern of the spatial filter is different for different L and DOAs of the direct sound. Moreover, the directivity pattern is essentially different from the desired response function $d_s(k, \varphi)$. In fact, it is not the aim of the spatial filter to resample the response function $d_s(k, \varphi)$ for all angles φ, but to provide the desired response for the DOAs of the L plane waves.

Adjusting the Control Parameters $\omega_{s,l}(k, n)$ in Practice

In many applications it is desired to capture a plane wave with low distortions if the wave is strong compared to the noise and diffuse sound. In this case, a strong attenuation of the noise and diffuse sound is less important. If, however, the plane wave is weak compared to the noise and diffuse sound, a strong attenuation of these components is desired. In the latter case, distortions of the plane wave signal can be assumed to be less critical.

To obtain such a behavior of the ISF, the parameters $\omega_{s,l}(k, n)$ can be made signal dependent. To ensure a computationally feasible algorithm, we compute all parameters $\omega_{s,1\ldots L}(k, n)$ independently, even though they jointly determine the signal distortion σ_l^2 for the lth plane. For the following example, let us first introduce the logarithmic input signal-to-diffuse-plus-noise ratio (SDNR) for the lth plane wave as

$$\xi_l(k, n) = 10 \log_{10} \left(\frac{\Psi_{s,l}(k, n)}{\Psi_d(k, n) + \Phi_n(k)} \right), \tag{7.24}$$

where $\Phi_n(k) = \frac{1}{M}\text{tr}\{\mathbf{\Phi}_n(k)\}$ is the mean noise power across the microphones. On the one hand, the parameter $\omega_{s,l}(k, n)$ should approach 0 (leading to the LCMV filter) if $\xi_l(k, n)$ is large. On the other hand, $\omega_{s,l}(k, n)$ should become 1 (leading to the MMSE filter) or larger than 1 (leading to a filter that is even more aggressive than the MMSE filter) if $\xi_l(k, n)$ is small. This behavior for $\omega_{s,l}(k, n)$ can be achieved, for instance, if we compute $\omega_{s,l}(k, n)$ via a sigmoid function that is monotonically decreasing with decreasing input SDNR, for example

$$\omega_{s,l}(k, n) = \text{sig}(\xi_l) \tag{7.25a}$$

$$= \frac{1}{2}\alpha_1 \left[1 + \tanh\left(\alpha_2 \frac{\alpha_3 - \xi_l(k, n)}{2} \right) \right], \tag{7.25b}$$

where $\text{sig}(\xi_l)$ denotes the sigmoid function and $\alpha_{1\ldots 3}$ are the sigmoid parameters that control the behavior of $\omega_{s,l}(k, n)$. In practice, the sigmoid parameters may need to be adjusted specifically for the given application, and also depending on the accuracy of the parameter estimators in Section 7.5, to obtain the best performance. Clearly, different functions to control $\omega_{s,l}(k, n)$ can be designed depending on the specific application

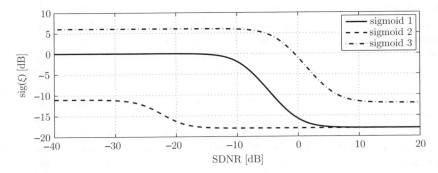

Figure 7.7 Example sigmoid functions depending on the SDNR. The parameters of the function "sigmoid 1" are: $a_1 = 1, a_2 = 0.5, a_3 = -9, a_4 = -18$ dB. Source: Thiergart 2015. Reproduced with permission of Oliver Thiergart.

and desired behavior of the filter. However, the sigmoid function in Equation (7.25b), with the associated parameters, provides high flexibility in adjusting the behavior of the spatial filters.

Figure 7.7 shows three examples of $\text{sig}(\xi)$ in Equation (7.25b). Note that the sigmoid functions are plotted in decibels. The "sigmoid 1" function may represent a suitable function for many applications. In general, with a_1 we control the maximum of the sigmoid function for low input SDNRs $\xi_l(k, n)$. Higher values for a_1 lead to more aggressive noise and diffuse sound suppression when the input SDNR is low. With a_2 and a_3 we control the slope and shift along the ξ_l axis, respectively. For instance, shifting the sigmoid function towards low input SDNRs and using a steep slope means that the plane waves are extracted with low undesired distortions unless the diffuse sound and noise become very strong. Accordingly, the "sigmoid 2" function in Figure 7.7 would yield a less aggressive filter, while the "sigmoid 3" function would yield a more aggressive filter. Note that all the parameters can be frequency dependent.

Influence of Early Reflections

The signal model in Equation (7.3) assumes that the direct sounds, the L plane waves, are mutually uncorrelated. This assumption greatly simplifies the derivation of the informed PMMW filter leading to the closed-form solution in Equation (7.22). Assuming mutually uncorrelated plane waves is reasonable for the direct sounds of different sources, but typically does not hold for the direct sound of a source and its early reflections.[3] Hence, we assume that the plane waves for a given time–frequency instant correspond to different sound sources (rather than one or more reflections of the same source). Analyzing the impact of mutually correlated direct sounds and reflections on the filter performance remains a topic for further research. In fact, it should be noted that not only the underlying assumption of the filter is violated, but also of the parameter estimators in Section 7.5.

---

[3] Note that early reflections of the sound sources are not directly considered by the model in Equation (7.3), but they can be represented as plane waves as well.

Influence of the Assumed Number of Plane Waves L

The number of plane waves L assumed in Equations (7.1) and (7.3) strongly influence the performance of the filter. If L is too small (smaller than the actual number of prominent waves), then the signal model in Equation (7.1) is violated. In this case, the filter in Equation (7.16) extracts fewer plane waves than desired, which leads to undesired distortions of the direct sound. The effects of such model violations are discussed, for example, in Thiergart and Habets (2012). On the other hand, if L is too high, the spatial filter has fewer degrees of freedom to minimize the noise and diffuse sound power in Equation (7.20). Therefore, the assumed value for L is a trade-off that depends on the number of microphones and desired filter performance. For spatial sound reproduction applications, using $L = 2$ or $L = 3$ usually represents a reasonable choice. Alternatively, the number of sources can be estimated, for example as discussed in Section 7.5.1.

7.4.2   Estimation of the Diffuse Signal $Y_d(k, n)$

State of the Art

In most applications, the diffuse sound is extracted from a single microphone using single-channel filters. For example in Uhle *et al.* (2007), the diffuse sound is extracted based on non-negative matrix factorization (NMF). In DirAC (Pulkki, 2007), the diffuse sound is extracted by designing a single-channel filter based on the diffuseness of the sound. In Del Galdo *et al.* (2012), it was shown that diffuseness-based sound extraction is a close approximation to a single-channel square-root Wiener filter.

Only a few approaches are available for the extraction of diffuse sounds from multiple microphones. Compared to the single-channel filters, these multi-channel filters have the advantage that they can extract the diffuse sound while at the same time attenuating the direct sound, which avoids direct sounds leaking into the diffuse sound estimate. To extract the diffuse sound from a microphone pair, most available approaches assume that the diffuse sound is uncorrelated between the microphones whereas the direct sound is correlated. The diffuse sound is extracted by removing the correlated sound components from the recorded microphone signals (Avendano and Jot, 2002; Irwan and Aarts, 2002; Faller, 2006; Usher and Benesty, 2007; Merimaa *et al.*, 2007). The drawback of these approaches is that only two microphones can be employed directly. Moreover, the diffuse sound at lower frequencies or for smaller microphones is typically correlated between the microphones such that removing correlated signal components cancels the diffuse sound as well. The approaches in Irwan and Aarts (2002), Faller (2006), Usher and Benesty (2007), and Merimaa *et al.* (2007) remove the coherent sound by computing the difference between the two microphone signals after properly delaying or equalizing the signals. This is essentially spatial filtering, where a spatial null is steered towards the DOA from which the direct sound arrives. In Thiergart and Habets (2013), this idea was applied to a linear array of omnidirectional microphones. Unfortunately, the resulting spatial filter was highly directional for higher frequencies, which is suboptimal when we aim to capture an isotropic diffuse field.

Informed Diffuse Sound Filtering

In the following, we use an ISF for estimating $Y_d(k, n)$ in Equation (7.2b), similarly to Thiergart and Habets (2013). As in the previous subsection, an estimate of the desired signal is obtained by a linear combination of the microphone signals $\mathbf{x}(k, n)$,

$$\hat{Y}_d(k, n) = \mathbf{w}_d^H(k, n)\,\mathbf{x}(k, n), \tag{7.26}$$

where $\mathbf{w}_{\mathrm{d}}(k, n)$ is a complex weight vector of length M. As shown later, all the $\mathbf{w}_{\mathrm{d}}(k, n)$ filters discussed can be decomposed as

$$\mathbf{w}_{\mathrm{d}}(k, n) = \mathbf{h}_{\mathrm{d}}(k, n) D_{\mathrm{d}}^*(k, n), \tag{7.27}$$

where $\mathbf{h}_{\mathrm{d}}(k, n)$ is an application-independent filter while $D_{\mathrm{d}}^*(k, n)$ is application dependent. Inserting the previous equation into Equation (7.26) yields

$$\widehat{Y}_{\mathrm{d}}(k, n) = D_{\mathrm{d}}(k, n)\widehat{P}_{\mathrm{d}}(k, n, \mathbf{r}_1), \tag{7.28}$$

where

$$\widehat{P}_{\mathrm{d}}(k, n, \mathbf{r}_1) = \mathbf{h}_{\mathrm{d}}^{\mathrm{H}}(k, n)\mathbf{x}(k, n) \tag{7.29}$$

is an estimate of the diffuse sound at the reference position. The filter $\mathbf{h}_{\mathrm{d}}(k, n)$ does not depend on the target response $D_{\mathrm{d}}(k, n)$. Hence, it is application independent and can be computed and applied at the recording side. Since we recompute the filters $\mathbf{w}_{\mathrm{d}}(k, n)$ and $\mathbf{h}_{\mathrm{d}}(k, n)$, respectively, for each time and frequency, we focus on filters with a closed-form expression.

Distortionless Response Filters

We first discuss spatial filters that estimate the desired diffuse sound $Y_{\mathrm{d}}(k, n)$ with a distortionless response. Distortionless response means that the filter $\mathbf{w}_{\mathrm{d}}(k, n)$ extracts the diffuse sound $P_{\mathrm{d}}(k, n)$ at the reference microphone with the target response $D_{\mathrm{d}}(k, n)$. Therefore, a distortionless filter must satisfy the constraint

$$\mathbf{w}_{\mathrm{d}}^{\mathrm{H}}(k, n)\mathbf{x}_{\mathrm{d}}(k, n) - D_{\mathrm{d}}(k, n)P_{\mathrm{d}}(k, n, \mathbf{r}_1) = 0. \tag{7.30}$$

An alternative expression is found by dividing Equation (7.30) by $P_{\mathrm{d}}(k, n, \mathbf{r}_1)$ and using Equation (7.14):

$$\mathbf{w}_{\mathrm{d}}^{\mathrm{H}}(k, n)\mathbf{u}(k, n) - D_{\mathrm{d}}(k, n) = 0. \tag{7.31}$$

The left-hand side in Equation (7.30) is the distortion of the target diffuse signal at the filter output, which is forced to zero by the filter. All filters that satisfy Equation (7.31) require $\mathbf{u}(k, n)$ and thus are denoted as $\mathbf{w}_{\mathrm{d}}(k, n, \mathbf{u})$. Unfortunately, the vector $\mathbf{u}(k, n)$ is unavailable in practice and must be considered as an unobservable random variable. As a consequence, we cannot compute the distortionless response filters in practice. However, the filters serve as a basis for deriving the distortionless response *average* filters at the end of this section.

Example Distortionless Response Filter

An optimal distortionless response filter for estimating the target diffuse sound $Y_{\mathrm{d}}(k, n)$ cancels out or reduces the direct sounds and noise while satisfying Equation (7.31). Such a filter is found, for example, by minimizing the residual noise power at the filter output while using L linear constraints for nulling out the L plane waves. In this case, the optimal filter is referred to as an LCMV filter and is computed as

$$\mathbf{w}_{\mathrm{dLCMV}}(k, n, \mathbf{u}) = \arg\min_{\mathbf{w}_{\mathrm{d}}} \mathbf{w}_{\mathrm{d}}^{\mathrm{H}}\mathbf{\Phi}_{\mathrm{n}}(k)\mathbf{w}_{\mathrm{d}}, \tag{7.32}$$

subject to Equation (7.31) and subject to

$$\mathbf{w}_{\mathrm{d}}^{\mathrm{H}}\mathbf{A}(k, n) = \mathbf{0}, \tag{7.33}$$

where $\mathbf{A}(k, n)$ is the array steering matrix. This filter satisfies $L + 1$ constraints and hence requires at least $M = L + 1$ microphones.

The optimization problem of this section has a well-known solution (Van Trees, 2002). The solution for the LCMV filter in Equation (7.32) is given by Equation (7.27), where $\mathbf{h}_d(k, n)$ is

$$\mathbf{h}_{\mathrm{dLCMV}}(k, n, \mathbf{u}) = \boldsymbol{\Phi}_{\mathrm{n}}^{-1}(k)\mathbf{C}(k, n)\left[\mathbf{C}^{\mathrm{H}}(k, n)\boldsymbol{\Phi}_{\mathrm{n}}^{-1}(k)\mathbf{C}(k, n)\right]^{-1}\mathbf{g}. \tag{7.34}$$

Here, $\mathbf{C}(k, n) = [\mathbf{A}(k, n), \mathbf{u}(k, n)]$ is the constraint matrix and $\mathbf{g} = [0, 1]^{\mathrm{T}}$ the corresponding response vector. We can see that computing the filters requires knowledge about $\mathbf{u}(k, n)$, which is unavailable in practice as explained before.

Distortionless Response Average Filters

Since we cannot compute distortionless response filters in practice, we investigate a second class of filters. To compute them, we consider the expectation of the distortionless response filters computed across the different realizations of $\mathbf{u}(k, n)$. This leads to the distortionless response average filters $\bar{\mathbf{w}}_d(k, n)$, which are computed as

$$\bar{\mathbf{w}}_d(k, n) = \mathrm{E}\{\mathbf{w}_d(k, n, \mathbf{u})\} \tag{7.35a}$$

$$= \int_{\mathbf{u}} f(\mathbf{u})\mathbf{w}_d(k, n, \mathbf{u})\, d\mathbf{u}, \tag{7.35b}$$

where $\mathbf{w}_d(k, n, \mathbf{u})$ is the distortionless response filter and $f(\mathbf{u})$ is the probability density function (PDF) of $\mathbf{u}(k, n)$. Unfortunately, no closed-form solution exists to compute the integral, and obtaining a numerical solution is computationally very expensive. As shown by Thiergart and Habets (2014), a close approximation of Equation (7.35a) is given by

$$\mathrm{E}\{\mathbf{w}_d(k, n, \mathbf{u})\} \approx \mathbf{w}_d(k, n, \mathrm{E}\{\mathbf{u}\}). \tag{7.36}$$

This equation means that the expected weights of the linearly constrained filters can be found approximately by computing the filter with the expectation of the linear constraint. The expectation of $\mathbf{u}(k, n)$ is given in Equation (7.15). The approximate distortionless response average filters, denoted by $\mathbf{w}_{dA}(k, n)$, can now be computed as

$$\mathbf{w}_{dA}(k, n) = \mathbf{w}_d(k, n, \boldsymbol{\gamma}_{d,1}), \tag{7.37}$$

which is an approximation of $\bar{\mathbf{w}}_d(k, n)$ in Equation (7.35a). An example filter that can be computed in practice is shown next.

Example Distortionless Response Average Filter

The distortionless response LCMV filter $\mathbf{w}_{\mathrm{dLCMV}}(k, n, \mathbf{u})$ introduced earlier minimizes the power of the noise at the filter output while satisfying a linear diffuse sound constraint and L additional linear constraints to attenuate the direct sounds. The corresponding approximate average LCMV filter is defined using Equation (7.37):

$$\mathbf{w}_{\mathrm{dALCMV}}(k, n) = \mathbf{w}_{\mathrm{dLCMV}}(k, n, \boldsymbol{\gamma}_{d,1}). \tag{7.38}$$

A closed-form solution for $\mathbf{w}_{\mathrm{dALCMV}}(k, n)$ is found when solving the optimization problem in Equation (7.32) subject to Equation (7.33) and subject to Equation (7.31) with

$\gamma_{d,1}(k)$ instead of $\mathbf{u}(k, n)$. The result is given by Equation (7.27), where $\mathbf{h}_d(k, n)$ is given in Equation (7.34) when substituting $\mathbf{u}(k, n)$ with $\gamma_{d,1}(k)$:

$$\mathbf{h}_{\text{dALCMV}}(k, n) = \mathbf{h}_{\text{dLCMV}}(k, n, \gamma_{d,1}). \tag{7.39}$$

Computing this filter requires the DOA of the L plane waves, to compute the array steering matrix $\mathbf{A}(k, n)$, and the noise PSD matrix $\mathbf{\Phi}_n(k)$. The filter requires $M > L + 1$ microphones to satisfy the $L + 1$ linear constraints and to minimize the residual noise.

Figure 7.8 shows the directivity pattern of the approximate average LCMV filter $\mathbf{h}_{\text{dALCMV}}(k, n)$ and the SOA spatial filter used in Thiergart and Habets (2013), denoted by $\mathbf{h}_{\Psi L}(k, n)$. The directivity patterns were computed for different frequencies assuming a ULA of $M = 8$ omnidirectional microphones with spacing $r = 3$ cm. A single plane wave was arriving from $\varphi = 72°$. We can see that for increasing frequencies, the filter $\mathbf{h}_{\Psi L}(k, n)$ (the plots on the left-hand side) became very directional such that the diffuse sound was captured mainly from one direction. In contrast, the proposed approximate average LCMV filter (right-hand side) provided the desired almost omnidirectional directivity even for high frequencies (besides the spatial null for the direction of the plane wave).

7.5 Parameter Estimation

This section deals with the estimation of the parameters that are required to compute the informed filters in Section 7.4.

7.5.1 Estimation of the Number of Sources

A good overview of approaches for estimating the number of sources forming a mixture can be found in Jiang and Ingram (2004). The most popular approaches consider the eigenvalues of the microphone input PSD matrix, in our case $\mathbf{\Phi}_x(k, n)$ defined in Equation (7.4). In general, these approaches are suited to our framework since the computationally expensive eigenvalue decomposition (EVD) of $\mathbf{\Phi}_x(k, n)$ is also required later, namely when estimating the DOAs of the direct sound (see Section 7.5.2).

The well-known approach proposed by Wax and Kailath (1985) models the microphone signals as a sum of L mutually uncorrelated source signals plus independent and identically distributed (iid) noise. Diffuse sound was not considered. Under these assumptions, $\mathbf{\Phi}_x(k, n)$ in Equation (7.4a) possesses $M - L$ eigenvalues which are equal to the noise power $\mathbf{\Phi}_n(k)$. Thus, the number of sources can be determined from the multiplicity of the smallest eigenvalue, that is, from the number of eigenvalues that are equal. In practice, determining this number is difficult since $\mathbf{\Phi}_x(k, n)$ needs to be estimated from the microphone signals and contains estimation errors. Hence, the $M - L$ smallest eigenvalues will be different. Therefore, the approach in Wax and Kailath (1985) determines the multiplicity of the smallest eigenvalue based on information theoretic criteria (ITC) such as the minimum description length (MDL) or Akaike information criterion (AIC).

A conceptually different approach, which also considers the eigenvalues of $\mathbf{\Phi}_x(k, n)$, was used by Markovich et al. (2009). Here, L was determined by considering the difference of the eigenvalues compared to the maximum eigenvalue and a fixed lower

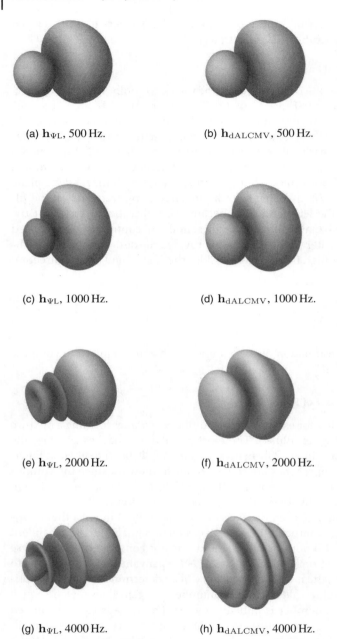

(a) $\mathbf{h}_{\Psi L}$, 500 Hz.

(b) $\mathbf{h}_{\mathrm{dALCMV}}$, 500 Hz.

(c) $\mathbf{h}_{\Psi L}$, 1000 Hz.

(d) $\mathbf{h}_{\mathrm{dALCMV}}$, 1000 Hz.

(e) $\mathbf{h}_{\Psi L}$, 2000 Hz.

(f) $\mathbf{h}_{\mathrm{dALCMV}}$, 2000 Hz.

(g) $\mathbf{h}_{\Psi L}$, 4000 Hz.

(h) $\mathbf{h}_{\mathrm{dALCMV}}$, 4000 Hz.

Figure 7.8 Directivity pattern of the SOA filter (left) and approximate average LCMV filter (right) for extracting diffuse sound in the presence of a single plane wave. Uniform linear array with $M = 8$ omnidirectional microphones ($r = 3$ cm). Source: Thiergart 2015. Reproduced with permission of Oliver Thiergart.

eigenvalue threshold. The main advantage of this approach is the almost negligible computational costs compared to the earlier method. Therefore, we consider this approach in the example application in Section 7.6.

7.5.2 Direction of Arrival Estimation

The DOA of the L narrowband plane waves represents a crucial parameter for computing the informed spatial filters. Once the DOAs are estimated, the elements of the array steering matrix $\mathbf{A}(k, n)$, which is required to compute the informed spatial filters, can be determined using Equation (7.6). The required DOA estimators are available for almost any microphone configuration, such as (non-)uniform linear arrays, planar arrays, or the popular B-format microphone.

The most popular multi-wave DOA estimators (which can estimate multiple DOAs per time and frequency) for arrays of omnidirectional microphones are ESPRIT (Roy and Kailath, 1989) and Root MUSIC (Rao and Hari, 1988). Both approaches can estimate the DOA of $L < M$ plane waves. ESPRIT requires an array that can be separated into two identical, rotationally invariant, subarrays. It is also extensible to spherical microphone arrays (Goossens and Rogier, 2009). Root MUSIC was initially derived for ULAs, but later extended to non-uniform linear arrays (NLAs) with microphones located on an equidistant grid (Mhamdi and Samet, 2011). Root MUSIC is also available for circular microphone arrays (Zoltowski and Mathews, 1992; Zoltowski *et al.*, 1993).

Both ESPRIT and Root MUSIC exploit the phase differences between the microphones to estimate the DOAs. Root MUSIC is generally more accurate than ESPRIT, but ESPRIT is computationally more efficient, especially its real-valued formulation Unitary ESPRIT (Haardt and Nossek, 1995). Due to the required EVD, both ESPRIT and Root MUSIC are computationally expensive. Moreover, none of these approaches considers diffuse sound in the signal model, which yields biased DOA estimates in reverberant environments.

7.5.3 Microphone Input PSD Matrix

All estimators required throughout this work operate (directly or indirectly) on the microphone input PSD matrix $\mathbf{\Phi}_x(k, n)$, or on some of its elements. To estimate $\mathbf{\Phi}_x(k, n)$, the expectation operator in Equation (7.4a) is usually replaced by temporal averaging. The averaging is often carried out as block averaging – see, for example, Van Trees (2002) and the references therein. Alternatively, $\mathbf{\Phi}_x(k, n)$ can be estimated using a recursive averaging filter,

$$\hat{\mathbf{\Phi}}_x(k, n) = \alpha_\tau \mathbf{x}(k, n)\mathbf{x}^H(k, n) + (1 - \alpha_\tau)\hat{\mathbf{\Phi}}_x(k, n - 1). \tag{7.40}$$

This approach is used in many applications of spatial sound processing, such as DirAC (Pulkki, 2007), as it requires less memory than block averaging. Here, $\alpha_\tau \in (0, 1]$ is the filter coefficient corresponding to a specific time constant τ. The filter coefficient depends on the parameters of the time–frequency transform used. For instance, for a short-time Fourier transform (STFT) with hop size R at a sampling frequency f_s, we have

$$\alpha_\tau = \min\left\{\frac{R}{\tau f_s}, 1\right\}. \tag{7.41}$$

Replacing the expectation operator in Equation (7.40) by a temporal averaging as in Equation (7.41) assumes that the underlying random processes are ergodic. In practice, there is always a trade-off between a low estimation variance (longer time averaging) and a sufficiently high temporal resolution to track changes in the acoustic scene (shorter time averaging). By using small values for τ (typically in the range $30\,\text{ms} \le \tau \le 60\,\text{ms}$), we obtain almost instantaneous estimates of $\mathbf{\Phi}_x(k, n)$, and, thus, of all the parameters in this section. This enables the informed filters in this work to adapt sufficiently fast to quick changes in the acoustic scene.

7.5.4 Noise PSD Estimation

Since the SOS of the noise are assumed to be time invariant or slowly time variant, we can estimate the noise PSD matrix $\mathbf{\Phi}_n(k)$ and noise PSD $\Phi_n(k)$ during time periods where the sound sources are inactive and where no diffuse sound is present. Several corresponding approaches for estimating $\mathbf{\Phi}_n(k)$ that can be used in our framework are discussed, for instance, in Habets (2010), Souden *et al.* (2011), and Gerkmann and Hendriks (2012).

7.5.5 Diffuse Sound PSD Estimation

This subsection discusses the estimation of the diffuse sound PSD $\Psi_d(k, n)$ when $L \ge 1$ sound sources are active per time and frequency. Once $\Psi_d(k, n)$ is estimated, we can compute the diffuse sound PSD matrix $\mathbf{\Phi}_d(k, n)$ with Equation (7.12). The diffuse PSD is estimated using spatial filters that suppress the direct sounds and capture the diffuse sound. The filters require an array with at least $M = L + 1$ microphones to suppress the L direct sounds and capture the diffuse sound.

State of the Art

Irwan and Aarts (2002), Faller (2006), Usher and Benesty (2007), and Merimaa *et al.* (2007) used $M = 2$ microphones to extract the diffuse sound while suppressing a single direct sound. From the extracted diffuse sound it is straightforward to estimate the diffuse PSD, as shown later. The same principle was applied by Thiergart and Habets (2013) to multiple plane waves. Here, a linearly constrained spatial filter was employed which attenuates L plane waves while directing the main lobe of the filter towards a direction from which no direct sound arrives. We summarize this approach here.

The weights $\mathbf{h}_{\Psi L}(k, n)$ of the linearly constrained spatial filter can be found by minimizing the noise power at the filter output,

$$\mathbf{h}_{\Psi L}(k, n) = \arg \min_{\mathbf{h}} \mathbf{h}^H \mathbf{\Phi}_n(k) \mathbf{h}, \tag{7.42}$$

subject to $\mathbf{h}^H \mathbf{A}(k, n) = 0$ and $\mathbf{h}^H \mathbf{a}_0(k, n) = 1$. The first constraint cancels out the L plane waves, while the second constraint ensures non-zero filter weights $\mathbf{h}_{\Psi L}(k, n)$. The propagation vector $\mathbf{a}_0(k, n)$ corresponds to a specific direction $\mathbf{n}_0(k, n)$ from which no direct sound arrives. The optimal direction $\mathbf{n}_0(k, n)$, towards which we direct the main lobe of the spatial filter, is the direction which maximizes the output diffuse-to-noise ratio (DNR). Unfortunately, no closed-form solution to compute this direction is available.

Therefore, we choose for $\mathbf{n}_0(k, n)$ the direction which has the largest distance to all $\mathbf{n}_l(k, n)$, that is, in the two-dimensional case,

$$\varphi_0(k, n) = \arg\max_{\varphi} \left(\min_l |\varphi - \varphi_l(k, n)| \right), \quad \varphi \in \left[-\frac{\pi}{2}, \frac{\pi}{2} \right]. \tag{7.43}$$

To estimate the diffuse PSD $\Psi_{\mathrm{d}}(k, n)$, we apply the weights $\mathbf{h}_{\Psi\mathrm{L}}(k, n)$ to the signal PSD matrix $\mathbf{\Phi}_x(k, n)$ in Equation (7.4a). For a filter $\mathbf{h}_\Psi$, this leads to

$$\mathbf{h}_\Psi^{\mathrm{H}} \mathbf{\Phi}_x(k, n) \mathbf{h}_\Psi = \Psi_{\mathrm{d}}(k, n) \mathbf{h}_\Psi^{\mathrm{H}} \mathbf{\Gamma}_{\mathrm{d}}(k) \mathbf{h}_\Psi + \mathbf{h}_\Psi^{\mathrm{H}} \mathbf{\Phi}_{\mathrm{n}}(k) \mathbf{h}_\Psi. \tag{7.44}$$

Rearranging this equation yields

$$\frac{\mathbf{h}_\Psi^{\mathrm{H}} \mathbf{\Phi}_x(k, n) \mathbf{h}_\Psi}{\mathbf{h}_\Psi^{\mathrm{H}} \mathbf{\Gamma}_{\mathrm{d}}(k) \mathbf{h}_\Psi} = \Psi_{\mathrm{d}}(k, n) + \epsilon_\Psi(k), \tag{7.45}$$

where

$$\epsilon_\Psi(k) = \frac{\mathbf{h}_\Psi^{\mathrm{H}} \mathbf{\Phi}_{\mathrm{n}}(k) \mathbf{h}_\Psi}{\mathbf{h}_\Psi^{\mathrm{H}} \mathbf{\Gamma}_{\mathrm{d}}(k) \mathbf{h}_\Psi}. \tag{7.46}$$

Assuming that the error term $\epsilon_\Psi(k)$ is small compared to $\Psi_{\mathrm{d}}(k, n)$, a reasonable estimate of the diffuse sound PSD is given by

$$\widehat{\Psi}_{\mathrm{d}}(k, n) = \frac{\mathbf{h}_\Psi^{\mathrm{H}} \mathbf{\Phi}_x(k, n) \mathbf{h}_\Psi}{\mathbf{h}_\Psi^{\mathrm{H}} \mathbf{\Gamma}_{\mathrm{d}}(k) \mathbf{h}_\Psi}. \tag{7.47}$$

It is clear from Equation (7.46) that the estimator is biased in the presence of noise, that is, we overestimate $\Psi_{\mathrm{d}}(k, n)$. Note that the filter $\mathbf{h}_{\Psi\mathrm{L}}(k, n)$ proposed above minimizes the numerator in Equation (7.46), but not the whole error term. Also, the direction φ_0 computed in Equation (7.43) does not necessarily minimize $\epsilon_\Psi(k)$. Therefore, the filter $\mathbf{h}_{\Psi\mathrm{L}}(k, n)$ is not optimal in the sense of providing the most accurate diffuse PSD estimate. To reduce the bias, one could subtract the error term $\epsilon_\Psi(k)$, which can be computed with Equation (7.46), from $\widehat{\Psi}_{\mathrm{d}}(k, n)$. In practice, this may lead to negative PSD estimates, namely when the involved quantities $\mathbf{\Phi}_x(k, n)$ and $\mathbf{\Phi}_{\mathrm{n}}(k)$ contain estimation errors. Therefore, we use Equation (7.47) to estimate $\Psi_{\mathrm{d}}(k, n)$.

Diffuse PSD Estimation Using a Quadratically Constrained Spatial Filter

This subsection discusses a second approach to estimating the diffuse PSD $\Psi_{\mathrm{d}}(k, n)$, from which we can compute the diffuse sound PSD matrix $\mathbf{\Phi}_{\mathrm{d}}(k, n)$ with Equation (7.12). As in the previous subsection, we use a spatial filter that suppresses the L plane waves and captures the diffuse sound. In contrast to the previous subsection, we consider a quadratically constrained spatial filter which maximizes the output DNR. This is equivalent to minimizing the error term $\epsilon_\Psi(k)$ in Equation (7.46). The proposed spatial filter was published by Thiergart *et al.* (2014b), and now a summary is provided.

To compute the quadratically constrained filter, we minimize the noise at the output of the filter,

$$\mathbf{h}_{\Psi\mathrm{Q}}(k, n) = \arg\min_{\mathbf{h}} \mathbf{h}^{\mathrm{H}} \mathbf{\Phi}_{\mathrm{n}}(k) \mathbf{h}, \tag{7.48}$$

subject to the constraints

$$\mathbf{h}^H \mathbf{A}(k,n)\mathbf{A}^H(k,n)\mathbf{h} = 0, \tag{7.49a}$$

$$\mathbf{h}^H \mathbf{\Gamma}_{\mathrm{d}}(k)\mathbf{h} = a. \tag{7.49b}$$

The constraint in Equation (7.49a) ensures that the power of the L plane waves is zero at the filter output. Note that a weight vector $\mathbf{h}$, which satisfies Equation (7.49a), also satisfies $\mathbf{h}^H \mathbf{A}(k,n) = \mathbf{0}$, which means that each individual plane wave is canceled out. With the constraint in Equation (7.49b) we capture the diffuse sound power with a specific factor a. The factor a is necessarily real and positive, and ensures non-zero weights $\mathbf{h}_{\psi Q}(k,n)$. For any $a > 0$, the error in Equation (7.46) is minimized, which is equivalent to maximizing the output DNR, subject to Equation (7.49a).

To compute $\mathbf{h}_{\psi Q}(k,n)$, we first consider the $M \times M$ matrix $\mathbf{A}(k,n)\mathbf{A}^H(k,n)$ in Equation (7.49a), which is Hermitian with rank L and thus has L non-zero real positive eigenvalues, as well as $N = M - L$ zero eigenvalues. We consider the N eigenvectors $v_{1\ldots N}(k,n)$ corresponding to the N zero eigenvalues. Any linear combination of these vectors $v_{1\ldots N}(k,n)$ can be used as weight vector $\mathbf{h}$ which would satisfy Equation (7.49a), that is,

$$\mathbf{h} = \Upsilon(k,n)\mathbf{c}(k,n), \tag{7.50}$$

where $\Upsilon(k,n) = [v_1(k,n) \ \ldots \ v_N(k,n)]$ is a matrix containing the N eigenvectors and $\mathbf{c}(k,n)$ is a vector of length N containing the (complex) weights for the linear combination. The optimal $\mathbf{c}(k,n)$, which yields the weights $\mathbf{h}_{\psi Q}(k,n)$ in Equation (7.50) that minimize the stationary noise, is denoted by $\mathbf{c}^{\mathrm{opt}}(k,n)$ and can be found by inserting Equation (7.50) into Equation (7.48):

$$\mathbf{c}^{\mathrm{opt}}(k,n) = \arg\min_{\mathbf{c}} \ \mathbf{c}^H \Upsilon^H(k,n)\mathbf{\Phi}_{\mathrm{n}}(k,n)\Upsilon(k,n)\mathbf{c}, \tag{7.51}$$

subject to Equation (7.49b). The cost function to be minimized is now

$$\begin{aligned}\mathcal{J}(k,n) = \ &\mathbf{c}^H \Upsilon^H(k,n)\mathbf{\Phi}_{\mathrm{n}}(k,n)\Upsilon(k,n)\mathbf{c} \\ &+ \eta[\mathbf{c}^H \Upsilon^H(k,n)\mathbf{\Gamma}_{\mathrm{d}}(k,n)\Upsilon(k,n)\mathbf{c} - a],\end{aligned} \tag{7.52}$$

where η is the Lagrange multiplier. Setting the complex partial derivative of $\mathcal{J}(k,n)$ with respect to $\mathbf{c}^H$ to zero, we obtain

$$\mathbf{D}\mathbf{c}(k,n) = -\eta \mathbf{E}\mathbf{c}(k,n), \tag{7.53}$$

where $\mathbf{D} = \Upsilon^H(k,n)\mathbf{\Phi}_{\mathrm{n}}(k)\Upsilon(k,n)$ and $\mathbf{E} = \Upsilon^H(k,n)\mathbf{\Gamma}_{\mathrm{d}}(k)\Upsilon(k,n)$. This is a generalized eigenvalue problem, and $\mathbf{c}^{\mathrm{opt}}(k,n)$ is the generalized eigenvector of $\mathbf{D}$ and $\mathbf{E}$ corresponding to the largest eigenvalue. From this $\mathbf{c}^{\mathrm{opt}}(k,n)$, the weights $\mathbf{h}_{\psi Q}(k,n)$ are found with Equation (7.50).

To finally estimate the diffuse PSD $\Psi_{\mathrm{d}}(k,n)$, we use the filter $\mathbf{h}_{\psi Q}(k,n)$ in Equation (7.47) similarly to the previous subsection. If the output DNR is high, then $\epsilon_\psi(k)$ in Equation (7.45) becomes small compared to $\Psi_{\mathrm{d}}(k,n)$, and $\hat{\Psi}_{\mathrm{d}}(k,n)$ in Equation (7.47) represents an accurate estimate of $\Psi_{\mathrm{d}}(k,n)$. Since the output DNR is maximized by the filter $\mathbf{h}_{\psi Q}(k,n)$, the error term $\epsilon_\psi(k)$ is minimized and, thus, $\mathbf{h}_{\psi Q}(k,n)$ is optimal for estimating the diffuse sound power with Equation (7.47).

7.5.6 Signal PSD Estimation in Multi-Wave Scenarios

This section discusses the estimation of the signal PSDs $\Psi_{s,1\ldots L}(k,n)$ of the L mutually uncorrelated plane waves in a multi-wave scenario.

Minimum Variance Estimate

The minimum variance approach (Capon, 1969) for estimating the signal PSDs is well known. This approach uses a spatial filter which is applied to the microphone signals $\mathbf{x}(k,n)$ and extracts the signal of the lth plane wave. The power of this signal then represents an estimate of $\Psi_{s,l}(k,n)$. Usually, the spatial filter extracts the lth plane wave with unit gain while minimizing the power of the $L-1$ remain plane waves plus the diffuse sound and noise. Alternatively, we can use a filter that extracts the lth plane wave and places spatial nulls towards the $L-1$ remaining plane waves while minimizing, for example, the noise plus diffuse sound. The corresponding filter is given by the lth column of the filter matrix $\mathbf{H}_{sLCMV}(k,n)$ discussed in Section 7.4.1. The drawback of using such filters for estimating the PSDs is the overestimation in the presence of noise and diffuse sound, since the filters can minimize only the power of the noise and diffuse sound, and this minimization is strongly limited when only a few microphones are used.

Minimum Mean Square Error Estimate

For the signal model in Section 7.3 it is straightforward to derive an estimator for the signal PSDs $\Psi_{s,1\ldots L}(k,n)$ that is optimal in the least squares (LS) sense. This approach was published by Thiergart et al. (2014b). The approach requires the estimation of the microphone input PSD matrix, diffuse PSD matrix, and noise PSD matrix in a pre-processing step. To determine the signal PSDs, we compute

$$\hat{\mathbf{\Phi}}_s(k,n) = \hat{\mathbf{\Phi}}_x(k,n) - \hat{\mathbf{\Phi}}_u(k,n), \tag{7.54}$$

where $\hat{\mathbf{\Phi}}_x(k,n)$ is an estimate of the microphone input PSD matrix $\mathbf{\Phi}_x(k,n)$ (see Section 7.5.3) and $\hat{\mathbf{\Phi}}_u(k,n) = \hat{\mathbf{\Phi}}_d(k,n) + \hat{\mathbf{\Phi}}_n(k)$ is the estimated input diffuse-plus-noise PSD matrix (see Sections 7.5.4 and 7.5.5). It follows from Equation (7.8) that $\hat{\mathbf{\Phi}}_s(k,n)$ is an estimate of

$$\hat{\mathbf{\Phi}}_s(k,n) = \mathbf{A}(k,n)\mathbf{\Psi}_s(k,n)\mathbf{A}^H(k,n) + \mathbf{\Delta}, \tag{7.55}$$

where $\mathbf{\Delta}$ is the estimation error of $\hat{\mathbf{\Phi}}_s(k,n)$. Since the L plane wave signals are mutually uncorrelated, $\mathbf{\Psi}_s(k,n)$ is a diagonal matrix. Therefore, Equation (7.55) can be written as

$$\hat{\mathbf{\Phi}}_s(k,n) = \sum_{l=1}^{L} \Psi_{s,l}(k,n)\mathbf{a}_l(k,n)\mathbf{a}_l^H(k,n) + \mathbf{\Delta}. \tag{7.56}$$

We estimate the signal PSDs $\psi_s(k,n) = [\Psi_{s,1}(k,n), \ldots, \Psi_{s,L}(k,n)]^T$ via an LS approach which minimizes the error $\mathbf{\Delta}$:

$$\hat{\psi}_s(k,n) = \arg\min_{\psi_s} \|\text{vec}\{\hat{\mathbf{\Phi}}_s(k,n)\} - \mathbf{\Lambda}(k,n)\psi_s\|^2, \tag{7.57}$$

where $\mathbf{\Lambda}(k,n) = \left[\text{vec}\{\mathbf{a}_1(k,n)\mathbf{a}_1^H(k,n)\}, \ldots, \text{vec}\{\mathbf{a}_L(k,n)\mathbf{a}_L^H(k,n)\}\right]$. The vector operator $\text{vec}\{\mathbf{\Phi}\}$ yields the columns of matrix $\mathbf{\Phi}$ stacked into one column vector. The solution to the minimization problem of Equation (7.57) is given by

$$\hat{\psi}_s(k,n) = [\mathbf{\Lambda}^H(k,n)\mathbf{\Lambda}(k,n)]^{-1}\mathbf{\Lambda}^H(k,n)\text{vec}\{\hat{\mathbf{\Phi}}_s(k,n)\}. \tag{7.58}$$

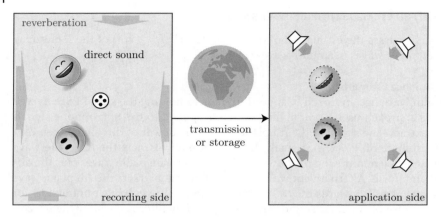

Figure 7.9 Block scheme of the parametric processing of spatial sound.

7.6 Application to Spatial Sound Reproduction

As discussed in Section 7.2.2, the compact description of the sound field in terms of direct sounds, diffuse sound, and sound field parameters, as shown in Figure 7.1, enables a huge variety of applications. In the following, we use the concepts presented in this work for the application of spatial sound recording and reproduction.

In spatial sound reproduction, we aim to capture the spatial sound on the recording side and reproduce it on the application side such that the listener perceives the sound with the original spatial impression. An example is depicted in Figure 7.9. Here, the spatial sound of two sources is recorded at the recording side, transmitted over a network, and then reproduced at the reproduction side using a loudspeaker setup that is unknown at the recording side. Clearly, this scenario requires an efficient processing scheme where only a few signals need to be transmitted while still being able to reproduce the sound on arbitrary reproduction setups.

7.6.1 State of the Art

The most popular approaches to efficient and flexible sound recording and reproduction are represented by DirAC (Pulkki, 2007) and HARPEX (Berge and Barrett, 2010a,b), which were derived for the so-called B-format microphone. As mentioned before, both approaches are based on a parametric sound field model. While DirAC assumes that the sound field for each time and frequency is composed of a single plane wave (direct sound) plus diffuse sound, HARPEX assumes two plane waves (and no diffuse sound). Both approaches aim to recreate the relevant features for the human perception of spatial sound. For instance, DirAC assumes that the interaural level difference (ILD) is correctly perceived when the direct sound is reproduced from the correct DOA, while a realistic rendering of the diffuse sound leads to a correct perception of the interaural coherence (IC). Both DirAC and HARPEX provide high flexibility, that is, the signals for any common loudspeaker setup – as well as the signals for binaural reproduction (Laitinen and Pulkki, 2009) – can be derived from the parametric description. The compact representation of the spatial sound enables efficient transmission and storage of the recorded audio scene, as shown, for example, by Herre *et al.* (2011).

Unfortunately, DirAC and HARPEX suffer from specific drawbacks: In DirAC, even though multiple microphones are used to estimate the required parameters, only single-channel filters are applied to extract the direct sound and diffuse sound, which limits the accuracy of the extracted sound components. Moreover, the single wave assumption can be violated easily in practice which impairs the direct sound reproduction, as shown by Thiergart and Habets (2012). Some of these drawbacks can be reduced with recently proposed algorithmic improvements, such as transient detection (Laitinen *et al.*, 2011) or virtual microphone processing (Vilkamo *et al.*, 2009). However, these improvements limit the flexibility and efficiency; for example, virtual microphone processing requires knowing the sound reproduction setup in advance or transmitting all the microphone signals. The higher-order extension to DirAC in Politis *et al.* (2015) reduces the model violations, but requires higher-order spherical harmonics as input signals, which can be obtained from signals measured with a spherical microphone array. HARPEX suffers from the drawback that no diffuse sound is considered in the model, which results in model violations when capturing reverberant or ambient sounds.

7.6.2 Spatial Sound Reproduction Based on Informed Spatial Filtering

We can use the parametric description of the sound field in Section 7.3 and the informed multi-channel filters in Section 7.4 to achieve efficient and flexible sound acquisition and reproduction. By assuming multiple plane waves per time and frequency in the signal model (that is, $L > 1$), and by extracting the signal components with the informed multi-channel filters that make use of all available microphone signals, some of the drawbacks of DirAC and HARPEX can be overcome.

The target signal $Y(k, n)$ we wish to obtain is given in Equation (7.2). This signal is computed at the application side in Figure 7.9 individually for each loudspeaker. In spatial sound reproduction, the direct sounds $P_{s,l}(k, n, \mathbf{r}_1)$ and diffuse sound $P_d(k, n, \mathbf{r}_1)$ represent the desired signals. Similarly to DirAC, the direct responses $D_{s,l}(k, n)$ are selected from panning functions $d_s(k, \varphi)$, which depend on the DOA of the direct sound and on the loudspeaker position. Figure 7.10 (both plots) shows the panning functions for a 5.1 loudspeaker setup, which were defined using the VBAP scheme (Pulkki, 1997). The target diffuse response $D_d(k, n)$ is set to $\sqrt{N^{-1}}$ for all N loudspeakers to reproduce the diffuse sound with the original power. Note that the target diffuse signals $Y_d(k, n)$ in Equation (7.2) for the different loudspeakers are decorrelated to ensure correct IC during sound reproduction (Pulkki, 2007).

Computing $Y(k, n)$ in Equation (7.2) requires the direct sounds $P_{s,l}(k, n, \mathbf{r}_1)$ and the diffuse sound $P_d(k, n, \mathbf{r}_1)$. These signals can be estimated at the recording side in Figure 7.9 using the spatial filters in Section 7.4. The parametric information required to compute these filters can be estimated with the approaches presented in Section 7.5. Once these signals are obtained at the recording side, they can be transmitted to the application side (together with the DOAs of the direct sounds), and $Y(k, n)$ can be computed for the desired loudspeaker setup.

Example: Direct Sound Extraction

In the following, we discuss an example of direct sound extraction using ISF. We carried out measurements in a reverberant room ($RT_{60} \approx 390$ ms). An NLA with $M = 6$ omnidirectional microphones (microphone spacings 12.8 cm–6.4 cm–3.2 cm–6.4 cm–12.8 cm) was placed in the room center. The spatial aliasing frequency was $f_{max} = 5.3$ kHz. Five

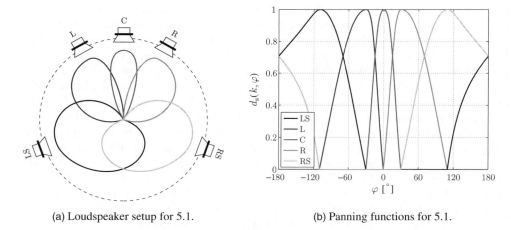

(a) Loudspeaker setup for 5.1.   (b) Panning functions for 5.1.

Figure 7.10 Left: Loudspeaker setup for a 5.1 configuration (L: left, R: right, C: center, LS: left surround, RS: right surround). Right: Panning functions for a 5.1 surround sound loudspeaker setup using the VBAP panning scheme (Pulkki, 1997). Source: Thiergart 2015. Reproduced with permission of Oliver Thiergart.

loudspeakers were located at the positions A–E at a distance of 1.7 m from the array center. The corresponding angles are depicted in Figure 7.11(a). Male speech was emitted from the center loudspeaker (denoted by the black square) and female speech was emitted from the other loudspeakers (one loudspeaker at a time, randomly changing). The dashed lines in Figure 7.11(b) indicate which loudspeaker was active when. This scenario represents a dynamic and rather challenging scenario where one source jumps to different positions, while at the same time another source is active from a fixed position. The sound was sampled at $f_s = 16\,\text{kHz}$ and transformed into the time–frequency domain using a 512-point STFT with 50% overlap.

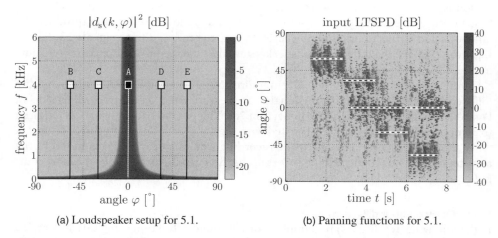

(a) Loudspeaker setup for 5.1.   (b) Panning functions for 5.1.

Figure 7.11 Left: Example target response functions to extract direct sounds. Right: Measured input long-term spatial power density.

In this example, we aim to extract the direct sounds with the direct sound response function $d_s(k, \varphi)$ depicted in Figure 7.11(a), that is, we aim to extract the male speaker from direction A (desired source) and attenuate the female speaker from the other directions (undesired interfering source). In source separation applications, the depicted response function could represent an arbitrary spatial window that extracts sounds from the center direction and attenuates sounds from other directions. In spatial sound reproduction applications, the depicted response function could be a loudspeaker panning window, for example for the center loudspeaker.

Parameter Estimation

The parameters required to compute the ISF were estimated as follows: The microphone input PSD matrix $\Phi_x(k, n)$ was estimated via the recursive averaging in Section 7.5.3 ($\tau = 40$ ms). The noise PSD matrix $\Phi_n(k)$ was computed from the signal beginning where no source was active. The number of sources L was estimated based on the eigenvalue ratios using the approach in Markovich *et al.* (2009) – see Section 7.5.1. The estimate of L was limited to a maximum of $L_{max} = 3$. The DOAs of the L direct sounds were estimated using Root MUSIC for non-linear arrays (Mhamdi and Samet, 2011). The array steering matrix $\mathbf{A}(k, n)$ was determined from the estimated DOAs using Equation (7.6). The diffuse PSD matrix $\Phi_d(k, n)$ and diffuse power $\Psi_d(k, n)$ were estimated with the approach in Section 7.5.5 assuming a spherically isotropic diffuse field. The signal PSDs in $\Psi_s(k, n)$ were obtained with the MMSE approach in Section 7.5.6.

Filter Computation

The L direct sounds $P_{s,l}(k, n, \mathbf{r}_1)$ were extracted from the microphone signals $\mathbf{x}(k, n)$ with Equation (7.19). We study the following filters:

- LCMV: The LCMV filter $\mathbf{H}_{sLCMV}(k, n)$ discussed in Section 7.4.1 and computed with Equation (7.22), where $\Omega_s(k, n) = \mathbf{0}$.
- MCW: The multi-channel Wiener filter $\mathbf{H}_{sMW}(k, n)$ discussed in Section 7.4.1 and computed with Equation (7.22), where $\Omega_s(k, n) = \mathbf{I}_L$.
- PMCW: The parametric multi-channel Wiener filter $\mathbf{H}_{sPMMW}(k, n)$ discussed in Section 7.4.1 and computed with Equation (7.22). The elements of $\Omega_s(k, n)$ were obtained with Equation (7.4.1). The logarithmic SDNR $\xi_l(k, n)$ was computed with Equation (7.24). The sigmoid function used is shown in Figure 7.7 (sigmoid 1).

For comparison, we computed the following fixed spatial filters, which did not exploit the instantaneous DOA information and multi-wave model:

- WNG: Maximum WNG filter. The filter is identical to the delay-and-sum filter (Doclo and Moonen, 2003) where the look direction corresponds to the direction of the (fixed) desired source (center loudspeaker in Figure 7.11(a)).
- SD: Robust superdirective beamformer (Cox *et al.*, 1987). This filter possesses a lower bound on the WNG which was set to -9 dB.

Finally, the target direct signal $Y_s(k, n)$ was computed with Equation (7.18) from the estimated direct sounds $\hat{P}_{s,l}(k, n, \mathbf{r}_1)$. The target direct responses $D_{s,l}(k, n)$ were selected from $d_s(k, \varphi)$ using the estimated DOAs (for WNG and SD we used the corresponding look directions).

Results

To study the performance of the parameter estimation, we consider the so-called input long-term spatial power density (LTSPD). This measure allows us to jointly evaluate the performance of the DOA and signal PSD estimation, that is, the estimation of $\Psi_{s,1...L}(k, n)$. The input LTSPD characterizes what direct power was localized for which direction (Thiergart *et al.*, 2014b). The input LTSPD is depicted in Figure 7.11(b). As mentioned before, the dashed lines indicate the loudspeaker positions and when which loudspeaker was used. The input LTSPD shows that most estimated direct power was localized near the direction of an active loudspeaker. This is true even during the double-talk periods. Nevertheless, some power was also localized in spatial regions where no source was active. This localized power resulted from the measurement uncertainties of the estimated DOAs and direct PSDs, mostly when the sound field was more diffuse.

A multiple stimuli with hidden reference and anchor (MUSHRA) listening test with the filters mentioned before was carried out to verify the perceptual quality of the presented approaches. The listening test results were published in Thiergart *et al.* (2014b). Note that the abrupt jumps of the undesired speaker during double talk with the desired speaker is a challenging scenario for spatial filters. The participants were listening to the output signal of the filters ($Y_s(k, n)$, after transforming back to the time domain), which were reproduced over headphones. The MUSHRA test was repeated five times and for each test the participants were evaluating different effects of the filters. The reference signal was the direct sound of the desired speaker plus the direct sound of the undesired speaker attenuated by 21 dB, both signals found from the windowed impulse responses. The lower anchor was an unprocessed and low-pass filtered microphone signal.

The MUSHRA scores are depicted in Figure 7.12. In the first MUSHRA test, denoted "interferer," the listeners were asked to grade the strength of the interferer attenuation, that is, the attenuation of the undesired speaker. Note that $m = 10$ in Figure 7.12 means that ten listeners passed the recommended post-screening procedure. We can see that the informed filters (LCMV, MCW, and PMMW) performed significantly better than the fixed SOA filters (WNG and DI). The rather aggressive MCW filter yielded the strongest interferer attenuation. In the second test (noise), the listeners were asked to grade the microphone noise reduction performance. The informed filters clearly outperformed the SOA filters, even the WNG filter which maximizes the WNG. The PMCW filter was as good as the MCW filter, and both were significantly better than the LCMV filter. The SD filter strongly amplified the noise, and hence was graded lowest. In the third test (dereverberation), the dereverberation performance was graded. Here, the PMCW filter was significantly better than the LCMV filter, but worse than the MCW filter. In the fourth test (distortion), the listeners were asked to grade the distortion of the direct sound of the desired speaker (high grades had to be assigned if the speech distortion was low). We can see that the MCW filter yielded strongly noticeable speech distortion, while the LCMV and PMCW filters resulted in a low distortion. The SOA filters (for which the correct look direction towards the desired source was provided) yielded the lowest speech distortion. Finally, the overall quality (listener's preference) was evaluated in the fifth test (quality). The best overall quality was provided by the PMCW and LCMV filters, which were significantly better than the MCW filter and the SOA filters. In general, the results show that the PMCW filter can provide a good trade-off between noise and diffuse sound reduction and speech distortion.

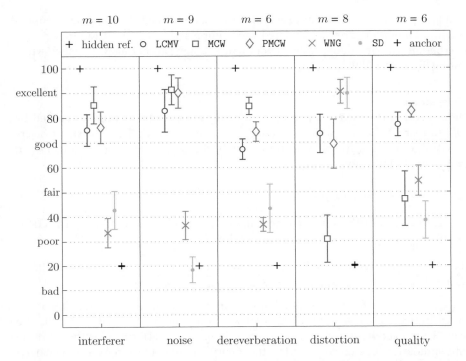

Figure 7.12 MUSHRA listening test results for direct sound extraction. The plot shows the average and 95% confidence intervals. Source: Thiergart 2014b. Reproduced with permission of IEEE.

Example: Diffuse Sound Extraction

In the following, we show an example of diffuse sound extraction using ISF. We simulated a reverberant shoebox with $RT_{60} = 390$ ms and an NLA with $M = 6$ omnidirectional microphones (microphone spacings 4 cm–2 cm–2 cm–2 cm–4 cm). Two sound sources were located at a distance of 1.6 m from the array center at the angles $\varphi_A = 39°$ and $\varphi_B = -7°$ (the array broadside was 0°). Source A was emitting speech while B was emitting transient castanet sound. Both sources were active simultaneously. The microphone signals were simulated with the image-source method (Allen and Berkley, 1979; Habets, 2008). Spatially white microphone noise was added to the microphone signals resulting in a segmental signal-to-noise ratio (SegSNR) of 39 dB. The sound was sampled at $f_s = 16$ kHz and transformed into the time–frequency domain using a 256-point STFT with 50% overlap.

Parameter Estimation

All the parameters required to compute the spatial filters were provided as accurate prior information. We were assuming $L = 2$. The DOAs of the L direct sounds were corresponding to the loudspeaker directions φ_A and φ_B. The microphone input PSD $\Phi_{x,11}(k,n)$ was computed from the reference microphone signals $X_1(k,n)$ using the recursive temporal averaging approach in Section 7.5.3 (with $\tau = 60$ ms, which yielded a typical temporal resolution). The diffuse sound PSD $\Psi_d(k,n)$ was computed directly from the diffuse sound at the reference position. Note that this signal is not directly observable in practice. Here, the diffuse sound was made available by applying an appropriate temporal window to the simulated room impulse responses (RIRs) to separate the reverberant part (including early reflections). The same averaging was applied when computing $\Psi_d(k, n)$ and $\Phi_{x,11}(k, n)$.

Filter Computation

The diffuse sound $P_d(k, n, \mathbf{r}_1)$ was estimated using a single-channel filter, denoted by $H_d(k, n)$, similarly to DirAC (which serves as reference), and with the approximate average LCMV filter $\mathbf{h}_{dALCMV}(k, n)$ in Section 7.4.2. In the case of the single-channel filter, $P_d(k, n, \mathbf{r}_1)$ was found with

$$\hat{P}_d(k, n, \mathbf{r}_1) = H_d(k, n)X_1(k, n), \tag{7.59}$$

whereas in the case of the multi-channel filter, $P_d(k, n, \mathbf{r}_1)$ was obtained with Equation (7.29). The two filters were computed as follows:

- $H_d(k, n)$: This filter represents the square-root Wiener filter for estimating the diffuse sound. The filter was computed as $H_d(k, n) = \sqrt{\Psi_d(k, n)\Phi_{x,11}^{-1}(k, n)}$.
- $\mathbf{h}_{dALCMV}(k, n)$: This filter is easy to compute in practice since only the DOA of the direct sound needs to be known. The filter was computed with Equation (7.39). The noise PSD matrix $\Phi_n(k)$ was not required in Equation (7.34), since for the iid noise in this simulation it can be replaced by the identity matrix. The filter was normalized such that the diffuse sound power is preserved at the filter output, that is, such that $\mathbf{h}_{dALCMV}^H \Gamma_d(k)\mathbf{h}_{dALCMV} = 1$.

Results

Figure 7.13 presents the effects of the single-channel filter $H_d(k, n)$ and the multi-channel filter $\mathbf{h}_{dALCMV}(k, n)$. The plot in Figure 7.13(a) shows the power of the unprocessed microphone signal $X_1(k, n)$, in which we can see the onsets of the castanets and the reverberant tails. Figure 7.13(b) shows the level difference between the true diffuse sound $P_d(k, n, \mathbf{r}_1)$ and $X_1(k, n)$. We can mainly see that in the diffuse sound, the onsets of the castanets were not present, as indicated by the high negative level differences. Figure 7.13(c) shows the level difference between the estimated diffuse sound $\hat{P}_d(k, n, \mathbf{r}_1)$ and $X_1(k, n)$ when using the filter $H_d(k, n)$. It can be observed that the onsets of the castanets were almost not attenuated, and hence were still present in the filtered output. Figure 7.13(d) depicts the level difference between the estimated diffuse sound and $X_1(k, n)$ when using the filter $\mathbf{h}_{dALCMV}(k, n)$. We can see that this filter strongly attenuated the onsets of the castanets, as desired, and we obtained almost the same result as in Figure 7.13(b).

In general, Figure 7.13 demonstrates the main advantage when using multi-channel filters for diffuse sound extraction compared to single-channel filters, namely the effective attenuation of the (in this case undesired) direct sounds. This is especially true for transient direct sounds or signal onsets. For these components, single-channel filters typically cannot provide the desired attenuation. This is problematic for spatial sound reproduction for two reasons: (i) The estimated diffuse sound (and hence also the contained direct sound) is reproduced from all directions. Therefore, the direct sound

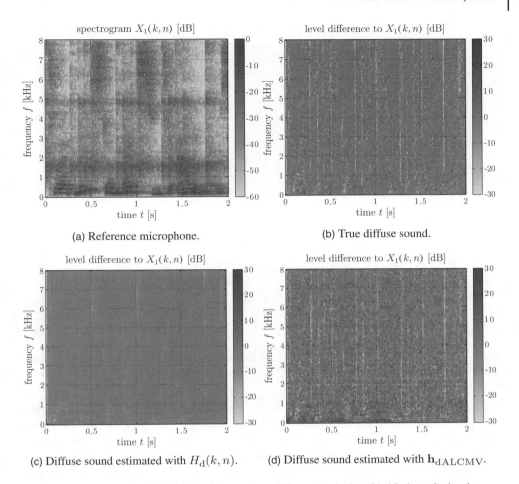

(a) Reference microphone.

(b) True diffuse sound.

(c) Diffuse sound estimated with $H_d(k, n)$.

(d) Diffuse sound estimated with $\mathbf{h}_{dALCMV}$.

Figure 7.13 Effect of the different filters for extracting diffuse sound. Plots (b)–(d) show the level difference between $X_1(k, n)$ and the extracted diffuse sound. Source: Thiergart 2014. Reproduced with permission of Oliver Thiergart.

contained in the diffuse estimate interferes with the desired direct sound, which is reproduced from the original direction. This impairs the localization of the desired direct sound. (ii) The diffuse sound is typically decorrelated before the sound is reproduced. Applying decorrelators to transient direct sounds yields unpleasant artifacts. This represents a major problem of SOA approaches for spatial sound reproduction, such as DirAC, which use single-channel filters for extracting the diffuse sound.

A MUSHRA listening test with eight participants was conducted to grade the perceptual quality achieved with the single-channel filter $H_d(k, n)$ and the multi-channel filter $\mathbf{h}_{dALCMV}(k, n)$. For this purpose, the diffuse sound extracted in the simulation was presented to the participants with a 5.1 loudspeaker setup. Note that depending on the filter performance, the extracted diffuse sound contained true diffuse sound but also direct sound components that were not accurately suppressed by the filter. The true diffuse part was decorrelated and reproduced from all loudspeakers, while the direct sound

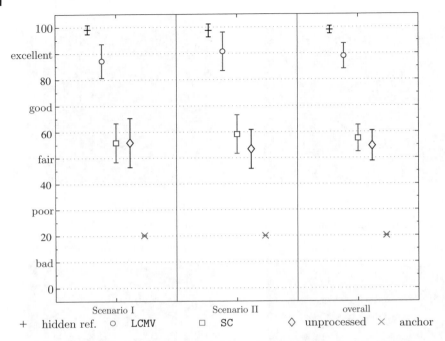

Figure 7.14 MUSHRA listening test results for the diffuse sound extraction. The plot shows the average and 95% confidence intervals. Source: Thiergart 2014. Reproduced with permission of Oliver Thiergart.

components were reproduced from the center loudspeaker to make them more audible. The participants were asked to grade the overall quality of the extracted diffuse sound compared to the true diffuse sound, which represented the reference. As lower anchor, we were using a low-pass filtered unprocessed microphone signal. In the listening test, we compared the output signal of the single-channel filter $H_d(k, n)$, the output signal of the multi-channel filter $\mathbf{h}_{dALCMV}(k, n)$, and an unprocessed microphone signal.

The results are depicted in Figure 7.14. In Scenario I, the two loudspeakers were emitting speech and castanet sounds, as explained before. In Scenario II, the same two loudspeakers were emitting two different speech signals at the same time. The multi-channel filter $\mathbf{h}_{dALCMV}(k, n)$ is denoted by LCMV and the single-channel filter $H_d(k, n)$ is denoted by SC. As can be seen, for both scenarios the multi-channel filter provided very good results, while the single-channel filter yielded fair results. The participants reported that for the single-channel filter, direct sound components were clearly audible. It was difficult to perceive a difference from the unprocessed signal in this dynamic scenario. For the multi-channel filter, the direct sound was very well attenuated and the filter output was very similar to the reference (true diffuse sound).

7.7 Summary

Parametric representations of spatial sound provide a complete yet compact description of the acoustic scene at the recording location independent of the microphone configuration. The parametric representation can be obtained with almost any multi-microphone configuration and enables a huge variety of applications, which can be defined and controlled freely by the user at the reproduction side. The flexible acquisition and processing of spatial sound is realized by modeling the sound field for each time and frequency as a sum of multiple plane waves and diffuse sound. In contrast to existing parametric approaches, we assume multiple plane waves per time and frequency, which enables us to accurately model complex acoustic scenes where multiple sources are active at the same time.

The direct sounds (plane waves) and diffuse sounds are extracted from the microphones by applying multi-channel filters. These filters are recomputed for each time and frequency using current information on the underlying parametric sound field model. The multi-channel filter for the direct sound extraction was derived using well-known spatial filtering approaches, which were applied to the proposed multi-wave signal model. This resulted in the parametric multi-channel Wiener filter that adapts to different recording conditions, for example to provide low signal distortions when the direct sounds are strong, a good dereverberation performance when the diffuse sound is prominent, or a high robustness against noise in noisy situations. To extract the diffuse sound, we derived the approximate average LCMV filter that can attenuate (direct) sounds from arbitrary directions, while capturing the sound with an almost omnidirectional directivity from all other directions.

To compute the filters, it is of paramount importance to obtain accurate estimates of the required parameters. This includes the DOA of the direct sounds, but also SOS such as the powers of the direct sounds and diffuse sound. Throughout this work, we provided an overview of appropriate estimators for omnidirectional microphone configurations.

Application of the filters presented was discussed for spatial sound reproduction, where we aimed to reproduce the recorded spatial sound with the original spatial impression. We showed that classical approaches for extracting the diffuse sound, which use single-channel filters, suffer from a leakage of transient direct sounds into the extracted diffuse sound. In contrast, when using the proposed multi-channel filters for the diffuse sound extraction, we can better attenuate the direct sounds.

References

Affes, S., Gazor, S., and Grenier, Y. (1996) An algorithm for multisource beamforming and multitarget tracking. *IEEE Transactions on Signal Processing*, **44**(6), 1512–1522.

Allen, J.B. and Berkley, D.A. (1979) Image method for efficiently simulating small-room acoustics. *Journal of the Acoustical Society of America*, **65**(4), 943–950.

Avendano, C. and Jot, J.M. (2002) Ambience extraction and synthesis from stereo signals for multi-channel audio up-mix. *IEEE International Conference on Acoustics, Speech, and Signal Processing (ICASSP)*, vol. 2, pp. II-1957–II-1960.

Benesty, J., Chen, J., and Huang, Y. (2008) *Microphone Array Signal Processing*, Springer-Verlag, Berlin.

Berge, S. and Barrett, N. (2010a) A new method for B-format to binaural transcoding. *40th International Audio Engineering Society Conference: Spatial Audio*, Tokyo, Japan.

Berge, S. and Barrett, N. (2010b) High angular resolution planewave expansion. *2nd International Symposium on Ambisonics and Spherical Acoustics*.

Bitzer, J. and Simmer, K.U. (2001) Superdirective microphone arrays, in *Microphone Arrays: Signal Processing Techniques and Applications* (ed. Brandstein, M. and Ward, D.), Springer, Berlin, chapter 2, pp. 19–38.

Braun, S., Thiergart, O., and Habets, E.A.P. (2014) Automatic spatial gain control for an informed spatial filter. *IEEE International Conference on Acoustics, Speech, and Signal Processing (ICASSP)*.

Capon, J. (1969) High-resolution frequency–wavenumber spectrum analysis. *Proceedings of the IEEE*, **57**(8), 1408–1418.

Cox, H., Zeskind, R., and Kooij, T. (1986) Practical supergain. *IEEE Transactions on Acoustics, Speech and Signal Processing*, **34**(3), 393–398.

Cox, H., Zeskind, R., and Owen, M. (1987) Robust adaptive beamforming. *IEEE Transactions on Acoustics, Speech and Signal Processing*, **35**(10), 1365–1376.

Del Galdo, G., Taseska, M., Thiergart, O., Ahonen, J., and Pulkki, V. (2012) The diffuse sound field in energetic analysis. *Journal of the Acoustical Society of America*, **131**(3), 2141–2151.

Doclo, S. and Moonen, M. (2002) GSVD-based optimal filtering for single and multimicrophone speech enhancement. *IEEE Transactions on Signal Processing*, **50**(9), 2230–2244.

Doclo, S. and Moonen, M. (2003) Design of far-field and near-field broadband beamformers using eigenfilters. *Signal Processing*, **83**(12), 2641–2673.

Doclo, S. and Moonen, M. (2005) On the output SNR of the speech-distortion weighted multichannel Wiener filter. *IEEE Signal Processing Letters*, **12**(12), 809–811.

Doclo, S. and Moonen, M. (2007) Superdirective beamforming robust against microphone mismatch. *IEEE Transactions on Audio, Speech, and Language Processing*, **15**(2), 617–631.

Doclo, S., Spriet, A., Wouters, J., and Moonen, M. (2005) Speech distortion weighted multichannel Wiener filtering techniques for noise reduction, in *Speech Enhancement* (ed. Benesty, J., Makino, S., and Chen, J.), Springer, Berlin, chapter 9, pp. 199–228.

Elko, G., West, J.E., and Kubli, R. (1998) An adaptive close-talking microphone array. *Conference Record of the Thirty-Second Asilomar Conference on Signals, Systems and Computers*, vol. 1, pp. 404–408.

Elko, G.W. (2000) Superdirectional microphone arrays, In *Acoustic Signal Processing for Telecommunication* (ed. Gay, S.L. and Benesty, J.), Kluwer Academic Publishers, Dordrecht, chapter 10, pp. 181–237.

Elko, G.W. (2001) Spatial coherence functions for differential microphones in isotropic noise fields, In *Microphone Arrays: Signal Processing Techniques and Applications* (ed. Brandstein, M. and Ward, D.), Springer, Berlin, chapter 4, pp. 61–85.

Faller, C. (2006) Multiple-loudspeaker playback of stereo signals. *Journal of the Audio Engineering Society*, **54**(11), 1051–1064.

Frost, III, O.L. (1972) An algorithm for linearly constrained adaptive array processing. *Proceedings of the IEEE*, **60**(8), 926–935.

Gannot, S. and Cohen, I. (2008) Adaptive beamforming and postfiltering, in *Springer Handbook of Speech Processing* (ed. Benesty, J., Sondhi, M.M., and Huang, Y.), Springer-Verlag, Berlin, chapter 47.

Gannot, S., Burshtein, D., and Weinstein, E. (2001) Signal enhancement using beamforming and nonstationarity with applications to speech. *IEEE Transactions on Signal Processing*, **49**(8), 1614–1626.

Gerkmann, T. and Hendriks, R. (2012) Unbiased MMSE-based noise power estimation with low complexity and low tracking delay. *IEEE Transactions on Audio, Speech, and Language Processing*, **20**(4), 1383–1393.

Goossens, R. and Rogier, H. (2009) Unitary spherical ESPRIT: 2-D angle estimation with spherical arrays for scalar fields. *IET Signal Processing*, **3**(3), 221–231.

Haardt, M. and Nossek, J. (1995) Unitary ESPRIT: How to obtain increased estimation accuracy with a reduced computational burden. *IEEE Transactions on Signal Processing*, **43**(5), 1232–1242.

Habets, E.A.P. (2008) Room impulse response generator. https://www.audiolabs-erlangen .de/fau/professor/habets/software/rir-generator/.

Habets, E.A.P. (2010) A distortionless subband beamformer for noise reduction in reverberant environments *Proceedings of the International Workshop on Acoustic Echo Control (IWAENC)*, Tel Aviv, Israel.

Herbordt, W. and Kellermann, W. (2003) Adaptive beamforming for audio signal acquisition, in *Adaptive Signal Processing: Applications to Real-World Problems* (ed. Benesty, J. and Huang, Y.), Springer-Verlag, Berlin, chapter 6, pp. 155–194.

Herre, J., Falch, C., Mahne, D., Del Galdo, G., Kallinger, M., and Thiergart, O. (2011) Interactive teleconferencing combining spatial audio object coding and DirAC technology. *Journal of the Audio Engineering Society*, **59**(12), 924–935.

Hoshuyama, O., Sugiyama, A., and Hirano, A. (1999) A robust adaptive beamformer for microphone arrays with a blocking matrix using constrained adaptive filters. *IEEE Transactions on Signal Processing*, **47**(10), 2677–2684.

Irwan, R. and Aarts, R.M. (2002) Two-to-five channel sound processing. *Journal of the Audio Engineering Society*, **50**(11), 914–926.

Jacobsen, F. and Roisin, T. (2000) The coherence of reverberant sound fields. *Journal of the Acoustical Society of America*, **108**(1), 204–210.

Jiang, J.S. and Ingram, M.A. (2004) Robust detection of number of sources using the transformed rotational matrix. *IEEE Wireless Communications and Networking Conference*, vol. 1, pp. 501–506. IEEE.

Krueger, A., Warsitz, E., and Haeb-Umbach, R. (2011) Speech enhancement with a GSC-like structure employing eigenvector-based transfer function ratios estimation. *IEEE Transactions on Audio, Speech, and Language Processing*, **19**(1), 206–219.

Laitinen, M.V. and Pulkki, V. (2009) Binaural reproduction for directional audio coding. *IEEE Workshop on Applications of Signal Processing to Audio and Acoustics*, pp. 337–340.

Laitinen, M.V., Kuech, F., Disch, S., and Pulkki, V. (2011) Reproducing applause-type signals with directional audio coding. *Journal of the Audio Engineering Society*, **59**(1/2), 29–43.

Markovich-Golan, S., Gannot, S., and Cohen, I. (2012) A weighted multichannel Wiener filter for multiple sources scenarios. *IEEE 27th Convention of Electrical Electronics Engineers in Israel (IEEEI)*, pp. 1–5.

Markovich, S., Gannot, S., and Cohen, I. (2009) Multichannel eigenspace beamforming in a reverberant noisy environment with multiple interfering speech signals. *IEEE Transactions on Audio, Speech, and Language Processing*, **17**(6), 1071–1086.

Merimaa, J., Goodwin, M.M., and Jot, J.M. (2007) Correlation-based ambience extraction from stereo recordings. *Audio Engineering Society Convention 123*. Audio Engineering Society.

Mhamdi, A. and Samet, A. (2011) Direction of arrival estimation for nonuniform linear antenna. *International Conference on Communications, Computing and Control Applications (CCCA)*, pp. 1–5.

Naylor, P.A, and Gaubitch, N.D. (2010) *Speech Dereverberation*. Springer, New York.

Nélisse, H. and Nicolas, J. (1997) Characterization of a diffuse field in a reverberant room. *Journal of the Acoustical Society of America*, **101**(6), 3517–3524.

Nordholm, S., Claesson, I., and Bengtsson, B. (1993) Adaptive array noise suppression of handsfree speaker input in cars. *IEEE Transactions on Vehicular Technology*, **42**(4), 514–518.

Politis, A., Vilkamo, J., and Pulkki, V. (2015) Sector-based parametric sound field reproduction in the spherical harmonic domain. *IEEE Journal of Selected Topics in Signal Processing*, **9**(5), 852–866.

Pulkki, V. (1997) Virtual sound source positioning using vector base amplitude panning. *Journal of the Audio Engineering Society*, **45**(6), 456–466.

Pulkki, V. (2007) Spatial sound reproduction with directional audio coding. *Journal of the Audio Engineering Society*, **55**(6), 503–516.

Rao, B. and Hari, K. (1988) Performance analysis of Root-MUSIC. *Twenty-Second Asilomar Conference on Signals, Systems and Computers*, vol. 2, pp. 578–582.

Reuven, G., Gannot, S., and Cohen, I. (2008) Dual-source transfer-function generalized sidelobe canceller. *IEEE Transactions on Audio, Speech, and Language Processing*, **16**(4), 711–727.

Roy, R. and Kailath, T. (1989) ESPRIT-estimation of signal parameters via rotational invariance techniques. *IEEE Transactions on Acoustics, Speech and Signal Processing*, **37**(7), 984–995.

Souden, M., Chen, J., Benesty, J., and Affes, S. (2011) An integrated solution for online multichannel noise tracking and reduction. *IEEE Transactions on Audio, Speech, and Language Processing*, **19**(7), 2159–2169.

Spriet, A., Moonen, M., and Wouters, J. (2004) Spatially pre-processed speech distortion weighted multi-channel Wiener filtering for noise reduction. *Signal Processing*, **84**(12), 2367–2387.

Talmon, R., Cohen, I., and Gannot, S. (2009) Convolutive transfer function generalized sidelobe canceler. *IEEE Transactions on Audio, Speech, and Language Processing*, **17**(7), 1420–1434.

Teutsch, H. and Elko, G. (2001) First- and second-order adaptive differential microphone arrays. *Proceedings of the 7th International Workshop on Acoustic Echo and Noise Control (IWAENC 2001)*.

Thiergart, O. (2015) *Flexible Multi-Microphone Acquisition and Processing of Spatial Sound Using Parametric Sound Field Representations*. PhD thesis, Friedrich-Alexander-Universitat Erlangen-Nurnberg, Erlangen, Germany.

Thiergart, O. and Habets, E.A.P. (2012) Sound field model violations in parametric spatial sound processing. *Proceedings of the International Workshop on Acoustic Signal Enhancement (IWAENC)*, Aachen, Germany.

Thiergart, O. and Habets, E.A.P. (2013) An informed LCMV filter based on multiple instantaneous direction-of-arrival estimates. *IEEE International Conference on Acoustics Speech and Signal Processing (ICASSP)*, pp. 659–663.

Thiergart, O. and Habets, E.A.P. (2014) Extracting reverberant sound using a linearly constrained minimum variance spatial filter. *IEEE Signal Processing Letters*, **21**(5), 630–634.

Thiergart, O., Kowalczyk, K., and Habets, E.A.P. (2014a) An acoustical zoom based on informed spatial filtering. *Proceedings of the International Workshop on Acoustic Signal Enhancement (IWAENC)*, Antibes, France.

Thiergart, O., Taseska, M., and Habets, E.A.P. (2013) An informed MMSE filter based on multiple instantaneous direction-of-arrival estimates. *21st European Signal Processing Conference (EUSIPCO 2013)*.

Thiergart, O., Taseska, M., and Habets, E.A.P. (2014b) An informed parametric spatial filter based on instantaneous direction-of-arrival estimates. *IEEE/ACM Transactions on Audio, Speech, and Language Processing*, **22**(12), 2182–2196.

Uhle, C., Walther, A., Hellmuth, O., and Herre, J. (2007) Ambience separation from mono recordings using non-negative matrix factorization. *30th International Audio Engineering Society Conference: Intelligent Audio Environments*.

Usher, J. and Benesty, J. (2007) Enhancement of spatial sound quality: A new reverberation-extraction audio upmixer. *IEEE Transactions on Audio, Speech, and Language Processing*, **15**(7), 2141–2150.

Van Trees, H.L. (2002) *Detection, Estimation, and Modulation Theory: Part IV: Optimum Array Processing*, vol. 1, John Wiley & Sons, Chichester.

Van Veen, B.D. and Buckley, K.M. (1988) Beamforming: A versatile approach to spatial filtering. *IEEE ASSP Magazine*, **5**(2), 4–24.

Vilkamo, J., Lokki, T., and Pulkki, V. (2009) Directional audio coding: Virtual microphone-based synthesis and subjective evaluation. *Journal of the Audio Engineering Society*, **57**(9), 709–724.

Wax, M. and Kailath, T. (1985) Detection of signals by information theoretic criteria. *IEEE Transactions on Acoustics, Speech, and Signal Processing*, **33**(2), 387–392.

Yilmaz, O. and Rickard, S. (2004) Blind separation of speech mixtures via time–frequency masking. *IEEE Transactions on Signal Processing*, **52**(7), 1830–1847.

Zoltowski, M. and Mathews, C.P. (1992) Direction finding with uniform circular arrays via phase mode excitation and beamspace Root-MUSIC. *IEEE International Conference on Acoustics, Speech, and Signal Processing*, vol. 5, pp. 245–248.

Zoltowski, M., Kautz, G., and Silverstein, S. (1993) Beamspace Root-MUSIC. *IEEE Transactions on Signal Processing*, **41**(1), 344–364.

8

Adaptive Mixing of Excessively Directive and Robust Beamformers for Reproduction of Spatial Sound

Symeon Delikaris-Manias[1] and Juha Vilkamo[2]

[1]*Department of Signal Processing and Acoustics, Aalto University, Finland*
[2]*Nokia Technologies, Finland*

8.1 Introduction

The spatial sound reproduction methods that utilize a parameter-based model describing some properties of the sound field reproduce sound scenes with a high perceptual accuracy when compared with non-parametric techniques. These techniques were discussed in Chapter 4; they usually employ a direction of arrival (DOA) as an estimated parameter, and are highly suitable as a transmission format since the sound field can be synthesized by using these parameters and only a few of the microphone signals. However, they can be subject to parameter estimation bias in cases where the applied model does not match the actual organization of the sound field. As an example, when reproducing a sound field with two simultaneously active sources in the same DOA analysis region using an active intensity-based model, the sound energy at the instances of spectral overlap is leaked to the loudspeakers between the source directions. This effect can occur in a temporally variant way with signals such as speech and music. Even if the effects of model mismatch may have only a small impact in an overall perceptual sense, the perceived quality of individual channels may be impaired when monitored at a close distance.

An alternative parametrization of the sound field is presented in this chapter that circumvents the modeling of the sound field with parameters such as DOA and deals with the aforementioned issues related to the perceived quality of the individual channels while retaining the high overall perceived quality of reproduction. The motivation behind the development of this technique was the use of compact microphone arrays. Utilizing such arrays for sound reproduction with signal-independent techniques is challenging since the microphone self-noise is either boosted in such a way that the beamforming signals are noisy, or attenuated in such a way that the beamforming signals become highly coherent.

The sound field is analyzed directly in terms of the covariance matrix in the domain of the target loudspeaker signals, which is a parametrization analogous to that commonly

Parametric Time–Frequency Domain Spatial Audio, First Edition. Edited by Ville Pulkki, Symeon Delikaris-Manias, and Archontis Politis.

applied in multi-channel audio coding. The parametrization and the processing with the proposed method is based on two sets of beamformers, both designed with respect to the target reproduction setup. These are the analysis beamformers, designed to provide high spatial selectivity, and the synthesis beamformers, designed to provide a high signal-to-noise ratio (SNR). The beamforming weights for the analysis are applied at the parameter estimation stage, and the beamforming weights for the synthesis provide the actual signals that are processed with least-squares optimized mixing and decorrelation, as proposed in Section 1.3.3, to obtain the estimated target parametric properties. In other words, the proposed technique employs the parametric approach to combine the high spatial selectivity of the analysis beamformers and the high signal quality of the broader beamformers having high SNR. Since the parametrization is based on signal-independent techniques, it can cope with multiple coherent or incoherent sound sources, providing high robustness for varying types of sound fields.

The fundamental difference from DirAC reproduction methods (Pulkki, 2007) is the higher bandwidth requirements for transmission since prior information about the output setup is necessary. Spatial aliasing will affect the proposed method, since it assumes that the loudspeaker or binaural beamforming signals can be designed using the microphone array in use. DirAC-based system are also affected by spatial aliasing and produce erroneous DOA estimates; however, the diffuse stream appears to be masking some of these effects. The biggest advantage of the proposed method is robustness to noise amplification, which is especially useful when using compact microphone arrays.

The remainder of this chapter is organized as follows. Section 8.2 provides the notation and definitions. Section 8.3 provides an overview of the adaptive time–frequency processing. Sections 8.4 and 8.5 provide details of the method and listening test results for loudspeaker and headphone implementations.

8.2 Notation and Signal Model

Matrices and vectors are denoted with bold-faced symbols, where upper case denotes a matrix. The entries of the vectors are expressed with the same symbols in lower case, appended with a subindex; for example, x_i denotes the ith element of vector $\mathbf{x}$. The frequency-band signal time index is n and the frequency-band index is k. Let us denote with $x_i(k, n)$ the signal originating from the ith microphone of the array and with $y_j(k, n)$ the signal produced with the jth loudspeaker. For M microphones and L loudspeakers, the input signal vector $\mathbf{x}(k, n) \in \mathbb{C}^{M \times 1}$ and the output signal vector $\mathbf{y}(k, n) \in \mathbb{C}^{L \times 1}$ are

$$\mathbf{x}(k, n) = \begin{bmatrix} x_1(k, n) \\ x_2(k, n) \\ \vdots \\ x_M(k, n) \end{bmatrix}, \mathbf{y}(k, n) = \begin{bmatrix} y_1(k, n) \\ y_2(k, n) \\ \vdots \\ y_L(k, n) \end{bmatrix}. \tag{8.1}$$

The parametric processing employs short-time estimates of the covariance matrices of the frequency-band signals. For stationary signals, the covariance matrices are defined by

$$\mathbf{C}_\mathbf{x} = E[\mathbf{x}(k, n)\mathbf{x}^H(k, n)],$$
$$\mathbf{C}_\mathbf{y} = E[\mathbf{y}(k, n)\mathbf{y}^H(k, n)], \tag{8.2}$$

where $\mathbf{C_x} \in \mathbb{C}^{M \times M}$, $\mathbf{C_y} \in \mathbb{C}^{L \times L}$, and $\mathrm{E}[\cdot]$ is the expectation operator. In a practical implementation, the covariance matrices are considered for a finite number of time frames assumed appropriate in a perceptual sense, typically in the range of tens of milliseconds, and estimated using an average operator across the frame.

For the microphone array, the steering vector for azimuthal angle $\phi \in [0, 2\pi)$ is denoted by $\mathbf{a}(\phi, k) \in \mathbb{C}^{(M \times 1)}$. Assuming R data points in which the steering vector $\mathbf{a}(\phi_r, k)$ for $r = 1, \ldots, R$ is known, the matrix of all steering vectors $\mathbf{A}(k) \in \mathbb{C}^{M \times R}$ is defined by

$$\mathbf{A}(k) = [\, \mathbf{a}(\phi_1, k) \quad \mathbf{a}(\phi_2, k) \quad \ldots \quad \mathbf{a}(\phi_R, k)\,]. \tag{8.3}$$

A beamforming weight matrix for a frequency-band signal from the microphones to all loudspeakers is denoted by $\mathbf{W}(k) \in \mathbb{C}^{L \times M}$. The beamformer output is given by $\mathbf{y_W}(k, n) = \mathbf{W}(k)\mathbf{x}(k, n)$. The present study utilizes two beamforming designs, the analysis and synthesis designs, which are differentiated by their subscripts as $\mathbf{W_a}(k)$ and $\mathbf{W_s}(k)$, respectively.

8.3 Overview of the Method

The proposed method applies the covariance rendering method, as discussed in Chapter 1, to perform the adaptive signal mixing. The block diagram of the parametric spatial sound reproduction method is shown in Figure 8.1. The fundamental idea of the proposed method is to generate three types of beamformers for the analysis and synthesis of a sound field:

- A set of high-directivity and potentially noisy *analysis beamformers* (organized as mixing weights to a matrix $\mathbf{W_a}$).

Figure 8.1 Block diagram of the proposed parametric spatial sound reproduction system. A noise-robust beamformer frequency-band signal ($\mathbf{W_s x}$) is processed to have the spatial energy distribution and the inter-channel dependencies according to the spatially more selective analysis beamformer frequency-band signal ($\mathbf{W_a x}$) which are normalized with the overall sound field energy, estimated with beamformer weights $\mathbf{w_o}$. The adaptive mixing and decorrelation technique is summarized in Section 8.4.3.

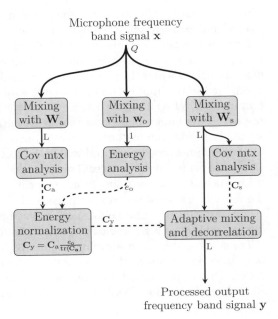

- A set of high-quality and potentially less directive *synthesis beamformers* (organized as mixing weights to a matrix $\mathbf{W}_s$).
- A set of *omnidirectional beamformers* (organized as mixing weights to a vector $\mathbf{w}_o$) to estimate the overall sound field energy.

The synthesis beamformer signals ($\mathbf{W}_s\mathbf{x}$) are then processed in order to obtain a target covariance matrix $\mathbf{C}_y$ defined as

$$\mathbf{C}_y = \gamma \mathbf{W}_a \mathbf{C}_x \mathbf{W}_a^H, \tag{8.4}$$

where γ is an energy normalization. The details of the technique are shown in the next section for loudspeaker and binaural rendering modes. More detail about the technique can be found in Vilkamo and Delikaris-Manias (2015) and Delikaris-Manias *et al.* (2015).

8.4 Loudspeaker-Based Spatial Sound Reproduction

The processing involves a parametric analysis and synthesis of the spatial sound. At the analysis stage, the perceptually relevant target stochastic properties for the loudspeaker frequency-band signals are estimated. At the synthesis stage, a regularized least-squares mixing technique is employed to produce from the microphone frequency-band signals a set of loudspeaker frequency-band signals with the target stochastic properties. The required processing blocks, as shown in the block diagram in Figure 8.1, are detailed in the following sections.

8.4.1 Estimation of the Target Covariance Matrix $\mathbf{C}_y$

The spatial parametric analysis in the proposed method is designed similarly to that of a signal-independent spatial sound reproduction system. It is based on the principles of vector base amplitude panning (VBAP; Pulkki, 1997). To determine the panning gains for a phantom source in a particular direction, VBAP first selects the pair or the triplet of loudspeakers where the phantom source resides and obtains the gains by formulating a weighted sum of unit vectors pointing towards the loudspeakers involved. In the current method, VBAP is applied as a design principle for the analysis beamformers with respect to the target loudspeaker setup, as illustrated in Figure 8.2. Assuming such a set of idealized beam patterns, a sound source or a group of sound sources arriving at the microphone array are reproduced at the receiver end at their directions of arrival as amplitude panned point sources. This reproduction principle extends to complex sound fields, such as those including prominent early reflections, since the phase relations of the reproduced point sources are preserved. Furthermore, the reproduction principle also covers diffuse sound fields, such as reverberation, since a typical expression for a diffuse sound field is a combination of an infinite set of plane waves (Jacobsen, 1979).

The set of analysis beamforming weights $\mathbf{W}_a(k)$ that approximate the amplitude panning function are designed with pattern matching optimization (Josefsson and Persson, 2006). This is performed by defining a frequency-independent gain vector $\mathbf{t}(\phi_r) \in \mathbb{R}^{L \times 1}$ determining the VBAP gains for all loudspeakers corresponding to the angle ϕ_r,

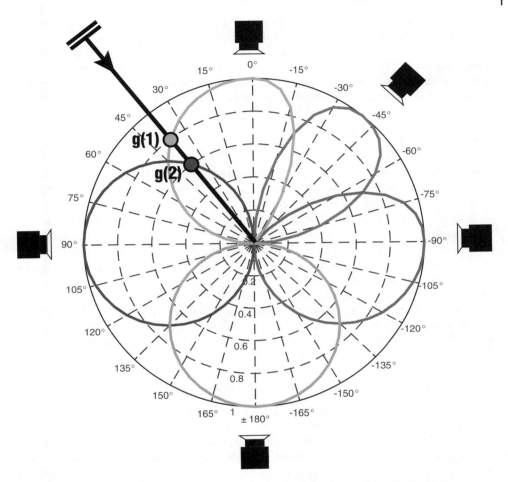

Figure 8.2 Arbitrary two-dimensional loudspeaker setup and its corresponding VBAP panning functions. A plane wave that arrives at the array from the illustrated angle is reproduced with two loudspeakers with gains $g(1)$ and $g(2)$, which are the panning gains for the corresponding directions. Such gain values form the entries of the target pattern matrix **T** in Equation (8.5).

determined for the R angular points in which the steering vectors in $\mathbf{A}(k)$ in Equation (8.3) are known. The matrix containing the gain vectors for all loudspeakers is

$$\mathbf{T} = [\, \mathbf{t}(\phi_1) \quad \mathbf{t}(\phi_2) \quad \dots \quad \mathbf{t}(\phi_R) \,], \tag{8.5}$$

where $\mathbf{T} \in \mathbb{R}^{L \times R}$. The analysis weight matrix $\mathbf{W}_a(k)$ is designed to realize the VBAP function when applied to the steering vectors: $\mathbf{W}_a(k)\mathbf{A}(k) = \mathbf{T}$. The least-squares solution is thus

$$\mathbf{W}_a(k) = \mathbf{T}\mathbf{A}^+(k), \tag{8.6}$$

where $\mathbf{A}^+(k)$ denotes the Moore–Penrose pseudoinverse of matrix $\mathbf{A}(k)$. This design of $\mathbf{W}_a(k)$ leads to large beamforming weights, especially for compact microphone arrays.

Thus, if applied directly to the microphone array frequency-band signals, the microphone noise is potentially amplified to the extent of being clearly audible.

In the proposed technique, and by omitting the frequency and time indices (k, n), the weight matrix $\mathbf{W_a}$ is not applied directly, but as part of the target parameter estimation. The covariance matrix of the beamformer output $(\mathbf{W_a x})$ is $\mathbf{C_a} = \mathbf{W_a C_x W_a}^H$. Through energy normalization, we define the target covariance matrix for the loudspeaker signals as

$$C_y = \frac{C_a e_o}{\text{tr}(C_a)}, \tag{8.7}$$

where $\text{tr}(\cdot)$ is the matrix trace and e_o is the estimated omnidirectional energy at the time–frequency estimation interval. The matrix $\mathbf{C_a}$ in Equation (8.7) determines the spatial energy distribution and the inter-channel dependencies, and the factor e_o determines the overall energy of the sound field. The parameter e_o can be estimated, for example, by using an additional omnidirectional microphone capsule, an omnidirectional beamformer, or by formulating a weighted sum of the energies of the microphones in the array. If the microphone noise covariance matrix $\mathbf{C_n} \in \mathbb{C}^{(M \times M)}$ is known, the target covariance matrix can also be estimated by $\mathbf{C_y} = \mathbf{W_a}(\mathbf{C_x} - \mathbf{C_n})\mathbf{W_a}^H$. Assuming orthogonality between the desired signal and the noise, the noise estimation can be performed, for example, by using the minimum statistics approach (Martin, 2001). The effect of the noise is partially taken into account in Equation (8.7) by the energy normalization, although at low signal energies the noise contributes to the spatial spread of the reproduced sound energy.

8.4.2 Estimation of the Synthesis Beamforming Signals $\mathbf{W_s}$

The loudspeaker signals with the target covariance matrix determined in Equation (8.7) are synthesized by applying parametric processing to signals that are characterized by a higher SNR than provided by the weight matrix $\mathbf{W_a}$. For this purpose, a set of noise-robust beamforming weights $\mathbf{W_s}$ are required. Potential candidates for a robust beamforming design are adaptive beamformers or pattern-matching algorithms based on regularized least squares. The regularization usually leads to lower directional selectivity, especially at low frequencies. Thus, if applied directly, these weights are not well suited for the overall spatial sound reproduction due to the excessive inter-channel coherence typically noticeable in the low-order signal-independent systems discussed in Chapter 4.

The design of the weights $\mathbf{W_s}$ is based on conventional target matching beamforming designs with Tikhonov regularization, as shown in Kirkeby *et al.* (1998) and Farina *et al.* (2010):

$$\mathbf{W_s} = \mathbf{T}[\mathbf{A}^H \mathbf{A} + \beta \mathbf{I}]^{-1} \mathbf{A}^H, \tag{8.8}$$

where $\mathbf{I}$ is the identity matrix and β is a frequency-dependent regularization parameter.

8.4.3 Processing the Synthesis Signals $(\mathbf{W_s x})$ to Obtain the Target Covariance Matrix $\mathbf{C_y}$

In this section the adaptive mixing of the two types of beamformers is described. The analysis beamformers $(\mathbf{W_a x})$ are characterized by a high directional selectivity and a low SNR, while the synthesis beamformers $(\mathbf{W_s x})$ are characterized by a high SNR and a lower directional selectivity. The goal of the parametric processing is thus to combine, in a perceptual sense, the noise-robust signal quality of the signals $(\mathbf{W_s x})$ with the high directional selectivity of the signals $(\mathbf{W_a x})$. This is realized by adaptive processing of the signal $(\mathbf{W_s x})$ to obtain the target covariance matrix $\mathbf{C_y}$ in Equation (8.7).

The proposed method adopts the optimized mixing and decorrelation technique described in Chapter 1. In the following, the notation of the technique is specified for the present task. An input–output relation is assumed:

$$y = M(\mathbf{W_s x}) + \mathbf{M_r} D[(\mathbf{W_s x})], \tag{8.9}$$

where $\mathbf{M} \in \mathbb{C}^{L \times L}$ is the primary mixing matrix that is solved with the aim of producing the target covariance matrix $\mathbf{C_y}$ for the output signal $\mathbf{y}$, which, however, is subject to regularization. The operator $D[\cdot]$ denotes a set of decorrelating processes that generate non-coherent signals with respect to their inputs and to each other. Ideally, the decorrelated signals should be perceptually as similar as possible to the corresponding input signals. $\mathbf{M_r} \in \mathbb{C}^{L \times L}$ is a secondary mixing matrix that is employed to process the decorrelated signal to obtain a covariance matrix that is complementary with respect to the effect of the regularization of $\mathbf{M}$. In other words, $\mathbf{M_r}$ is formulated such that the signal $\mathbf{M_r} D[(\mathbf{W_s x})]$ obtains the covariance matrix

$$C_r = C_y - M(\mathbf{W_s C_x W_s}^H)M^H, \tag{8.10}$$

where $\mathbf{C_r} \in \mathbb{C}^{L \times L}$.

With the temporary assumption that the matrix $\mathbf{K_s}$, defined in the following, is full rank and that $\mathbf{M_r} = \mathbf{0}$, the set of mixing solutions generating the target covariance matrix $\mathbf{C_y}$ for the output signal $\mathbf{y}$ is shown to be (Vilkamo *et al.*, 2013)

$$M = K_y P K_s^{-1}, \tag{8.11}$$

where $\mathbf{K_s} \in \mathbb{C}^{L \times L}$ and $\mathbf{K_y} \in \mathbb{C}^{L \times L}$ are matrices with the properties $\mathbf{K_s K_s}^H = (\mathbf{W_s C_x W_s}^H)$ and $\mathbf{K_y K_y}^H = \mathbf{C_y}$, for example as the result of the Cholesky decomposition, and $\mathbf{P} \in \mathbb{C}^{L \times L}$ is any unitary matrix. An optimized solution is found by minimizing the square difference of the processed signal and the energy-normalized input signal:

$$e = \|y - G(\mathbf{W_s x})\|^2 = \|M(\mathbf{W_s x}) - G(\mathbf{W_s x})\|^2, \tag{8.12}$$

where $\mathbf{G} \in \mathbb{R}^{L \times L}$ is a non-negative diagonal matrix that normalizes the energies of $(\mathbf{W_s x})$ to those of $\mathbf{y}$. This normalization ensures that the error measure is weighted with the channel energies. The solution is found by minimizing the error in Equation 8.12, and is shown in Chapter 1. The secondary mixing matrix $\mathbf{M_r}$ is solved with an equivalent procedure, such that the decorrelated signal $\mathbf{M_r} D[(\mathbf{W_s x})]$ obtains $\mathbf{C_r}$. When the primary and the secondary signals are mixed as in Equation (8.9), the summed signal has the target covariance matrix $\mathbf{C_y}$ by definition.

8.4.4 Spatial Energy Distribution

Four simulated sound field conditions were generated to compare signal-dependent and signal-independent reproduction modes in terms of the spatial energy distribution. These included cases with a source in the direction of a loudspeaker, with a source between the loudspeakers, with two simultaneous incoherent sources with the same magnitude spectrum, and with only a diffuse sound field. As shown in Figures 8.3

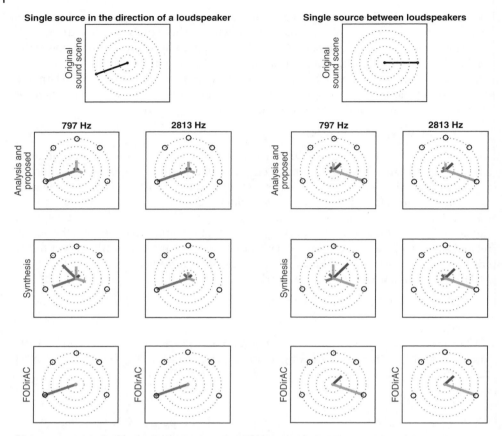

Figure 8.3 Comparison of the spatial energy distribution of the tested reproduction modes for two different frequencies for a single sound source at the position of a loudspeaker and a single source between the loudspeakers. The vector length corresponds to the reproduced amplitude. The square sum of the vector lengths, that is, the total energy, is normalized to a constant in each case.

and 8.4, the state-of-the-art model performs particularly well if the model matches the sound field; however, the bias is large in the case of two sources. The proposed method replicates the spatial energy distribution of the analysis patterns, which is more consistent regardless of the sound field type. The synthesis patterns, commonly utilized in signal-independent spatial sound reproduction systems, spread the sound energy at low frequencies due to their regularized design.

8.4.5 Listening Tests

A listening test was performed to illustrate the performance of the proposed method in terms of the perceived quality of reproduction. The designed item set was chosen so that it characterized a range of acoustic conditions: a single source, two sources imitating the effect of a prominent early reflection and the fully diffuse reverberation. The scenarios were all simulated so that a reference could be utilized. Four different methods were

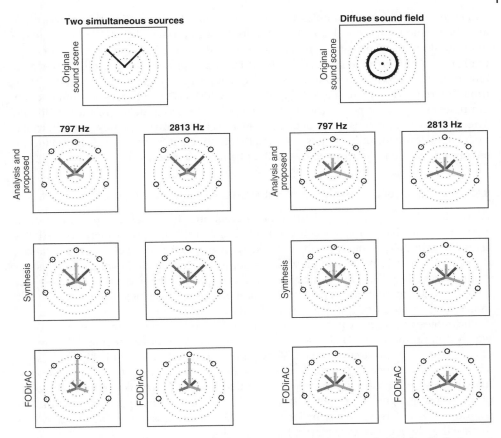

Figure 8.4 Comparison of the spatial energy distribution of the tested reproduction modes for two different frequencies for two non-coherent simultaneous sources with an equal magnitude spectrum, and a diffuse sound field. The vector length corresponds to the reproduced amplitude. The square sum of the vector lengths, that is, the total energy, is normalized to a constant in each case.

evaluated in two sessions with a multiple-stimulus test using the reference. The methods were two signal-independent methods, analysis and synthesis, which consisted of narrow and noise-robust beamformers, approximating the loudspeaker setup, respectively, FO-DirAC (Pulkki, 2007), and the proposed method. The results showed that the proposed method was rated closer to the reference when compared to the analysis and synthesis. FO-DirAC was rated similarly in the cases when all channels were active and when the sound field matched the model. A more detailed description of the listening test setup and statistical analysis can be found in Vilkamo and Delikaris-Manias (2015).

8.5 Binaural-Based Spatial Sound Reproduction

Similarly to loudspeaker reproduction, the binaural version of the algorithm is designed with respect to the left and right ear head-related transfer functions (HRTFs) $H_l(\theta, \phi, k)$

and $H_r(\theta, \phi, k)$, which are complex values as functions of source azimuth θ, elevation ϕ, and the frequency band index k. The amplitudes and phases of the HRTFs define the inter-aural level and phase differences, as well as the overall sum energy. The perceptually relevant information in a pair of head-related impulse responses (HRIRs) is contained in their energies as a function of frequency and a single wide-band inter-aural time-difference (ITD) parameter (Plogsties *et al.*, 2000). The ITD can be estimated, for example, as the difference of the median values of the group delays of the HRIR pair. The spectra can be estimated by decomposing the HRIRs into frequency bands in which their energies are formulated. It is assumed that a sufficient frequency resolution for the energy estimation is determined by the Bark bands (Zwicker, 1961).

8.5.1   Estimation of the Analysis and Synthesis Beamforming Weight Matrices

Parametrized processing is applied to the microphone signals x based on three sets of static beamforming weights: the analysis matrix $\mathbf{W_a}(k)$ that approximates the HRTFs spatially accurately, being thus subject to excessive microphone self-noise amplification, the synthesis matrix $\mathbf{W_s}(k)$ that only loosely approximates the HRTFs and produces signals with a higher SNR, and the sum row vector $\mathbf{w_0}(k)$ that approximates the sum spatial energy capture pattern of the left and right HRTFs with a single beam pattern. The indices (k, l) are omitted in the following for notational brevity.

The processing is performed in time–frequency areas in terms of covariance matrices, which contain information about the perceptually relevant binaural cues. The covariance matrix $\mathbf{C_a}$ of the analysis signal $\mathbf{W_a}x$ is first estimated by $\mathbf{C_a} = \mathrm{E}[(\mathbf{W_a}x)](\mathbf{W_a}x)^H]$. The target covariance matrix $\mathbf{C_y}$ is then obtained by normalizing $\mathbf{C_a}$ to have the estimated energy e_o of the sum signal $\mathbf{w_0}x$ as in Equation (8.7). The covariance matrix $\mathbf{C_s}$ of the synthesis signal $\mathbf{W_s}x$ is also estimated. The same least-squares optimized mixing solution is formulated, as described in Section 8.5.3, to process the synthesis signal $\mathbf{W_s}x$ to have the target covariance matrix $\mathbf{C_y}$.

The matrices $\mathbf{W_a}$, $\mathbf{W_s}$, and vector $\mathbf{w_0}$ are derived with different static beamforming designs. These designs are based on conventional least-squares beamforming, in which the complex target HRTF beam patterns, $\mathbf{H}$, are approximated by utilizing the steering vectors $\mathbf{A}$ of the microphone array as in the case of loudspeakers. $\mathbf{H}$ consists of the diffuse-field equalized HRTFs for the same set of directions (θ, ϕ) as the steering vectors. Vector $\mathbf{h}_o$ consists of the gains corresponding to the sum energy of the HRTFs, also for the same set of directions.

8.5.2   Diffuse-Field Equalization of HRTFs

Having parametrized the set of HRIRs to a set of wide-band ITDs and energies as a function of frequency, the parameters at arbitrary data points can be obtained with the following steps by interpolation. Here, the CIPIC database is utilized (Algazi *et al.*, 2001). The position of a target data point (θ, ϕ) is first translated to the CIPIC coordinate system, and the ITD and energies are then linearly interpolated separately, first along the elevation axis, then azimuth. As a result, the energy and time-difference parameters $E_l(\theta, \phi, k)$, $E_r(\theta, \phi, k)$, and $\mathrm{ITD}(\theta, \phi)$ at the target data point are obtained. The ITD parameter is translated into an IPD parameter between $\pm\pi$ by

$$\mathrm{IPD}(\theta, \phi, k) = [(\mathrm{ITD}(\theta, \phi) \cdot 2\pi f_b(k) + \pi) \bmod 2\pi] - \pi, \tag{8.13}$$

where $f_b(k)$ is the band center frequency. The interpolated HRTFs are then

$$H_l(\theta, \phi, k) = e^{i \cdot \text{IPD}/2} \sqrt{E_l(\theta, \phi, k)} G(k),$$

$$H_r(\theta, \phi, k) = e^{-i \cdot \text{IPD}/2} \sqrt{E_r(\theta, \phi, k)} G(k),$$

(8.14)

where $G(k)$ is a diffuse-field equalizing gain that is formulated such that

$$\frac{1}{A} \sum_{a=0}^{A-1} [\|H_l(\theta_a, \phi_a, k)\|^2 + \|H_r(\theta_a, \phi_a, k)\|^2] = 1$$

(8.15)

for a set of data points (θ_a, ϕ_a) selected to be uniformly distributed across a sphere, where $a = 0, \ldots, (A - 1)$ and A is the total number of data points.

8.5.3 Adaptive Mixing and Decorrelation

In the binaural version, dual-band processing is applied. At low frequencies ($f < 2700$ Hz), in which human hearing is more sensitive to the inter-aural phase difference and coherence, the adaptive mixing block employs the technique proposed by Vilkamo *et al.* (2013). The technique solves mixing matrices $\mathbf{M}$ and $\mathbf{M_r}$ in a least-squares sense using an input–output relation as described in Equation (8.9). At high frequencies ($f \geq 2700$ Hz), human hearing is less sensitive to the IPD and the coherence. Thus, at these frequencies only gain modulation is applied to synthesize the overall energy and the ILD by $\mathbf{M} = \mathbf{G}$ and $\mathbf{M_r} = \mathbf{0}$.

8.5.4 Subjective Evaluation

A listening test was performed using a multiple-stimulus test with a known reference to evaluate the performance of a binaural rendering system. Five different modes were evaluated: the proposed technique, three signal-independent rendering modes based on target pattern matching with different regularization approaches, and a monophonic anchor. The noiseless, analysis, and synthesis methods utilized pattern-matching beamforming techniques to approximate the HRTF panning functions. The noiseless method utilized the same beamforming technique as the analysis, but no noise was added to the microphone signals. The results, depicted in Figure 8.5, show that the proposed

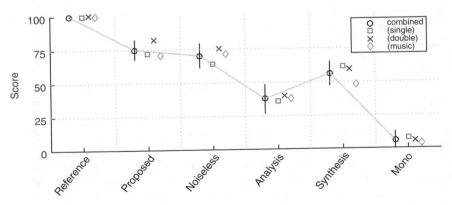

Figure 8.5 Means and 95% confidence intervals of the overall results, and the means with each reference scene separately. Source: Delikaris-Manias 2015. Reproduced with permission of IEEE.

technique provides higher perceptual quality than that obtainable by the signal-independent techniques. The proposed mode performed equally well to the noiseless mode. More details on the listening test, the different sound scenes, and the analysis can in be found in Delikaris-Manias *et al.* (2015).

8.6 Conclusions

In this chapter we proposed an alternative method to parametrize the sound field using time–frequency-dependent covariance matrices related to the multi-channel setup targeted in reproduction. In the method, two types of beamforming signals are generated: a set of narrow and potentially noisy beamformers that approximate with high accuracy the target reproduction setup, and a set of more noise-robust beamformers that spatially approximate the same setup only loosely. The narrow beamformers are used to estimate the target covariance matrices, which in turn are used to adaptively mix the broad beamformers to output the multi-channel signal. In loudspeaker reproduction, the proposed method provides superior performance when compared with signal-independent techniques. The overall quality has been found to be on the same scale as FO-DirAC, and the performance depends less on the source material. In binaural reproduction, listening tests showed the noise robustness of the method for a compact microphone array where the overall quality was higher than signal-independent approaches.

References

Algazi, V.R., Duda, R.O., Thompson, D.M., and Avendano, C. (2001) The CIPIC HRTF database. *IEEE Workshop on the Applications of Signal Processing to Audio and Acoustics*, pp. 99–102.

Delikaris-Manias, S., Vilkamo, J., and Pulkki, V. (2015) Parametric binaural rendering utilizing compact microphone arrays. *IEEE International Conference on Acoustics, Speech, and Signal Processing (ICASSP)*, pp. 629–633. IEEE.

Farina, A., Capra, A., Chiesi, L., and Scopece, L. (2010) A spherical microphone array for synthesizing virtual directive microphones in live broadcasting and in post production. *40th International Audio Engineering Society Conference: Spatial Audio – Sense the Sound of Space*. Audio Engineering Society.

Jacobsen, F. (1979) *The Diffuse Sound Field: Statistical Considerations Concerning the Reverberant Field in the Steady State*. Technical report, The Acoustics Laboratory, Technical University of Denmark.

Josefsson, L. and Persson, P. (2006) *Conformal Array Antenna Theory and Design*, vol. 29, John Wiley & Sons, Chichester.

Kirkeby, O., Nelson, P.A., Hamada, H., and Orduna-Bustamante, F. (1998) Fast deconvolution of multichannel systems using regularization. *IEEE Transactions on Speech and Audio Processing*, **6**(2), 189–194.

Martin, R. (2001) Noise power spectral density estimation based on optimal smoothing and minimum statistics. *IEEE Transactions on Speech and Audio Processing*, **9**(5), 504–512.

Plogsties, J., Minnaar, P., Olesen, S.K., Christensen, F., and Møller, H. (2000) Audibility of all-pass components in head-related transfer functions. *Audio Engineering Society Convention 108*. Audio Engineering Society.

Pulkki, V. (1997) Virtual sound source positioning using vector base amplitude panning. *Journal of the Audio Engineering Society*, **45**(6), 456–466.

Pulkki, V. (2007) Spatial sound reproduction with directional audio coding. *Journal of the Audio Engineering Society*, **55**(6), 503–516.

Vilkamo, J. and Delikaris-Manias, S. (2015) Perceptual reproduction of spatial sound using loudspeaker-signal-domain parametrization. *IEEE/ACM Transactions on Audio, Speech, and Language Processing*, **23**(10), 1660–1669.

Vilkamo, J., Bäckström, T., and Kuntz, A. (2013) Optimized covariance domain framework for time–frequency processing of spatial audio. *Journal of the Audio Engineering Society*, **61**(6), 403–411.

Zwicker, E. (1961) Subdivision of the audible frequency range into critical bands (*frequenzgruppen*). *Journal of the Acoustical Society of America*, **33**(2), 248–248.

9

Source Separation and Reconstruction of Spatial Audio Using Spectrogram Factorization

Joonas Nikunen and Tuomas Virtanen

Department of Signal Processing, Tampere University of Technology, Finland

9.1 Introduction

Natural sound scenes often consist of multiple sound sources at different spatial locations. When such a scene is captured with a microphone, a signal with a mixture of the sources is obtained, and the application of signal processing operations to signals from individual sources within the mixture is difficult. Processing of spatial audio through source and object separation allows altering the rendition of audio, such as changing the spatial position of sound sources. Alternatively, in many applications the separation of sources from the mixture is of great interest itself, that is, separating essential content (performing artist, speech, and so forth) from interfering sources in the recording scenario. These applications include assisted listening, robust content analysis of audio, modification of the sound scene for augmented reality, and three-dimensional audio in general. Spectrogram factorization aided with spatial analysis abilities can be used to parameterize spatial audio to achieve these tasks of modifying the audio content by sound objects.

Audio signals consist of sound events that repeat over time, such as individual phonemes in speech and notes from musical instruments. The magnitude spectrogram, that is, the absolute value of the short-time Fourier transform (STFT), of such audio signals also consists of spectral patches that have a repetitive nature. This can be visually confirmed from Figure 9.1, which depicts a few different notes played on a piano in repeating sequence. In this chapter we introduce methods for factorizing the spectrogram of multichannel audio into repetitive spectral objects and apply the introduced models to the analysis of spatial audio and modification of spatial sound through source separation.

The purpose of decomposing an audio spectrogram using spectral templates is to learn the underlying structures (audio objects) from the observed data. The non-negative matrix factorization (NMF) of audio spectrograms has been utilized in a wide range of applications due to its ability to represent recurrent spectral events using a set of

Parametric Time–Frequency Domain Spatial Audio, First Edition. Edited by Ville Pulkki, Symeon Delikaris-Manias, and Archontis Politis.
Companion Website: www.wiley.com/go/pulkki/parametrictime-frequency

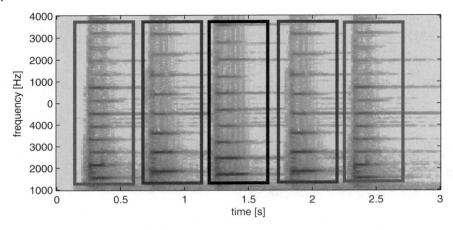

Figure 9.1 Example of magnitude spectrogram redundancy of piano notes. Repeating events are denoted by rectangles of the same shades of gray; the note in the middle occurs only once.

spectral basis functions and their activations. The representation of the magnitude spectrogram using such components allows, for example, easy labeling and classification of the spectral templates for separation (Virtanen, 2007; Ozerov and Fevotte, 2010; Sawada *et al.*, 2013), acoustic pattern classification (Raj *et al.*, 2010), and automatic music transcription (Smaragdis and Brown, 2003; Benetos *et al.*, 2012).

Extending the spectrogram factorization to multiple channels by also modeling channel-wise level and time differences allows the representation of multichannel audio signals and extraction of sound objects with spatial location-specific characteristics. Further, grouping spectral events with similar spatial attributes leads to separation of entire sound sources (Févotte and Ozerov, 2011; Mitsufuji and Roebel, 2014; Nikunen and Virtanen, 2014a). This allows modification of the spatial sound by, for example, repositioning sound sources in the mixture intended for listening through a surround-sound loudspeaker setup or headphones (binaural synthesis). The two main scenarios discussed in this chapter are parameterization of multichannel surround sound and parameterization of microphone array signals. These two share similar characteristics from the spectrogram factorization perspective: each sound object, present in multiple channels, is modeled by a single spectral template with a channel-dependent gain and delay.

This chapter is organized as follows. First, in Section 9.2, we introduce spectrogram modeling tools, the non-negative matrix and tensor factorization and their combination with spatial covariance matrix models. Additionally, we explain the principles of source separation by time–frequency filtering using separation masks constructed from the spectrogram models. The introduction of the theoretical models is followed by Section 9.3, which explains the use of spectrogram factorization with microphone array signals. We introduce a spatial covariance matrix model based on the directions of arrival of sound events and spectral templates, and discuss its relationship to conventional spatial audio signal processing. The use of spectrogram models in applications related to modified reconstruction of spatial audio is presented in Section 9.4. Examples include multichannel surround sound upmixing in Section 9.4.1, sound source separation using a compact microphone array in Section 9.4.2, and reconstruction of binaural sound through sound source separation in Section 9.4.3. The chapter concludes by discussing the strengths, prospects, and current shortcomings of the methods introduced here, and their possible future improvement.

9.2   Spectrogram Factorization

Audio spectrogram factorization is based on learning the redundant spectral patterns occurring over time in the spectrogram. The redundancy of sound events in the frequency domain can be utilized by a model that represents an audio spectrogram as the sum of a few spectral templates and their time-varying activations. The degree of the model, that is, the number of spectral templates, restricts the modeling accuracy, and thus similar yet slightly different sound events become modeled using the same spectral template. Such a model reduces the redundancy in representing the overall spectral content at the cost of introducing a small approximation error. Since spectrogram models are usually used for separation of different sounds from a mixture through filtering and not to actually reconstruct signals from the model parameters, some modeling error is allowed.

The family of algorithms known as non-negative matrix factorization is based on a simple model that consists of a linear combination of basis functions and their activations. The non-negative constraint on the parameters makes the model purely additive, and efficient algorithms for estimating optimal parameters have been developed (Lee and Seung, 2002). The NMF decomposition of an audio signal magnitude spectrogram produces a dictionary of spectral templates that model redundant parts of the audio signal and thus is able to utilize long-term redundancy in representing the signal.

The difference in decomposing monaural and multichannel spectrograms is that the latter includes inter-channel information that is related to the spatial positions of sound objects. The inter-channel information can be used in the decomposition algorithm instead of processing each channel individually. The spatial properties of audio objects are referred to as spatial cues, namely inter-channel level and time differences, ICLD and ICTD, respectively. The spectrogram factorization methods can utilize one of these (FitzGerald *et al.*, 2005, 2006; Nikunen *et al.*, 2011), or both (Sawada *et al.*, 2013; Nikunen and Virtanen, 2014a,b).

9.2.1   Mixtures of Sounds

The field of sound source separation studies the estimation of the original source signals from an observed mixture. The mixture signal $x_m(t)$ consisting of $q = 1, \ldots, Q$ sources captured using $m = 1, \ldots, M$ microphones (equivalent to audio with M channels) in the sampled time domain can be modeled as

$$x_m(t) = \sum_{q=1}^{Q} \sum_{\tau} h_{mq}(\tau) s_q(t - \tau), \tag{9.1}$$

where t is the time-domain sample index. Single-channel source signals $s_q(t)$ are convolved with their associated spatial impulse responses $h_{mq}(\tau)$. The above convolutive mixing can be approximated in the time–frequency domain by instant mixing at each frequency bin individually:

$$x_{il} \approx \sum_{q=1}^{Q} \mathbf{h}_{iq} s_{ilq} = \sum_{q=1}^{Q} \mathbf{y}_{ilq}, \tag{9.2}$$

where $x_{il} = [x_{il1}, \ldots, x_{ilM}]^T$ is the STFT of the multichannel signal. The frequency bins are indexed with $i = 1, \ldots, I$ and the time frame index is $l = 1, \ldots, L$. The window length is $N = 2I - 1$. The mixing of each source q is denoted at each frequency bin i by the spatial frequency response $\mathbf{h}_{iq} = [\mathrm{h}_{ik1}, \ldots, \mathrm{h}_{ikM}]^T$, which is the discrete Fourier transform of h_{mq}, and the source signals are represented by their STFTs, denoted by s_{ilq} at each time–frequency point (i, l). The source signals as observed by the microphones are given as $\mathbf{y}_{ilq} = \mathbf{h}_{iq} s_{ilq}$. The mixing filters h_{mq} and $\mathbf{h}_{iq}$ can also originate from the processing done at the audio production stage, that is, they correspond to the gain and delay for each source in each produced channel.

9.2.2 Magnitude Spectrogram Models

Single Channel Audio

Considering the single-channel magnitude spectrogram x_{il} (that is, the absolute value of the STFT), its NMF model $\hat{x}_{il}$ is given as

$$x_{il} \approx \hat{x}_{il} = \sum_{k=1}^{K} b_{ik} g_{kl}, \quad b_{ik}, g_{kl} \geq 0. \tag{9.3}$$

The parameters of the above model can be interpreted as follows: each component k has a fixed spectrum b_{ik}, $i = 1, \ldots, I$, and time-varying gain g_{kl}, $l = 1, \ldots, L$. Hereafter, the term "NMF component" refers to an entity consisting of a single spectral basis and its time-varying activation. The total number of components is denoted by K. The estimation of the parameters b_{ik} and g_{kl} is discussed in Section 9.2.2.

One NMF component with a fixed spectrum can only model parts of actual sound sources with varying spectral characteristics. However, in most cases the NMF components represent meaningful entities, and multiple components can be combined to represent each source in the mixture. An example of a three-component decomposition of piano notes is illustrated in Figure 9.2, where it can be seen that each NMF component represents the harmonic structure of one note and that the components are active during the presence of associated spectral features observed in the mixture.

The required degree of approximation denoted by K depends highly on the application and the data to be processed. In the case of the NMF utilized in audio signal processing in a blind manner (without any training material to estimate the templates for individual sources), the appropriate value varies from several to tens of components. In speech processing with supervised NMF (Raj *et al.*, 2010; Hurmalainen *et al.*, 2011), where the spectral templates of known source types are learned from a large dataset of training material, the number of components can be several thousands up to tens of thousands. However, in some supervised NMF applications only one component per

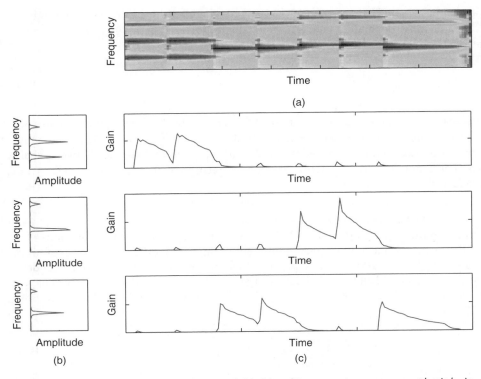

Figure 9.2 Illustration of NMF spectrogram model $\hat{x}_{il}$ (a) and its parameters: component basis b_{ik} in the left row (b), and activations in the right row g_{kl} (c).

separated source is used, as in the separation of drums from music (Helén and Virtanen, 2005).

Multichannel Audio

When dealing with multichannel data, there are two alternatives for defining the spectrogram model based on matrix and tensor factorization. The first one is non-negative tensor factorization (NTF; FitzGerald *et al.*, 2006) applied to a rank-3 tensor ($I \times L \times M$) obtained by combining the magnitude spectrograms of each channel along the third dimension. The NTF model has been used, for example, in source separation (FitzGerald *et al.*, 2008) and estimation of the spatial position of spectral components (Parry and Essa, 2006). The second approach is factorization of the entire complex-valued STFT spectrogram in the covariance domain incorporating both the magnitude and phase of the spectrogram elements. These types of spectrogram factorization algorithms differ greatly from the magnitude-only models and thus are discussed separately in Section 9.2.3.

We denote the magnitude spectrogram of a multichannel signal consisting of $m = 1, \ldots, M$ channels as x_{ilm}. The NTF model $\hat{x}_{ilm}$ consists of the same basic linear model, that is, the NMF model introduced in Equation (9.3), composed of several fixed spectral bases and their corresponding gain. The added third dimension is modeled using a channel-dependent gain a_{km} (spatial cue) for each component k, resulting in the NTF model

$$x_{ilm} \approx \hat{x}_{ilm} = \sum_{k=1}^{K} b_{ik} g_{kl} a_{km}, \quad b_{ik}, g_{kl}, a_{km} \geq 0. \tag{9.4}$$

The spatial cue parameter a_{km} denotes the gain of the NTF component k in each input channel m and it is constrained to unity scale as $\sum_{m=1}^{M} a_{km} = 1$. In the above formulation the mixing is only considered to occur via level difference between channels, corresponding to the ICLD. The phase relation determined by ICTD is not considered.

Estimating the Parameters of Spectrogram Models

Estimating the parameters of the NMF and NTF models introduced in Equations (9.3) and (9.4) can be done with respect to different optimization criteria. The original work by Lee and Seung (2002) considers two different cost functions, squared Euclidean distance (SED) and Kullback–Leibler divergence (KLD). In the context of this chapter the most relevant cost criterion is the SED for the NTF model in Equation (9.4), defined as

$$c_{SED}(x_{ilm}, \hat{x}_{ilm}) = \sum_{m=1}^{M} \sum_{i=1}^{I} \sum_{l=1}^{L} (x_{ilm} - \hat{x}_{ilm})^2. \tag{9.5}$$

The algorithm for estimating the NMF model parameters proposed by Lee and Seung (2002) consists of multiplicative updates that are repeated in an iterative manner. The update equations of each parameter in the model are evaluated in turn, while keeping the rest fixed. The update equations monotonically decrease the value of the cost function measuring the fit of the model to the observed data. In the vicinity of a local or global minimum, the decrease in the cost function becomes small between iterations and the algorithm is considered to have converged. No globally optimal solution can be guaranteed to be obtained, since multiple local minima exist even in the simplest NMF model formulations.

The update equations for the NTF model with the SED criterion in Equation (9.5) were originally proposed by FitzGerald *et al.* (2005), and are given as

$$b_{ik} \leftarrow b_{ik} \frac{\sum_m \sum_l g_{kl} a_{km} x_{ilm}}{\sum_t \sum_m g_{kl} a_{km} \hat{x}_{ilm}}, \tag{9.6}$$

$$g_{kl} \leftarrow g_{kl} \frac{\sum_m \sum_i b_{ik} a_{km} x_{ilm}}{\sum_m \sum_i b_{ik} a_{km} \hat{x}_{ilm}}, \tag{9.7}$$

$$a_{km} \leftarrow a_{km} \frac{\sum_i \sum_l b_{ik} g_{kl} x_{ilm}}{\sum_i \sum_l b_{ik} g_{kl} \hat{x}_{ilm}}. \tag{9.8}$$

The process for estimating the parameters of NMF-based spectrogram models consists of a few common steps that can be used to describe all the cost function variants

---

Algorithm 9.1 Estimating NMF model parameters.

Input: x_{ilm}

$b_{ik}, g_{kl}, a_{km} \leftarrow$ Initialize spectrogram model parameters with random positive values uniformly distributed in the range $[0, 1]$

maxiter $\leftarrow$ Set appropriate value for maximum number of iterations

iter $= 1$

while: iter $<$ maxiter **do**

$\quad\quad \hat{x}_{ilm} \leftarrow$ update NTF model (9.4)

$\quad\quad b_{ik} \leftarrow$ update component spectral basis (9.6)

$\quad\quad \hat{x}_{ilm} \leftarrow$ update NTF model (9.4)

$\quad\quad g_{kl} \leftarrow$ update component activations (9.7)

$\quad\quad \hat{x}_{ilm} \leftarrow$ update NTF model (9.4)

$\quad\quad a_{km} \leftarrow$ update component channel-wise activations (9.8)

$\quad\quad$ Scaling numerical range for stability:

$\quad\quad \eta_k = \left(\sum_{i=1}^{I} b_{ik}^2 \right)^{1/2}, \; b_{ik} \leftarrow b_{ik}/\eta_k, \; g_{kl} \leftarrow g_{kl}\eta_k$

$\quad\quad a_{km} \leftarrow a_{km} / \sum_{m=1}^{M} a_{km}$

$\quad\quad$ iter $\leftarrow$ iter $+ 1$

end while

return: $b_{ik}, g_{kl},$ and a_{km}

---

and different formulations of the model. The process is described in Algorithm 9.1. A Matlab implementation of the algorithm is given in Section 9.6.

9.2.3 Complex-Valued Spectrogram Models

The multichannel audio signal encoding the spatial position of the sound sources by the time difference between the channels is not considered in the magnitude-based models discussed in Section 9.2.2. This is because the phase of the STFT x_{il} is not retained in the magnitude-only decomposition models defined in Equations (9.3) and (9.4). Considering the phase of the STFT in spectrogram factorization models requires the processing of complex-valued input data, as well as consideration of the additivity of factorized components with different phases.

Spatial Covariance Domain

When incorporating both ICLD and ICTD into spectrogram factorization of spatial audio, it is useful to operate with spatial covariance matrices (SCMs) calculated from the input signal STFT (Arberet *et al.*, 2010; Sawada *et al.*, 2011). The benefit of operating with SCMs is that they represent channel-wise properties of the input signal in relation to other channels, for example the phase difference between two channels instead of the absolute signal phase. The spatial covariance matrices in the context of this chapter are calculated from the square-root of the STFT:

$$\tilde{\mathbf{x}}_{il} = [|x_{il1}|^{1/2} \, \text{sign}(x_{il1}), \ldots, |x_{ilM}|^{1/2} \, \text{sign}(x_{ilM})]^{\mathsf{T}}, \tag{9.9}$$

where $\text{sign}(z) = z/|z|$, $z \in \mathbb{C}$ is the signum function for complex numbers. The covariance matrix of a single time–frequency point (i, l) is obtained as the outer product

$$\mathbf{X}_{il} = \tilde{\mathbf{x}}_{il}\tilde{\mathbf{x}}_{il}^{\text{H}}, \tag{9.10}$$

where ^{H} stands for Hermitian transpose.

The elements of matrix $\mathbf{X}_{il} \in \mathbb{C}^{M \times M}$ for each time–frequency point (i, l) encode the spatial behavior of the captured signal in the form of the amplitude and phase differences with respect to each channel pair. The use of the square-root of the STFT $\tilde{\mathbf{x}}_{il}$ means that the diagonal of each matrix $[\mathbf{X}_{il}]_{n,m}$, $n = m$, contains the STFT magnitudes $|\mathbf{x}_{il}| = [|x_{il1}|, \ldots, |x_{ilM}|]^{\text{T}}$. The off-diagonal values $[\mathbf{X}_{il}]_{n,m}$, $n \neq m$, represent the magnitude correlation and phase difference $|x_{iln}x_{ilm}|^{1/2} \text{sign}(x_{iln}x_{ilm}^*)$ between each channel pair (n, m).

The mixing defined in Equation (9.2) can be approximated in the spatial covariance domain as

$$\mathbf{X}_{il} \approx \sum_{q=1}^{Q} \mathbf{H}_{iq}\hat{s}_{ilq}, \tag{9.11}$$

where $\mathbf{H}_{iq} \in \mathbb{C}^{M \times M}$ is the spatial covariance matrix for each source q at each frequency i, and $\hat{s}_{ilq} = (s_{ilq}\overline{s_{ilq}})^{(1/2)} = |s_{ilq}|$ denotes the magnitude spectrogram of source q. The spatial covariance matrices $\mathbf{H}_{iq}$ for all frequencies $i = 1, \ldots, I$ define the mixing of the qth source in the spatial covariance domain. The source spectrogram $\hat{s}_{ilq}$ is real and non-negative, and estimation of its absolute phase is not required. Equivalence between Equations (9.11) and (9.2) is achieved by defining $\mathbf{H}_{iq} = \dfrac{\mathbf{h}_{iq}\mathbf{h}_{iq}^{\text{H}}}{||\mathbf{h}_{iq}\mathbf{h}_{iq}^{\text{H}}||_{\text{F}}}$ and assuming that $||\mathbf{h}_{iq}||_1 = 1$. In the context of this chapter the covariance mixing defined in Equation (9.11) is referred to as the spatial covariance domain.

Complex-Valued Non-Negative Matrix Factorization

The NMF framework can be used to model the spatial covariance observations and the spatial covariance domain mixing defined in Equation (9.11) as follows. The NMF magnitude model defined in Equation (9.3) is used to represent the mixture magnitude spectrogram $x_{il} = \sum_q \hat{s}_{ilq} \approx \sum_k b_{ik}g_{kl}$, and the spatial properties $\mathbf{H}_{ik}$ are estimated separately for each NMF component k. This strategy is hereafter referred to as a component-wise SCM model. With the described substitutions to Equation (9.11), the complex-valued NMF model with component-wise SCMs is defined as

$$\mathbf{X}_{il} \approx \hat{\mathbf{X}}_{il} = \sum_{k=1}^{K} \mathbf{H}_{ik}b_{ik}g_{kl}, \tag{9.12}$$

where $\mathbf{H}_{ik}$ are the SCMs for each NMF component k at each frequency index i.

Several NMF components are needed to represent one actual acoustic sound source $\hat{s}_{ilq}$, and a separate linking strategy between NMF components and real sound sources is needed. Two or more components modeling the same sound source will ideally end up having similar SCM properties determined by the spatial position of the source. An explicit parameterization of this underlying spatial property has been proposed in Ozerov *et al.* (2011) and Févotte and Ozerov (2011), and utilized by Sawada *et al.* (2011)

by introducing a component-to-source linking parameter to the model and estimating only a single set of SCMs for a group of NMF components. The component linking can be estimated in the same way as any other non-negative parameter of the model, and thus no separate clustering of NMF components for separation is needed.

The complex-valued NMF model with source-wise SCMs is defined as

$$\mathbf{X}_{il} \approx \hat{\mathbf{X}}_{il} = \sum_{q=1}^{Q} \mathbf{H}_{iq} \underbrace{\sum_{k=1}^{K} c_{qk} b_{ik} g_{kl}}_{\hat{s}_{ilq}}, \tag{9.13}$$

where c_{qk} denotes the association of component k to source q. The association parameter c_{qk} is a non-negative scalar and the association is soft instead of binary, which means that one NMF component can be associated with multiple sources with different weights. Also, note that the SCMs denoted by $\mathbf{H}_{iq}$ are defined for the actual number of sources $q = 1, \ldots, Q$.

Using the source-wise SCM estimation as in Equation (9.13), the number of SCMs needed to be estimated at each frequency index decreases to the number of sources present in the mixture, which needs to be known in advance. Such models are somewhat sensitive to the source count initialization, since excess SCMs and sources in the model may end up representing the spectral and spatial properties of the same actual sound source. This is due to no restriction or regularization being imposed on SCMs with respect to the spatial location they represent. This problem is explicitly avoided by the SCM model based on direction of arrival introduced in Section 9.3.2.

Relation of the Model to Acoustic Quantities

The interpretation of the complex-valued NMF model in Equation (9.13) based on the covariance mixing defined in Equation (9.11) can be described as follows. The diagonal entries of $\mathbf{H}_{iq}$ represent the ICLD of each source q with respect to the input channels, and are not time dependent due to the assumption of stationary sources and mixing. Furthermore, the NMF magnitude model $\hat{s}_{ilq} \approx \sum_k c_{qk} b_{ik} g_{kl}$ for a single source denotes its overall spectrogram in all channels. The ICLD at the diagonal of $\mathbf{H}_{iq}$ represents frequency-dependent acoustical amplification or attenuation caused by the source positioning with respect to the capturing device. The off-diagonal values of $\mathbf{H}_{iq}$ model the cross-channel magnitude and phase difference properties, of which the latter represents the ICTD between the input channels.

The main difference between the magnitude-based NTF model in Equation (9.4) and the complex-valued NMF in Equation (9.12) is the modeling of the ICTD between channels. Additionally, the ICLD in the former is defined component-wise, whereas the complex-valued NMF model also allows frequency-dependent level differences between input channels. Frequency-dependent ICLD and ICTD modeling is more accurate when considering real recordings done using microphone arrays, where frequency-dependent attenuation may exist, especially at high frequencies which are easily absorbed by physical obstacles.

Estimating the Parameters of the NMF Model with Covariance Matrices

The estimation of the parameters of complex-valued NMF models was proposed by Sawada *et al.* (2013). The derivation of the update equations makes use of the auxiliary function technique, as in the expectation maximization algorithm (Dempster *et al.*, 1977). The optimization criterion used is the squared Frobenius norm between the observed covariance matrices and the model, given as

$$c_{FRO}(\mathbf{X}_{il}, \hat{\mathbf{X}}_{il}) = \sum_{i=1}^{I} \sum_{l=1}^{L} ||\mathbf{X}_{il} - \hat{\mathbf{X}}_{il}||_F^2. \tag{9.14}$$

Additionally, model parameter estimation using Itakura–Saito divergence as a cost function was proposed by Sawada *et al.* (2013), but it is not covered in this chapter.

The parameter updates for the model in Equation (9.13) can be derived using the framework proposed in Sawada *et al.* (2013), which is extensible for different magnitude spectrograms and covariance matrix parameterizations. The cost function in Equation (9.14) with the model of Equation (9.13) can be given as

$$f(c_{qk}, b_{ik}, g_{kl}, \mathbf{H}_{iq}) = \sum_{i,l} \left[\left(\sum_{k,q} \mathbf{H}_{iq} c_{qk} b_{ik} g_{kl} \right) \left(\sum_{k,q} \mathbf{H}_{iq} c_{qk} b_{ik} g_{kl} \right)^H \right]$$
$$- \sum_{i,l,k} c_{qk} b_{ik} g_{kl} \text{tr} \left(\mathbf{X}_{il} \mathbf{H}_{iq}^H \right) - \sum_{i,l,k} c_{qk} b_{ik} g_{kl} \text{tr} \left(\mathbf{H}_{iq} \mathbf{X}_{il}^H \right). \tag{9.15}$$

With the introduction of auxiliary variables $\mathbf{R}_{ilkq} = \hat{\mathbf{X}}_{il}^{-1} \mathbf{H}_{iq} c_{qk} b_{ik} g_{kl}$ (which are positive-definitive Hermitian and satisfy $\sum_{k,q} \mathbf{R}_{ilkq} = \mathbf{I}$), it has been shown (Sawada *et al.*, 2013) that the minimization of auxiliary function

$$f^+(c_{qk}, b_{ik}, g_{kl}, \mathbf{H}_{iq}, \mathbf{R}_{ilkq}) = \sum_{i,l,k,q} c_{qk}^2 b_{ik}^2 g_{kl}^2 \mathbf{H}_{iq} \mathbf{R}_{ilkq} \mathbf{H}_{iq}^H$$
$$- \sum_{i,l,k,q} c_{qk} b_{ik} g_{kl} \text{tr} \left(\mathbf{X}_{il} \mathbf{H}_{iq}^H \right) - \sum_{i,l,k,q} c_{qk} b_{ik} g_{kl} \text{tr} \left(\mathbf{H}_{iq} \mathbf{X}_{il}^H \right) \tag{9.16}$$

is equivalent to indirect minimization of Equation (9.15), that is, the squared Frobenius norm between observations and the complex-valued NMF model. The update equations can be obtained by partial differentiation of f^+ with respect to each parameter in the model, setting the derivative to zero and solving for each parameter to be updated. After substituting the auxiliary variable by its definition, we obtain the following updates:

$$c_{qk} \leftarrow c_{qk} \frac{\sum_{i,l} b_{ik} g_{kl} \text{tr}(\mathbf{X}_{il} \mathbf{H}_{iq})}{\sum_{i,l} b_{ik} g_{kl} \text{tr}(\hat{\mathbf{X}}_{il} \mathbf{H}_{iq})}, \tag{9.17}$$

$$b_{ik} \leftarrow b_{ik} \frac{\sum_{l,q} c_{qk} g_{kl} \text{tr}(\mathbf{X}_{il} \mathbf{H}_{iq})}{\sum_{l,q} c_{qk} g_{kl} \text{tr}(\hat{\mathbf{X}}_{il} \mathbf{H}_{iq})}, \tag{9.18}$$

$$g_{kl} \leftarrow g_{kl} \frac{\sum_{i,q} c_{qk} b_{ik} \text{tr}(\mathbf{X}_{il} \mathbf{H}_{iq})}{\sum_{i,q} c_{qk} b_{ik} \text{tr}(\hat{\mathbf{X}}_{il} \mathbf{H}_{iq})}, \tag{9.19}$$

$$\mathbf{H}_{iq} \leftarrow \mathbf{H}_{iq} \left(\sum_{l,k} c_{qk} b_{ik} g_{kl} \hat{\mathbf{X}}_{il} \right)^{-1} \left(\sum_{l,k} c_{qk} b_{ik} g_{kl} \mathbf{X}_{il} \right). \tag{9.20}$$

To prevent a subtractive model and negative values in the diagonals of $\mathbf{H}_{iq}$, each (i, q) matrix is forced to be Hermitian positive semidefinite. This can be done as proposed by

Sawada *et al.* (2011) by calculating its eigenvalue decomposition $\mathbf{H}_{iq} = \mathbf{VDV}^{H}$, setting all the negative eigenvalues to zero,

$$[\hat{\mathbf{D}}]_{m,m} = \begin{cases} [\mathbf{D}]_{m,m} & \text{if } [\mathbf{D}]_{m,m} > 0, \\ 0 & \text{otherwise,} \end{cases} \tag{9.21}$$

and reconstructing each matrix for all i, q with the modified eigenvalues as $\mathbf{H}_{iq} \leftarrow \mathbf{V}\hat{\mathbf{D}}\mathbf{V}^{H}$.

9.2.4   Source Separation by Time–Frequency Filtering

Source separation using spectrogram factorization models is achieved via time–frequency filtering of the original observation STFT by a generalized Wiener filter obtained from the spectrogram model parameters.

When the NMF model in Equation (9.3) or the NTF model in Equation (9.4) is used to represent the magnitude spectrogram of a multichannel audio signal, the required information for separating the sources from the mixture is the association of NTF components to entire sound sources. In the literature, there are attempts at using the spatial cue a_{km} as an indicator variable of the NTF components belonging to the same sound source (Févotte and Ozerov, 2011; Mitsufuji and Roebel, 2014). In supervised separation with spectrogram models (Weninger *et al.*, 2012), separate training material from each speaker is used to learn the association between estimated spectral templates and each speaker to be separated. The problem in supervised approaches is then to represent each speaker exclusively by its own dictionary of spectral templates, and adaptation of noise models for unseen acoustic material (Hurmalainen *et al.*, 2013).

In the following, we introduce a generic approach for reconstructing source signals by assuming that the association of NTF component k belonging to an audio source q is known or can be estimated as in the complex-valued NMF of Equation (9.13). The reconstruction of the individual source magnitude spectrogram $\hat{s}_{ilq}$ consisting of a subset of the spectral components denoted by the clustering variable (soft or binary) $c_{qk} \in [0, 1]$ is given as

$$\hat{s}_{ilq} = \sum_{m=1}^{M} \sum_{k=1}^{K} c_{qk} b_{ik} g_{kl} a_{km} = \sum_{k=1}^{K} c_{qk} b_{ik} g_{kl}. \tag{9.22}$$

Note that after the introduction of the component-to-source variable c_{qk} and taking into consideration the scaling $\sum_{m} a_{km} = 1$, the NTF spectrogram model becomes identical to the spectrogram model used in the complex-valued NMF, Equation (9.13). Estimate of the source signals $\hat{y}_{ilq}$, using the generalized Wiener filter, for all spectrogram the models in Equations (9.3), (9.4), and (9.13) can be defined as

$$\hat{y}_{ilq} = \frac{\hat{s}_{ilq}}{\sum_{q=1}^{Q} \hat{s}_{ilq}} x_{il} = \frac{\sum_{k=1}^{K} c_{qk} b_{ik} g_{kl}}{\sum_{q=1}^{Q} \sum_{k=1}^{K} c_{qk} b_{ik} g_{kl}} x_{il}. \tag{9.23}$$

The time-domain source signals are acquired by the inverse STFT of $\hat{y}_{ilq}$.

Alternatively, in the case of the complex-valued NMF model, Equation (9.13), a multichannel Wiener filter can be used to reconstruct the source spectrograms, which also utilizes the estimated spatial covariance information. The multichannel Wiener filter is defined as

$$\mathbf{y}_{ilq} = \sum_{k=1}^{K} c_{qk} b_{ik} g_{kl} \mathbf{H}_{iq} \hat{\mathbf{X}}_{il}^{-1} \mathbf{x}_{il}, \tag{9.24}$$

where $\hat{\mathbf{X}}_{il}^{-1}$ is the matrix inverse of the model in Equation (9.13) at each time–frequency point (i, l). The structure of the multichannel Wiener filter corresponds to a minimum variance distortionless response beamformer followed by a single channel post-filter (Simmer *et al.*, 2001).

9.3 Array Signal Processing and Spectrogram Factorization

Decomposition of the spectrogram of an array signal capture requires modeling *both* the spatial properties, ICLD and ICTD, of the learned spectral templates. In the spatial covariance domain the magnitude spectrum and spatial properties of the sources are represented as separate entries, a real-valued source spectrogram and its SCMs. As stated in Section 9.2.3, it is a useful property when combining spatial processing and spectrogram factorization, since the absolute phase of the sources need not be estimated. Additionally, we have seen the use of the complex-valued NMF model for analysis of the spatial covariance domain observations in Section 9.2.3. However, estimating $\mathbf{H}_{iq}$ in such a way that it corresponds to a single source at all frequencies requires an algorithm that operates jointly over frequencies and thus ties together possible aliased phase differences. This is not guaranteed by the models presented in Equations (9.12) and (9.13), which require frequency-wise estimation of the SCMs [update Equation (9.20)].

In conventional separation approaches combining the NMF magnitude model with covariance domain observations (Arberet *et al.*, 2010; Sawada *et al.*, 2011, 2013), the estimation of source SCMs assumes that the NMF magnitude model enforces $\hat{s}_{ilq}$ to correspond to a single source, and thus estimating $\mathbf{H}_{iq}$ yields an estimate that corresponds to a single source over frequency. However, it is not guaranteed that each NMF component models spatially coherent spectrogram components. For example, two sources having similar spectral characteristics may become modeled by the same NMF component even though they reside at different spatial locations. A direct investigation of whether the assumption is violated is difficult.

The remaining question is the estimation of the spatially coherent source SCMs and their parameterization via direction-dependent quantities, that is, the delay (phase difference) caused by a certain direction of arrival (DOA). In this section we introduce a DOA-based model for spatial covariance matrices, and the estimation of its parameters. The applications of such a model include sound source separation using small compact microphone arrays (Section 9.4.2) and reconstruction of binaural sound by using the DOA parameterized SCMs for retrieving the correct head-related transfer function of each separated source for binaural synthesis (Section 9.4.3).

9.3.1 Spaced Microphone Arrays

Frequency-domain signal processing with spaced microphone arrays is based on observing time delays in terms of phase differences between the array elements. In the following sections we assume far-field propagation and use of spaced microphone arrays that are composed of omnidirectional microphones with no assumption being made on the direction-dependent level difference between the microphones. In practice, the body of the array (where the microphones are attached) causes an acoustical shade, particularly at higher frequencies, leading to amplitude differences between micro-phones. In general, the methods for modeling source SCMs introduced in the follow-ing sections are not restricted to the use of spaced microphone arrays. The methods can be extended to account for the time and the amplitude difference of any direc-tional microphone arrangement of which the spatial response with respect to DOA is known.

Recording an auditory scene with one or several microphones captures the source signals $s_q(t)$ convolved with their spatial impulse response $h_{mq}(\tau)$, as described by the mixing model in Equation (9.1). The spatial impulse response is the transfer function for each sound source q at location $s_q \in \mathbb{R}^3$ to each microphone m at location $\mathbf{m} \in \mathbb{R}^3$. The spatial impulse response $h_{mq}(\tau)$ incorporates all the propagation attributes of a spe-cific source, which are the radiation pattern of the sound source and its position with respect to the microphone array and the surrounding environment. In room acoustics, the spatial response is referred to as the room impulse response (RIR).

9.3.2 Model for Spatial Covariance Based on Direction of Arrival

Considering the separation based on the covariance domain mixing model of Equa-tion (9.11), the problem in estimation of the spatial mixing of a source concentrates on the fact that the observed evidence, the phase difference, is dependent on frequency and begins to alias at rather low frequencies with microphone arrays of practical size. Our aim is to estimate $\mathbf{H}_{iq}$ in such a way that it corresponds to a single source at all frequencies. This requires an algorithm that operates jointly over frequencies and thus ties together possible aliased phase differences.

The problem can be contrasted with sound source separation by independent com-ponent analysis (ICA) applied frequency-wise on the mixture signal STFT and solving the source permutation problem based on the estimated mixing filter time delays at dif-ferent frequencies (Sawada *et al.*, 2007). Attempts at regularizing ICA parameter esti-mation over frequencies and avoiding the permutation problem altogether have been considered by Nesta *et al.* (2011), assuming no spatial aliasing occurring.

Time Difference of Arrival with Known Array Geometry

The wavefront DOA corresponds to a set of time difference of arrival (TDOA) values between each microphone pair, and the TDOAs depend on the geometry of the array. The geometry of an array consisting of two microphones n and m located on the xy-plane at locations $\mathbf{n} \in \mathbb{R}^3$ and $\mathbf{m} \in \mathbb{R}^3$ is illustrated in Figure 9.3(a). In the illustration, a look direction vector $\mathbf{k}_o$ is pointing towards the source at location $\mathbf{s} \in \mathbb{R}^3$ from the geo-metrical center $\mathbf{p} \in \mathbb{R}^3$ of the array. The geometrical center of the array is at the origin of the Cartesian coordinate system, $\mathbf{p} = [0, 0, 0]^{\mathrm{T}}$, and the norm of the look direction vector is $\|\mathbf{k}_o\| = 1$. Any given array geometry can be translated and rotated in such a way that its geometrical center is located at the origin of the coordinate system, which is done to simplify the TDOA calculations. The different look directions are indexed by o and their direction is given in a spherical coordinate system using elevation $\theta_o \in [0, \pi]$, azimuth

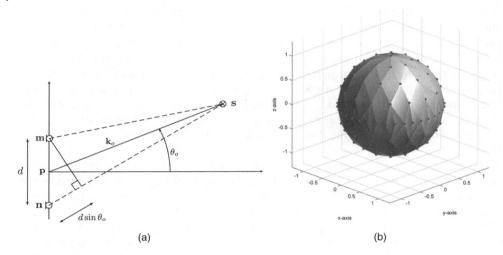

Figure 9.3 (a): Geometry of an array consisting of microphones at locations **n** and **m** and look direction vector $\mathbf{k}_o$ pointing towards a source located at **s**. (b): Approximately uniformly distributed points (sparsely sampled grid) on the surface of a sphere.

$\varphi_o \in [0, 2\pi]$, and a fixed radius of $r = 1$. The reference axis of the array $(\theta = 0, \varphi = 0)$ can be chosen arbitrarily.

With the above definitions and use of basic Euclidean geometry, the TDOA between microphones n and m in seconds is defined for look direction $\mathbf{k}_o$ as

$$\tau_{nm}(\mathbf{k}_o) = (\mathbf{k}_o) = \frac{-\mathbf{k}_o^{\mathrm{T}}(\mathbf{n} - \mathbf{m})}{v}, \tag{9.25}$$

where v is the speed of sound. The TDOA corresponds to a phase difference in the frequency domain, which can be expressed as the argument of $\exp(-j2\pi f_i \tau_{nm}(\mathbf{k}_o))$, where $f_i = (i - 1)F_s/N$ is the frequency of the ith STFT bin in Hertz, F_s is the sampling frequency, and N is the STFT window length.

Further, we denote the phase differences of all microphone pairs $n = 1, \ldots, M$ and $m = 1, \ldots, M$ using matrices of size $M \times M$ defined for each frequency index i and for all look directions $o = 1, \ldots, O$ as

$$[\mathbf{W}_{io}]_{n,m} = \exp\left(-j2\pi f_i \tau_{nm}(\mathbf{k}_o)\right). \tag{9.26}$$

Hereafter we refer to the $\mathbf{W}_{io} \in \mathbb{C}^{M \times M}$ as DOA kernels, since they denote ideal phase differences for the hypothesis of a source being at look direction o.

Spatial Covariance Matrix Model by Superposition of DOA Kernels

Based on the previous section, we can state that the direct path TDOA explains the observed phase difference even with aliasing, and the connection between source DOAs and phase differences can be found by the aid of TDOAs. This concept can be adapted to the SCM estimation in the complex-valued NMF framework by finding which direct-path TDOA explains the observed phase difference evidence the best.

The DOA-based SCM model considered in this section is based on superposition, that is, a weighted linear combination, of the DOA kernels defined in Equation (9.26). The look direction vectors $\mathbf{k}_o$ are set to approximately uniformly sample the surface of a unit sphere set around the geometrical center $\mathbf{p}$ of the array. A sparsely sampled grid used in experiments given later in this chapter is illustrated in Figure 9.3(b). The SCM of a point source in anechoic capturing conditions could be described by a single DOA kernel, which is analogous to the direct path propagation. Due to the echoes and diffractions from surfaces and objects in regular capturing conditions, a combination of several direct paths for modeling one sound source is used and a non-negative weight for each look direction is estimated. The weights of each DOA kernel describe the signal power originating from the corresponding direction.

Given the DOA kernels $\mathbf{W}_{io}$ as specified in Equation (9.26), the proposed SCM model based on their superposition is given as

$$\mathbf{H}_{iq} = \sum_{o=1}^{O} \mathbf{W}_{io} z_{qo},$$

(9.27)

where z_{qo} are the direction weights corresponding to each look direction $o = 1, \ldots, O$. The direction weights are scaled to satisfy $\sum_{o=1}^{O} z_{qo}^2 = 1$, which corresponds to each source's spatial power being equal. This scaling ensures that the spatial model and the covariance matrices $\mathbf{H}_{iq}$ have an approximately constant Frobenius norm, modeling the channel-wise properties only, so that the actual signal energies of different sound sources become modeled by spectrogram parameters.

The direction weights z_{qo} in the proposed model are independent of frequency, which makes the estimation of the entire SCM $\mathbf{H}_{iq}$ optimized over all frequencies. The frequency dependencies of the phase differences are taken into account in the definition of the DOA kernels, and the estimation can be seen as finding the most probable DOA in terms of how well the TDOA of each look direction explains the observations.

9.3.3   Complex-Valued NMF with the Spatial Covariance Model

The complex-valued NMF model with the DOA-based SCM model in Equation (9.27) is simply obtained from the source-wise SCM model defined in Equation (9.13) by replacing the source SCMs $\mathbf{H}_{iq}$ with the proposed model $\sum_{o=1}^{O} \mathbf{W}_{io} z_{qo}$. The entire model can be then written as

$$\mathbf{X}_{il} \approx \hat{\mathbf{X}}_{il} = \sum_{q=1}^{Q} \Big[\underbrace{\sum_{o=1}^{O} \mathbf{W}_{io} z_{qo}}_{\mathbf{H}_{iq}} \Big] \underbrace{\sum_{k=1}^{K} c_{qk} b_{ik} g_{kl}}_{\hat{s}_{ilq}} \cdot$$

(9.28)

Comparing the above model with the covariance mixing in Equation (9.11), it can be seen that the NMF magnitude model represents source spectra $\hat{s}_{ilq} \approx \sum_k c_{qk} b_{ik} g_{kl}$, and all the spatial properties are encoded by the kernels $\mathbf{W}_{io}$ and the direction weights z_{qo}.

The update equations for optimizing the parameters of the model in Equation (9.28) can be derived in a similar way to that described in Section 9.2.3. The process results in the updates

$$
c_{qk} \leftarrow c_{qk} \frac{\sum_{i,l,o} z_{qo} b_{ik} g_{kl} \text{tr}(\mathbf{X}_{il} \mathbf{W}_{io})}{\sum_{i,l,o} z_{qo} b_{ik} g_{kl} \text{tr}(\hat{\mathbf{X}}_{il} \mathbf{W}_{io})}, \qquad
b_{ik} \leftarrow b_{ik} \frac{\sum_{l,q,o} z_{qo} c_{qk} g_{kl} \text{tr}(\mathbf{X}_{il} \mathbf{W}_{io})}{\sum_{l,q,o} z_{qo} c_{qk} g_{kl} \text{tr}(\hat{\mathbf{X}}_{il} \mathbf{W}_{io})},
$$

$$
g_{kl} \leftarrow g_{kl} \frac{\sum_{i,q,o} z_{qo} c_{qk} b_{ik} \text{tr}(\mathbf{X}_{il} \mathbf{W}_{io})}{\sum_{i,q,o} z_{qo} c_{qk} b_{ik} \text{tr}(\hat{\mathbf{X}}_{il} \mathbf{W}_{io})}, \qquad
z_{qo} \leftarrow z_{qo} \frac{\sum_{i,l,k} c_{qk} b_{ik} g_{kl} \text{tr}(\mathbf{X}_{il} \mathbf{W}_{io})}{\sum_{i,l,k} c_{qk} b_{ik} g_{kl} \text{tr}(\hat{\mathbf{X}}_{il} \mathbf{W}_{io})}. \qquad (9.29)
$$

The requirement for also updating the DOA kernels $\mathbf{W}_{io}$ results from the fact that the real-valued entries in the diagonal of each $\mathbf{W}_{io}$ determine the amplitude difference between the channels, and the absolute values of the off-diagonal entries determine the cross-channel magnitude correlation (for each frequency at each direction). If we assumed that no amplitude differences occurred between the microphones in the array, then the amplitudes of all channels and their cross-correlations would be constant. However, all practical microphone arrays deviate from the ideal omnidirectional response by having direction-dependent attenuation at high frequencies due to the body or casing of the array. Because of this, and the possibility of using directional microphones, we need to update the parameters of the model representing the ICLD, that is, the absolute values of the entries of $\mathbf{W}_{io}$.

In the update of the DOA kernels, the argument of each complex-valued entry of $\mathbf{W}_{io}$ needs to be kept fixed. This is done to maintain the original phase difference, that is, the original TDOA caused by a certain look direction, while the magnitudes are subject to updating. First, a preliminary update for the complex-valued entries of DOA kernels is done:

$$
\hat{\mathbf{W}}_{io} \leftarrow \hat{\mathbf{W}}_{io} \left(\sum_{l,k,q} z_{qo} c_{qk} b_{ik} g_{kl} \hat{\mathbf{X}}_{il} \right)^{-1} \left(\sum_{l,k,q} z_{qo} c_{qk} b_{ik} g_{kl} \mathbf{X}_{il} \right), \qquad (9.30)
$$

which is based on the SCM update procedure introduced in Section 9.2.3. Similarly, each (i, o) matrix $\hat{\mathbf{W}}_{io}$ is forced to be Hermitian positive semidefinite. The final update maintaining the argument of each element of the DOA kernels but modifying the absolute values based on the update in Equation (9.30) is achieved as

$$
\mathbf{W}_{io} \leftarrow |\hat{\mathbf{W}}_{io}| \exp(i \arg(\mathbf{W}_{io})). \qquad (9.31)
$$

The proposed update of the DOA kernels is done frequency-wise, but it only updates the ICLD cues associated with each direction o. This can be interpreted as learning the ICLD properties of each look direction from the observed data $\mathbf{X}_{il}$. These properties can, for example, be the frequency-dependent attenuation caused by the scattering of the array and its surrounding environment. The actual spatial locations of the sources are estimated via updating of the spatial weights z_{qo} in Equation (9.29), which is independent of frequency.

The iterative process for estimating the parameters of the model is a straightforward extension of the generic NMF algorithm described in Algorithm 9.1. Each parameter is updated in turn, and the required scaling is applied between updates.

9.4   Applications of Spectrogram Factorization in Spatial Audio

At this point of this chapter we have presented the spectrogram factorization tools to analyze spatial audio of different forms: multichannel surround sound and microphone array captures. In this section we present applications of these spectrogram models in the tasks of multichannel audio upmixing and sound source separation with spatial synthesis.

9.4.1   Parameterization of Surround Sound: Upmixing by Time–Frequency Filtering

Multichannel spectrogram factorization applied to a surround-sound recording provides a compact representation of the spectral content of each channel. This representation can be used to achieve a downmix–upmix operation, where multiple channels are recovered from one or several downmixed channels. The setting is similar to the parametric coding of spatial audio (Schuijers *et al.*, 2004; Faller, 2004), though the clustering of the NTF components based on their spatial cues or other attributes also allows separation of sources from the surround sound mixture, as demonstrated by Nikunen *et al.* (2012).

In the simplest case, the spatial positioning of sound sources is determined only by the level difference between channels. A typical setting of surround sound with two sound sources positioned in the front channels is illustrated in Figure 9.4. The NTF model as given in Equation (9.4) can be used to approximate the magnitude spectrogram of such multichannel audio signal. The parameter a_{km} denotes the spectral object occurrence with respect to each channel determined by the ICLD. Comparing the NTF model with conventional SAC approaches, which estimate only a single set of spatial cues for each time–frequency block, the NTF model estimates the cues over frequency for each spectral template and allows upmixing of spectrally overlapping sound objects.

Figure 9.4 Surround sound and positioning of sound sources in the front channels.

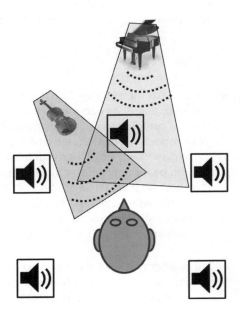

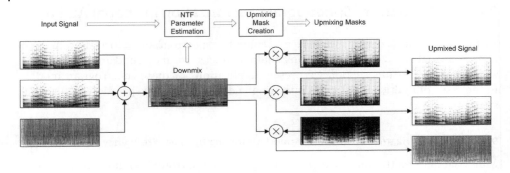

Figure 9.5 Process of upmixing based on the NTF model of a multichannel spectrogram and a single channel downmix. Source: Nikunen 2012. Reproduced with permission of the Audio Engineering Society.

The process of upmixing using the NTF model and a single downmix channel is illustrated in Figure 9.5. The first column depicts the magnitude spectrograms of each individual channel. In the second column is the magnitude spectrogram of the downmix. The third column illustrates upmixing masks generated from the NTF model, and the masks denote a gain in the range [0, 1] for each time–frequency point in each channel to be upmixed. The combination of the second and the third columns by element-wise multiplication realizes upmixing (time–frequency filtering), and the upmixed channels can be seen in the last column.

Upmixing by Multichannel Spectrogram Factorization

In the general upmixing setting, the NTF model in Equation (9.4) is estimated from the STFT magnitudes of a multichannel signal $x_{ilm} = |\mathbf{x}_{il}|$. The recovery of multiple channels (upmixing) by time–frequency filtering is applied to the downmix signal (the combined content of the multiple input channels) as

$$y_{ilm} = \frac{\sum_{k=1}^{K} b_{ik}g_{kl}a_{km}}{\sum_{m'=1}^{M}\sum_{k=1}^{K} b_{ik}g_{kl}a_{km'}}d_{il} = A_{ilm}d_{il}, \qquad (9.32)$$

where y_{ilm} are the recovered channels and $d_{il} \approx \sum_{m} \mathbf{x}_{il}$ is the STFT of the downmix signal. Note that d_{il} is complex valued, that is, it has a phase resulting from the combination of complex-valued input channels. The term in the numerator corresponds to the NTF model for channel m. In the denominator, one finds the sum of the NTF model over all channels m', which corresponds to the STFT of the downmix. The time-domain signals can be directly reconstructed from the upmixed STFT y_{ilm} by inverse DFT and overlap–add synthesis. The method of upmixing by spectrogram factorization has also been considered for the task of informed source separation (ISS), where the goal is to reconstruct individual source signals from a downmix while observing the source signals at the parameter estimation stage (Liutkus *et al.*, 2012; Ozerov *et al.*, 2012).

Taking into Account the Downmix Procedure

Minimizing a cost criterion between the observed multichannel signal and the NTF model $c(x_{ilm}, \hat{x}_{ilm})$ is not guaranteed to minimize the same cost criterion between the observation and the upmixed signal $c(x_{ilm}, y_{ilm})$. This is due to the downmix d_{il} being used to reconstruct y_{ilm}, and overlapping spectral content from different channels makes d_{il} less sparse in the time–frequency domain than the original separate channels. The spectral content of the individual channels being only closely separated in time and frequency causes crosstalk in upmixed channels if the NTF model is estimated based only on the observed spectrogram.

The approximation accuracy of the NTF model in representing the multichannel input signal spectrogram determines the time and frequency selectivity of the upmixing process. Errors in representing the input signal cause upmixing to assign false signal content to the upmixed channels. In order to minimize the upmixing cost criterion $c(x_{ilm}, y_{ilm})$, a modification to the parameter estimation was proposed by Nikunen *et al.* (2011). The accuracy of the model needs to be increased at time–frequency points where the downmix contains overlapping spectral content. This can be done by giving higher error weight (see the weighted SED in Blondel *et al.*, 2007) to time–frequency points where the downmix STFT has a high magnitude with respect to the NTF channel sum in the denominator of the Wiener mask in Equation (9.32). The experiments in Nikunen *et al.* (2011) indicated that the proposed weighting decreases the upmixed signal error.

Results of the Downmix–Upmix Operation using the NTF Model

In the upmixing example illustrated in Figure 9.5, the first two channels contain structured spectral content (piano), whereas the third channel contains a wideband noise-like signal (water fountain). The downmix magnitude spectrogram contains all of this spectral content overlapped. The upmixing masks A_{ilm} denote gains in the range [0, 1], and the overall mask shape resembles the original multichannel data, but with the exception of additionally being able to handle rejection of spectrally overlapping content. This is especially evident in the upmixing mask of the third channel, which contains the spectral shape of the piano with reversed gain, that is, rejecting it. In the last column the upmixed magnitude spectrograms before time-domain reconstruction can be seen. There are only faint remarks from the upmix processing, namely some noise of the third channel added to the first two and vice versa. This equates to some crosstalk between the channels, which in general have little or no perceptually degrading effect due to the masking effect caused by the correctly upmixed content.

9.4.2 Source Separation Using a Compact Microphone Array

In this section we present separation evaluation results of complex-valued NMF from Nikunen and Virtanen (2014a) and Pertilä and Nikunen (2015). The separation quality is contrasted against independent component analysis (ICA) and other complex-valued NMF formulations in Nikunen and Virtanen (2014a) and Pertilä and Nikunen (2015), includes reference methods based on conventional beamforming and neural network mask estimation using spatial features.

The evaluation of source separation quality consists of energy-based objective metrics proposed by Vincent *et al.* (2006): signal-to-distortion ratio (SDR), signal-to-interference ratio (SIR), and signal-to-artefact ratio (SAR). The SDR measures how much the separated signal resembles the original signal from the energy perspective, SIR measures the interference between separated sources, and SAR denotes how many artefacts are added in the separation process. Separation evaluation in Pertilä and Nikunen

(2015) also includes the short-time objective intelligibility measure (STOI; Taal *et al.*, 2011) and frequency-weighted segmental signal-to-noise ratio (fwSNR; Hu and Loizou, 2008).

Initialization of Direction Weights as a Preprocessing Step

The complex-valued NMF with DOA-based SCM model in Equation (9.28) allows initialization of the direction weights z_{qo} towards directions where sources are assumed to be located. This can be done by estimating the DOAs of the sources before optimizing the NMF separation model parameters. Calculating steered response power (SRP; Tashev, 2009), averaging it over the whole signal, and searching for its maxima produces estimates of the DOAs, that is, peaks in spatial energy.

The estimated DOAs denoted by azimuth angles $\varphi_q \in [0°, 360°]$ are used to set the weights of the look direction indices o of z_{qo} in the vicinity of φ_q to one, and all the other direction weights of the source are set to zero. We use a spatial window of $\pm 25°$, and the chosen spatial window width acts as a tradeoff between the spatial selectivity and assumed error in the preliminary DOA estimation. An example of the initialized direction weights is illustrated in Figure 9.7, along with final estimated direction weights z_{qo} for three sources with approximately a 90° angle between them.

Dealing with Reverberation

Signal reconstruction within the complex-valued NMF is based on generalized Wiener filtering, Equation (9.23), which divides the entire mixture signal energy to the separated sources and thus does not perform any dereverberation initially. The separation method based on complex-valued NMF and the DOA-based SCM model can be equipped with dereverberation properties with the following addition. The look directions o that are not assigned to any source in the preliminary DOA analysis stage can be used as initial directions for a background/ambient source. The initialization of ambient source spatial weights and an example result are illustrated in Figure 9.8. The complex-valued

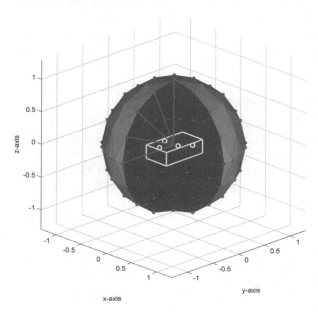

Figure 9.6 Illustration of the microphone array inside a sphere which is defined by the look directions $\mathbf{k}_o$; a few such vectors originating from [0,0,0] are illustrated at the edge of the cut-out area.

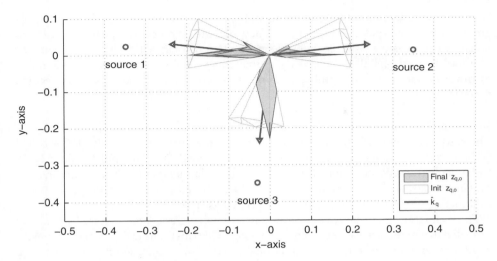

Figure 9.7 Initialization of spatial weights based on DOA analysis (contour plot) and final estimated spatial weights z_{qo} (gray surface) depicted as looking from above the array. The annotated source positions are illustrated as circles.

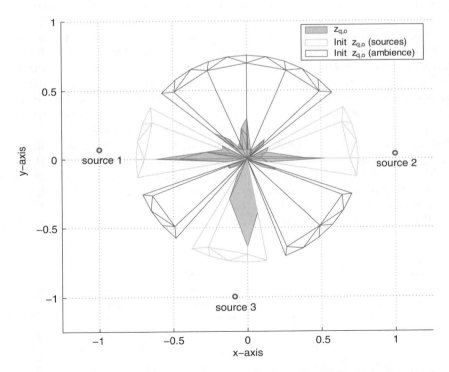

Figure 9.8 Initialization of spatial weights based on DOA analysis (light contour) and background source initial weights (darker contour). The final estimated direction weights are illustrated as a gray surface.

NMF model of Equation (9.28) with the ambient source results to building a spectrogram model of the spatial energy emitted between the target sources, which consists of noise and reverberated source components. Source reconstruction by time-frequency filtering then assigns all the correctly estimated spectral components corresponding to the reverberation to the ambient source and actual sound source signals become dereverberated.

Separation Quality Evaluation

The separation methods to be evaluated are: complex-valued NMF with DOA-based SCM model [as in Equation (9.28)], abbreviated as "CNMF DOA"; complex-valued NMF with unconstrained estimation of spatial covariances [as in Equation (9.13)], abbreviated as "CNMF baseline"; and finally conventional ICA-based separation with permutation alignment from Sawada *et al.* (2007), abbreviated as "ICA."

The evaluation material was generated by convolving anechoic material with RIRs recorded in a moderately reverberant room (T_{60} = 350 ms). The array placed at a height of 1.08 m consisted of four omnidirectional microphones enclosed in a metal casing (30 mm × 60 mm × 115 mm). A loudspeaker at a height of 1.40 m was placed at a distance of 1.50 m from the array at angles with spacing of 45°. The geometry of the array used in the recording is illustrated in Figure 9.6, and the exact locations of the microphones were [0.0, −45.8, 6.0; −22.0, −7.7, 6.0; 22, −7.7, 6.0; 0.0, 61.2, −18.0] mm. Mixtures of 10 s duration consisting of two and three simultaneous sound sources were generated. The source types included male and female speech, pop music, and various everyday noise sources. The algorithm processing details were a sampling frequency of F_s = 24 kHz and a window length of N = 2048 with 50% overlap. The number of NMF components was set to K = 60, and the algorithms were run for 500 iterations. The true number of sources was given to the methods. In total, 110 look directions sampling the spatial space around the array were used for the spatial covariance model in Equation (9.27). The azimuthal resolution at zero elevation was 10°, and the resolution was decreased close to the poles of the unit sphere, as can be seen in Figure 9.3(b).

The separation quality scores averaged over the test set of two and three simultaneous sources are given in Table 9.1. The scores indicate an increase in all separation metrics for the DOA-based SCM model over the unconstrained covariance matrix estimation with both two and three simultaneous sources. Additionally, the "CNMF DOA" method significantly improves over the conventional ICA-based separation. In Nikunen and Virtanen (2014b) the DOA-based covariance model is shown to increase its performance

Table 9.1 Separation quality evaluation with two and three simultaneous sources.

Method	CNMF DOA		CNMF baseline		ICA	
# of sources	2	3	2	3	2	3
SDR (dB)	**5.6**	**3.0**	3.7	2.0	2.0	0.5
SIR (dB)	**6.8**	**2.6**	4.5	0.4	4.5	1.3
SAR (dB)	**13.1**	**10.7**	12.7	9.9	8.2	5.6

as the angle between the two sound sources increases, while the performance of the unconstrained SCM estimation does not benefit from increasing the angle. The DOA-based approach outperforms unconstrained SCM estimation at all tested angles, with the smallest angle between the sources being 15°.

Comparison to Beamforming and Neural Network Based Separation

In many real-world applications the number of microphones on a single device is limited. The spatial covariance model based on complex-valued NMF with DOA was developed to work well with a low number of microphones, and in this section we present results comparing conventional beamforming with eight microphones while the NMF-based method uses only four in the same setting. Additionally, mask estimation based on a neural network (NN) utilizing spatial features is used as a contrast method. The computational complexity of estimating the parameters of the complex-valued NMF models in Equations (9.13) and (9.28) increases quadratically with respect to the number of microphones, and processing with all eight microphones was not considered computationally feasible.

The evaluation material was recorded using eight microphones in a circular arrangement with radius of 10 cm. Two rooms with low and moderate reverberation ($T_{60} = 260$ ms and 370 ms) were used to capture isolated speech at distances of 1.1 m (near) and 1.7 m (far). Mixtures of two and three speakers with 90° spacing were generated from the isolated speech recordings. The complex-valued NMF with DOA SCM model is denoted in the results as "CNMF," whereas the results with the ambient source extension (see "Dealing with reverberation" above) are abbreviated as "CNMF+A." The contrast methods include delay-and-sum beamforming (DSB), minimum variance distortionless (MVDR) beamformer, and separation using an NN with spatial features as proposed by Pertilä and Nikunen (2015), abbreviated as "spatial-NN." The separation metrics include the previously mentioned SDR and SIR, with the addition of STOI and fwSNR. A more detailed description of the recording conditions and contrast method implementation can be found in Pertilä and Nikunen (2015).

The average separation scores for the room with low reverberation are given in Table 9.2, and the results can be summarized as follows. The separation by complex-valued NMF achieves better performance than conventional beamforming with eight microphones. Additionally, when measured by SDR and SIR, the complex-valued NMF outperforms spatial-NN separation; however, the spatial-NN method achieves higher intelligibility and fwSNR, especially when the sources are farther away from the array. Both CNMF methods have significantly lower fwSNR performance, arguably due to them not being able to learn the difference in spectral characteristics of the far-field array signal and the close-field lavalier microphone which was used as a reference. Additionally, the different sensitivity of far and close-field microphones for capturing noise at extremely low frequencies, and it not being sufficiently rejected by the separation method, may lead to poor SNR performance. The modeling of the ambience and reverberation with the complex-valued NMF (CNMF+A) increases the intelligibility measure (STOI), and the SDR performance is also increased with three simultaneous sources. The method-wise results of the high reverberation case, reported in Pertilä and Nikunen (2015), are similar to those presented in Table 9.2, with a small relative improvement for the spatial-NN method.

Table 9.2 Separation results in low reverberation ($T_{60} = 260$ ms).

Method	DSB (8ch)		MVDR (8ch)		spatial-NN (8ch)		CNMF (4ch)		CNMF+A (4ch)	
	near	far	near	far	near	far	near	far	near	far
Two sources										
SDR (dB)	1.2	0.5	6.3	**5.2**	5.8	4.0	7.4	4.6	**7.5**	4.5
SIR (dB)	2.0	1.9	12.9	10.6	12.9	**11.5**	**13.6**	10.4	11.4	8.8
STOI (dB)	0.75	0.68	0.76	0.69	**0.82**	**0.78**	0.79	0.70	0.81	0.70
fwSNR (dB)	−0.7	0.0	3.5	2.4	**5.8**	**5.5**	−1.0	−1.3	−1.9	−1.8
Three sources										
SDR (dB)	1.4	2.2	5.9	2.6	3.7	1.7	6.0	2.9	**6.4**	**3.3**
SIR (dB)	0.9	1.1	9.8	5.8	9.8	7.5	**11.4**	**8.0**	9.7	6.7
STOI (dB)	0.65	0.63	0.69	0.65	**0.77**	**0.75**	0.71	0.66	0.76	0.69
fwSNR (dB)	−1.2	−0.6	1.3	0.7	**5.2**	**4.5**	−0.2	−1.2	−1.5	−2.0

9.4.3 Reconstruction of Binaural Sound Through Source Separation

Natural human perception of sounds is binaural, and sound scenes are perceived through direction- and location-dependent characteristics of sounds, that is, binaural cues caused by the anatomy of ears and head (Musicant and Butler, 1984; Faller and Merimaa, 2004). Binaural hearing is an important element in understanding speech in difficult conditions (Bronkhorst, 2000). However, binaural cues are missing in audio signals recorded by the stereo microphones found in conventional hand-held devices, and in order to provide the sensation of spatial sound for any recorded audio material, a spatial reconstruction of the sound scene is required. In this part of the chapter we present a method for synthesis and reconstruction of binaural sound through sound source separation using the complex-valued NMF and its direction of arrival estimation properties when using the spatial covariance model presented in Section 9.3.

Assisted Listening Scenario and System Overview

Lately, the term "assisted listening" has been associated with improved sound perception for persons with normal hearing in environments that are difficult for communication (Valimaki *et al.*, 2015). The goal of assisted listening is to provide a natural-sounding spatial reproduction from a multichannel audio signal recorded using a hand-held or wearable device, and to provide users with the possibility to enhance or concentrate on the desired content.

The general scenario in assisted listening by reconstruction of binaural sound through source separation is illustrated in Figure 9.9. The separation algorithm is used to obtain isolated source signals, and their DOAs need to be estimated for the binaural synthesis. Filtering anechoic source signals with a head-related transfer function (HRTF) can be used to generate the sensation of spatial sound through headphones. Head-related transfer functions consist of direction- and frequency-dependent level and time

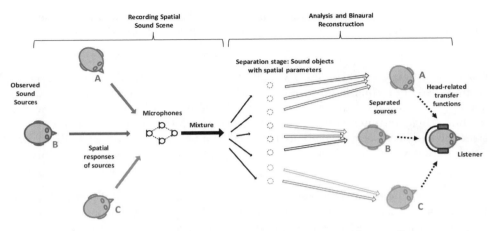

Figure 9.9   Assisted listening scenario using a compact microphone array for recording.

differences between the ears. Measured databases of HRTFs from various subjects exist (Algazi *et al.*, 2001; Warusfel, 2003).

Recent studies have considered assisted listening using blind source separation (Doclo *et al.*, 2015; Sunder *et al.*, 2015) and parametric spatial sound scene models (Kowalczyk *et al.*, 2015) accompanied with spatial synthesis. Additionally, the perceived quality of a modified sound scene is of great interest. The modification and reconstruction of spatial sound through source separation has one very favorable characteristic: the audibility of separation artifacts is greatly reduced when sources with added spatial cues are presented simultaneously to the listener (spatial sound reconstruction), in comparison to listening to individual separated sources. This is due to the simultaneous masking phenomenon of our hearing, which limits the audibility of wrongly assigned spectral elements from interfering sources when the interfering source is also present in the mixture but at a different spatial location.

Extracting a Single Direction of Arrival for Binaural Synthesis

In order to analyze a single DOA for the sound sources separated by the complex-valued NMF, the following procedure was originally proposed by Nikunen *et al.* (2015). The direction weights z_{qo} estimated for the spatial part of the spectrogram model in Equation (9.28) can be interpreted as probabilities of the source q originating from each direction o. Weighting the look direction vectors $\mathbf{k}_o = [x_o, y_o, z_o]$ by the spatial weights z_{qo} and summing over all directions gives the average direction vector

$$\hat{\mathbf{k}}_q = \sum_{o=1}^{O} \mathbf{k}_o z_{qo}, \tag{9.33}$$

which is considered as the overall DOA estimate of each source q. By simple Euclidean geometry the vector can be transformed into azimuth and elevation, denoted by (φ_q, θ_q). The low azimuthal resolution of the model, 10° at the lateral plane, is effectively increased by the weighted sum, and therefore the final DOA estimate might be located in between the discretized look directions. An example of the direction weights z_{qo} after estimation of the spectrogram model parameters is shown in Figure 9.7, and the final source direction, retrieved using Equation (9.33), is illustrated as a red arrow.

Binaural Synthesis

Given the estimated source signals $\mathbf{y}_{ilq}$ and their DOAs (φ_q, θ_q), the binaural synthesis is carried out by filtering the source signals using the left and right HRTFs $h_{\text{left},q}(t)$ and $h_{\text{right},q}(t)$ corresponding to the estimated direction (φ_q, θ_q) of each source. In order to obtain single-channel estimates of the sources, an arbitrary channel from the reconstruction can be chosen, or alternatively a DSB can be applied to the separation algorithm output $\mathbf{y}_{ilq}$ obtained through Equation (9.23). The DSB can be described as $y_{il} = \mathbf{w}_{iq}^{\text{H}} \mathbf{y}_{ilq}$, where the DSB weights are obtained as $\mathbf{w}_{iq}^{\text{H}} = [e^{-j\omega_i \tau_{11}(\hat{\mathbf{k}}_q)} \ \dots \ e^{-j\omega_i \tau_{1m}(\hat{\mathbf{k}}_q)}]$, which correspond to enhancing signals coming from the estimated direction $\hat{\mathbf{k}}_q$. The time-domain signals after applying DSB are denoted as $\tilde{y}_q(t)$. The binaural synthesis for the left and right channels, intended for headphone playback, are obtained as

$$b_{\text{left}}(t) = \sum_{q=1}^{Q} \tilde{y}_q(t) * h_{\text{left},q}(t), \tag{9.34}$$

$$b_{\text{right}}(t) = \sum_{q=1}^{Q} \tilde{y}_q(t) * h_{\text{right},q}(t), \tag{9.35}$$

where $*$ denotes convolution.

In practice, the separated source signals are not anechoic and contain some amount of environmental reverberation. Most of the reverberated parts of the sources are reflections that have different DOAs, or they are diffuse. Thus, we assume that only a small amount of reverberation remains after separation and beamforming, and fairly accurate binaural reconstruction can be achieved by positioning the entire reverberated source at the estimated direction $\hat{\mathbf{k}}_q$.

Speech Intelligibility with Spatial Reconstruction

To evaluate the effect of the binaural reconstruction of separated sources, a speech intelligibility test was conducted; its detailed description can be found in Nikunen *et al.* (2015). Sentences spoken by three simultaneous speakers were recorded using the same compact microphone array mentioned in Section 9.4.2. A true binaural reference was recorded using a Brüel & Kjær head and torso simulator (HATS). Both captures were recorded simultaneously by placing the microphone array on top of the HATS. Three speakers positioned themselves around the HATS (and the array) at approximately 1 m distance with a 90° angle between them. Six different source positions were used, including cases where sources were located at the same time on both sides, front and back, and at two indirect directions (that is, intercardinal directions).

Separation using the complex-valued NMF with the above-described binaural synthesis was considered as the system under test (CNMF+HRTF). Other methods tested were omnidirectional stereo recording (Stereo) using two channels from the compact microphone array (distance 4.4 cm, which corresponds to the size of a conventional hand-held device).[1] The reference separation method was MVDR beamforming (MVDR+HRTF)

---

[1] It should be noted that directional stereo microphone arrangements with a spacing corresponding to the average distance between the ears would produce a better stereo recording from a binaural perspective.

with the same estimated DOAs as used for complex-valued NMF initialization. Also, separated source signals (Separated) were included in the intelligibility evaluation, and the HATS recordings acted as an upper reference (True binaural).

The textual data of the test consisted of Finnish sentences from Vainio *et al.* (2005), where a carefully prepared list of phonetically balanced sentences is proposed. In the listening test, the subjects were instructed to listen to the speech content of the target speakers, and the target was indicated by an initialization sentence, "Listen to my voice," spoken by the target person from the same position as the test sentence. The subjects were instructed to write down what they heard, and the word recognition accuracy was calculated from the answers. Due to the restricted number of unique sentences available from Vainio *et al.* (2005), and the need to have a representative subset of test samples from each speaker position for each tested method to be listened to by the same listening-test subject, the five methods were tested in two different listening tests with different listeners. Separation by complex-valued NMF acted as an anchor between the two groups. All of the test sentences were unique, and only listened to once by the test subjects to prevent learning and remembering the lingual content.

The mean word recognition accuracies over all the test sentences and all listening test subjects are presented in Figure 9.10. In the second listening test group the CNMF+HRTF method with the exact same sentence mixtures and target speakers achieved 6.7 percentage points better recognition accuracy, indicating that the second listening test group were more concentrated on the task or simply were more experienced at recognizing overlapping speech. Statistical significance between all methods in both tests was achieved according to ANOVA applied to the method-wise averages from each test subject.

In summary, the results indicated that source separation and simple binaural synthesis by HRTFs improved the intelligibility of speech over omnidirectional stereo, but did not achieve the performance of the true binaural recording done using an artificial head. Additionally, the proposed separation algorithm based on spectrogram factorization was evaluated to be superior to MVDR beamforming for spatial sound reconstruction through source separation. Arguably, the high separation in terms of source crosstalk rejection is favorable for the developed spatial synthesis when considering the fact that separation artifacts become masked in spatial reconstruction where all signal content is presented to the listener simultaneously. Details of the statistical analysis and more in-depth analysis of the results, for example position-wise recognition accuracies, can be found in Nikunen *et al.* (2015).

Perceptual Quality of Proposed Spatial Reconstruction

To evaluate the overall perceptual quality and perceived spaciousness of the different methods for producing binaural signals, a subjective listening test was performed with the second listening test group. The participants listened to 60 mixture sentences (12 from each method). The mean opinion score (MOS) was used to evaluate the subjective annoyance of artifacts and the spaciousness of the signal. A description of each grade (1–5) for both evaluated quantities can be seen from the x-axis of the result graphs (Figure 9.11).

The results (mean scores with 95% confidence intervals) of the subjective listening test are illustrated in Figure 9.11. In summary, the true binaural recording achieves the best subjective quality in both tested quantities by a fair margin. The binaural synthesis does not improve the perceived spaciousness over the stereo recording, while it achieved

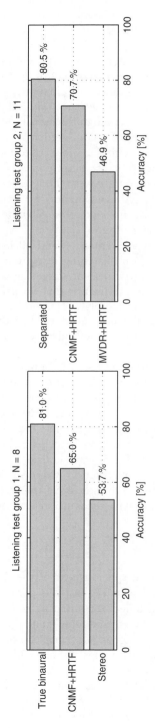

Figure 9.10 Word recognition accuracies for each tested method.

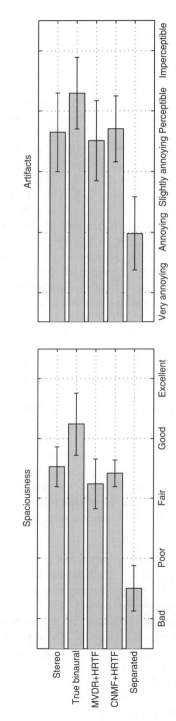

Figure 9.11 Results of the subjective listening test.

significantly higher intelligibility. Presumably, the lack of natural reverberation in the binaurally synthesized signals may decrease its spaciousness score. The overall perceptual quality (artifacts/unnaturalness) is not degraded by the separation and binaural processing, which is due to the separation artefacts becoming masked by the presence of other sources. While the separated signals achieved very high intelligibility score, the subjective quality of the separated signals in both quantities is significantly less preferred. In general, the listeners had a high dispersion in grading of the artifacts and the overall perceptual quality, leading to large confidence intervals for all the tested methods.

9.5 Discussion

This chapter has considered the separation and modification of spatial audio using spectrogram factorization combined with spatial analysis abilities. The applications included source separation using a compact array with few microphones, upmixing of surround sound, and synthesizing binaural audio by positioning separated sources at their estimated direction. The separation performance of conventional beamforming even with a higher number of microphones was exceeded with the methods based on spectrogram factorization. Additionally, comparison with the state-of-the-art separation methods with various objective separation quality criteria indicated better or comparable performance.

The strengths of the complex-valued NMF models for source separation include the capability to operate with a small number of microphones, the ability to analyze spatial parameters over multiple frequency bins, and, most importantly, good separation performance. The shortcomings include high computational complexity in the estimation of spatial covariance parameters, that is, updating the DOA kernels and recalculation of the spatial covariance model after each update of the spatial weights. The effect of updating channel-wise level differences of the DOA kernels when using compact arrays has only a small effect in the separation performance. Additionally, a good initial DOA estimate for the sources would allow initialization of the spatial weights directly to a close vicinity of the true direction, and requires no updating of the spatial weights. These modifications would greatly decrease the computational complexity of the model, reducing to only estimation of the spectrogram model parameters.

The synthesis procedure used in the reconstruction of binaural sound in this chapter is rather simplified, and better perceptual quality and speech intelligibility can be expected by incorporating individualized HRTFs, modeling reverberation, and taking into account the source distance in the synthesis procedure. The recent increased interest in assisted listening and the expected revolution in augmented reality applications with realistic audio makes the methods considered very timely and relevant for future research. The obvious question of moving sound sources is left open here. The presented complex-valued NMF with DOA-based covariance matrix model is extensible to time-variant mixing, and thus modeling the movement of sound sources. The association of source parameters in adjacent frames also needs to be solved in such a case.

9.6 Matlab Example

Listing 9.1 shows example Matlab code for using the NTF model in Equation (9.4) and the algorithm shown in Algorithm 9.1 to learn the spectral basis, activations, and panning of five notes in a pentatonic scale based on a random sequence of 100 notes with a 3% tuning error.

Listing 9.1: Non-Negative Tensor Factorization Example

```
1   function NTFexample()
2   %% Generate signal
3   % Center frequencies for L=5 notes in pentatonic scale (A=440 Hz)
4   L = 5; note_freq = 440*(2^(1/12)).^([0 3 5 7 10]);
5   note_pan = rand(1,L); % Random left/right panning [0 1] for notes
6
7   % Generate random sequence of 100 notes
8   N = 100; seq = randi(L,1,N); % N random integers between 1..L
9   seq_freq = note_freq(seq); % Note frequencies corresponding to seq
10  seq_pan = note_pan(seq); % Panning of notes corresponding to seq
11
12  % Random +-3% tuning error to simulate realistic conditions
13  seq_freq = seq_freq.*1+(rand(size(seq_freq))*0.06)-0.03;
14  % Random amplitude [0.3 1.0] for each note
15  seq_ampl = rand(size(seq_freq))*0.7+0.3;
16
17  % Generate 200ms sawtooth wave for each note
18  Fs = 32000; % Sampling frequency
19  Ts = 0.2*Fs; % Length of 200ms note in samples
20  x = zeros(N*Ts,2); % Empty array for 2ch signal with N notes
21  t = 0:1:Ts-1; % Time vector for one note
22  Ldecay = linspace(1,0,Ts); % Linear decay of note amplitude
23
24  % Loop for generating sawtooth waves by additive synthesis
25  for i = 1:N
26      f0 = seq_freq(i); note = zeros(size(t));
27      for k = 1:floor((Fs/2)/f0);
28          note = note + (-1)^k*sin(2*pi*k*f0*(t/Fs))/k.*Ldecay;
29      end
30      note = (2/pi)*note*seq_ampl(i); % apply amplitude
31      x(t+Ts*(i-1)+1,1) = note.*seq_pan(i); % panning for left
32      x(t+Ts*(i-1)+1,2) = note.*(1-seq_pan(i)); % panning for right
33  end
34
35  Nplot = 1500; Tplot = linspace(0,(Nplot/Fs)*1e3,Nplot);
36  figure(1); plot(Tplot,x(1:Nplot,1)); hold on; grid on;
37  plot(Tplot,x(1:Nplot,2),'r'); legend('Left ch','Right ch');
38  axis([0 Tplot(end) -1 1]); xlabel('Time [ms]'); ylabel('Amplitude');
39
40  %% Spectrogram: window length = 1024, 50% overlap, fft length = 1024;
41  X1 = spectrogram(x(:,1),hanning(1024),512,1024);
42  X2 = spectrogram(x(:,2),hanning(1024),512,1024);
43  X = cat(3,abs(X1),abs(X2)); clear X1 X2
44  F = size(X,1); T = size(X,2); lim = [5 40];
45  Fvec = linspace(0,Fs/2,F); Tvec = linspace(0,N*Ts/Fs,T);
46
```

```
47  figure(2); subplot(2,2,1); imagesc(Tvec,Fvec,20*log10(X(:,:,1)),lim);
48  set(gca,'Ydir','normal'); title('|X| Left ch');
49  xlabel('time [s]'); ylabel('Frequency [Hz]');
50  subplot(2,2,2); imagesc(Tvec,Fvec,20*log10(X(:,:,2)),lim);
51  set(gca,'Ydir','normal'); title('|X| Right ch');
52  xlabel('time [s]'); ylabel('Frequency [Hz]');
53
54  %% NTF factorization
55  K = 5; % Learning 5 components
56  C = size(X,3); % Number of channels in input
57  max_iter = 150; % Maximum number of iterations
58
59  B = 1 + rand(F,K); % Spectral basis
60  G = 1 + rand(K,T); % Time activations
61  A = 1 + rand(K,C); % Channel gain
62  mX = calcModel(B,G,A); % Calculate model mX:
63
64  iter = 1; fprintf('NTF parameter estimation \n');
65  while iter < max_iter;
66      % Update B
67      GA = permute(bsxfun(@times,G,permute(A,[1 3 2])),[4 1 2 3]);
68      num = bsxfun(@times,GA,permute(X,[1 4 2 3]));
69      num = sum(sum(num,4),3);
70      denom = bsxfun(@times,GA,permute(mX,[1 4 2 3]));
71      denom = sum(sum(denom,4),3);
72      B = B.* (num./(denom + 1e-12));
73
74      figure(3); imagesc(1:K,linspace(0,Fs/2,F),20*log10(B),[-40 0]);
75      set(gca,'Xtick',1:K); set(gca,'Ydir','normal');
76      title(['Spectral basis at iteration:' num2str(iter)]);
77      ylabel('Frequency [Hz]'); xlabel('NTF component #'); drawnow;
78
79      % Apply numerical range scaling
80      norm_B = sqrt(sum(B.^2,1));
81      B = B./repmat(norm_B,size(B,1),1);
82      G = G.*repmat(norm_B',1,size(G,2));
83
84      mX = calcModel(B,G,A); % Calculate model mX:
85
86      % Update G
87      BA = permute(bsxfun(@times,B,permute(A,[3 1 2])),[1 2 4 3]);
88      num = bsxfun(@times,BA,permute(X,[1 4 2 3]));
89      num = squeeze(sum(sum(num,4),1));
90      denom = bsxfun(@times,BA,permute(mX,[1 4 2 3]));
91      denom = squeeze(sum(sum(denom,4),1));
92      G = G.* (num./(denom + 1e-12));
93
94      mX = calcModel(B,G,A); % Calculate model mX:
95
96      % Update A
97      BG = permute(bsxfun(@times,B,permute(G,[3 1 2])),[1 2 3 4]);
```

```
98         num = bsxfun(@times,BG,permute(X,[1 4 2 3]));
99         num = squeeze(sum(sum(num,3),1));
100        denom = bsxfun(@times,BG,permute(mX,[1 4 2 3]));
101        denom = squeeze(sum(sum(denom,3),1));
102        A = A.* (num./(denom + 1e-12));
103
104        % Gain scaling
105        A = A./repmat(sum(A,2),1,size(A,2));
106
107        mX = calcModel(B,G,A); % Calculate model mX:
108
109        cost = sum(sum(sum(abs(X-mX).^2)));
110        fprintf('Cost criterion at iteration %i : %.2f \n',iter,cost);
111        iter = iter + 1;
112    end
113    fprintf('Done -> plotting results \n\n');
114
115    %% Plotting results
116    figure(2); subplot(2,2,3);imagesc(Tvec,Fvec,20*log10(mX(:,:,1)),lim);
117    set(gca,'Ydir','normal'); title('NTF model of X Left ch');
118    xlabel('time [s]'); ylabel('Frequency [Hz]');
119    subplot(2,2,4); imagesc(Tvec,Fvec,20*log10(mX(:,:,2)),lim);
120    set(gca,'Ydir','normal'); title('NTF model of X Right ch');
121    xlabel('time [s]'); ylabel('Frequency [Hz]');
122
123    % Finding NTF component f0 by assuming it has largest amplitude and
124    % sorting to increasing order (DFT has only 1/(F-1)*Fs/2 resolution)
125    [~,max_i] = max(B,[],1); [~,sort_i] = sort(max_i);
126    B = B(:,sort_i); G = G(sort_i,:); A = A(sort_i,:);
127    fprintf('Original note f0: \n');
128    fprintf('f0: %3.0f Hz \n',note_freq);
129    fprintf('Learned note f0: \n');
130    fprintf('f0: %3.0f Hz \n',Fvec(max_i(sort_i)));
131
132    fprintf('\nOriginal note panning: \n');
133    fprintf('Note #%i: %0.4f \n',[1:L; note_pan]);
134    fprintf('Learned note panning: \n');
135    fprintf('Note #%i: %0.4f \n',[1:K; A(:,1)']);
136
137    figure(4); imagesc(1:K,linspace(0,Fs/2,F),20*log10(B),[-40 0]);
138    set(gca,'Xtick',1:K); set(gca,'Ydir','normal');
139    xlabel('NTF component #'); ylabel('Frequency [Hz]');
140    title('Learned NTF basis functions');
141
142    figure(5); plot(repmat(1:100,K,1)',G(:,1:100)');
143    xlabel('STFT frame'); ylabel('Activation amplitude'); grid on;
144    title({'Learned component activations'; ...
145        ['Original sequence ' num2str(seq(1:16))]});
146    legend('comp #1','comp #2','comp #3','comp #4','comp #5');
147    end
148
```

```
149  function mX = calcModel(B,G,A)
150  % Calculate model mX:
151  F = size(B,1); T = size(G,2); C = size(A,2); K = size(B,2);
152  GA = bsxfun(@times,G,permute(A,[1 3 2]));
153  mX = reshape(B*reshape(GA,K,[]),[F T C]);
154  end
```

References

Algazi, V., Duda, R., Thompson, D., and Avendano, C. (2001) The CIPIC HRTF database. *IEEE Workshop on the Applications of Signal Processing to Audio and Acoustics (WASPAA)*, pp. 99–102.

Arberet, S., Ozerov, A., Duong, N.Q., Vincent, E., Gribonval, R., Bimbot, F., and Vandergheynst, P. (2010) Nonnegative matrix factorization and spatial covariance model for under-determined reverberant audio source separation. *Proceedings of the International Conference on Information Sciences, Signal Processing, and their Applications (ISSPA)*, pp. 1–4.

Benetos, E., Klapuri, A., and Dixon, S. (2012) Score-informed transcription for automatic piano tutoring. *Proceedings of the 20th European Signal Processing Conference (EUSIPCO)*, pp. 2153–2157, Bucharest, Romania.

Blondel, V., Ho, N.D., and van Dooren, P. (2007) Weighted nonnegative matrix factorization and face feature extraction. *Image and Vision Computing* (submitted).

Bronkhorst, A.W. (2000) The cocktail party phenomenon: A review of research on speech intelligibility in multiple-talker conditions. *Acta Acustica united with Acustica*, **86**(1), 117–128.

Dempster, A.P., Laird, N.M., and Rubin, D.B. (1977) Maximum likelihood from incomplete data via the EM algorithm. *Journal of the Royal Statistical Society*, **39**(1), 1–38.

Doclo, S., Kellermann, W., Makino, S., and Nordholm, S.E. (2015) Multichannel signal enhancement algorithms for assisted listening devices: Exploiting spatial diversity using multiple microphones. *IEEE Signal Processing Magazine*, **32**(2), 18–30.

Faller, C. (2004) Parametric coding of spatial audio. *Proceedings of the 7th International Conference on Audio Effects (DAFx)*, Naples, Italy.

Faller, C. and Merimaa, J. (2004) Source localization in complex listening situations: Selection of binaural cues based on interaural coherence. *Journal of the Acoustical Society of America*, **116**(5), 3075–3089.

Févotte, C. and Ozerov, A. (2011) Notes on nonnegative tensor factorization of the spectrogram for audio source separation: Statistical insights and towards self-clustering of the spatial cues. *Proceedings of the 7th International Symposium on Exploring Music Contents*, pp. 102–115. Springer, New York.

FitzGerald, D., Cranitch, M., and Coyle, E. (2005) Non-negative tensor factorisation for sound source separation. *Proceedings of the Irish Signals and Systems Conference*, Dublin, Ireland.

FitzGerald, D., Cranitch, M., and Coyle, E. (2006) Sound source separation using shifted non-negative tensor factorisation. *Proceedings of the 31st IEEE International Conference on Acoustics, Speech and Signal Processing (ICASSP)*, Toulouse, France.

FitzGerald, D., Cranitch, M., and Coyle, E. (2008) Extended nonnegative tensor factorisation models for musical sound source separation. *Computational Intelligence and Neuroscience*, **2008**, 872425.

Helén, M. and Virtanen, T. (2005) Separation of drums from polyphonic music using non-negative matrix factorization and support vector machine. *Proceedings of the 13th European Signal Processing Conference (EUSIPCO)*.

Hu, Y. and Loizou, P.C. (2008) Evaluation of objective quality measures for speech enhancement. *IEEE Transactions on Audio, Speech, and Language Processing*, **16**(1), 229–238.

Hurmalainen, A., Gemmeke, J., and Virtanen, T. (2011) Non-negative matrix deconvolution in noise-robust speech recognition. *Proceedings of the 36th International Conference on Acoustics, Speech, and Signal Processing (ICASSP)*, pp. 4588–4591, Prague, Czech Republic.

Hurmalainen, A., Gemmeke, J., and Virtanen, T. (2013) Modelling non-stationary noise with spectral factorisation in automatic speech recognition. *Computer Speech & Language*, **27**(3), 763–779.

Kowalczyk, K., Thiergart, O., Taseska, M., et al. (2015) Parametric spatial sound processing: A flexible and efficient solution to sound scene acquisition, modification, and reproduction. *IEEE Signal Processing Magazine*, **32**(2), 31–42.

Lee, D.D. and Seung, H.S. (2002) Algorithms for non-negative matrix factorization. *Advances in Neural Information Processing Systems*, **13**, 556–562.

Liutkus, A., Pinel, J., Badeau, R., Girin, L., and Richard, G. (2012) Informed source separation through spectrogram coding and data embedding. *Signal Processing*, **92**(8), 1937–1949.

Mitsufuji, Y. and Roebel, A. (2014) On the use of a spatial cue as prior information for stereo sound source separation based on spatially weighted non-negative tensor factorization. *EURASIP Journal on Advances in Signal Processing*, **2014**(1), 1–9.

Musicant, A.D. and Butler, R.A. (1984) The influence of pinnae-based spectral cues on sound localization. *Journal of the Acoustical Society of America*, **75**(4), 1195–1200.

Nesta, F., Svaizer, P., and Omologo, M. (2011) Convolutive BSS of short mixtures by ICA recursively regularized across frequencies. *IEEE Transactions on Audio, Speech, and Language Processing*, **19**(3), 624–639.

Nikunen, J., Diment, A., Virtanen, T., and Vilermo, M. (2015) Binaural rendering of microphone array captures based on source separation. *Speech Communication*, submitted.

Nikunen, J. and Virtanen, T. (2014a) Multichannel audio separation by direction of arrival based spatial covariance model and non-negative matrix factorization. *Proceedings of the 39th International Conference on Acoustic, Speech, and Signal Processing (ICASSP)*, pp. 6727–6731, Florence, Italy.

Nikunen, J. and Virtanen, T. (2014b) Direction of arrival based spatial covariance model for blind sound source separation. *IEEE/ACM Transactions on Audio, Speech, and Language Processing*, **22**(3), 727–739.

Nikunen, J., Virtanen, T., and Vilermo, M. (2011) Multichannel audio upmixing based on non-negative tensor factorization representation. *Proceedings of the IEEE Workshop on Applications of Signal Processing to Audio and Acoustics (WASPAA)*, pp. 33–36.

Nikunen, J., Virtanen, T., and Vilermo, M. (2012) Multichannel audio upmixing by time–frequency filtering using non-negative tensor factorization. *Journal of the Audio Engineering Society*, **60**(10), 794–806.

Ozerov, A. and Févotte, C. (2010) Multichannel nonnegative matrix factorization in convolutive mixtures for audio source separation. *IEEE Transactions on Audio, Speech, and Language Processing*, **18**(3), 550–563.

Ozerov, A., Févotte, C., Blouet, R., and Durrieu, J.L. (2011) Multichannel nonnegative tensor factorization with structured constraints for user-guided audio source separation. *Proceedings of the 36th International Conference on Acoustics, Speech, and Signal Processing (ICASSP)*, pp. 257–260, Prague, Czech Republic.

Ozerov, A., Liutkus, A., Badeau, R., and Richard, G. (2012) Coding-based informed source separation: Nonnegative tensor factorization approach. *IEEE Transactions on Audio, Speech and Language Processing*, **21**(8), 1699–1712.

Parry, R.M. and Essa, I.A. (2006) Estimating the spatial position of spectral components in audio. *Proceedings of International Conference on Independent Component Analysis and Blind Signal Separation*, pp. 666–673.

Pertilä, P. and Nikunen, J. (2015) Distant speech separation using predicted time–frequency masks from spatial features. *Speech Communication*, **68**, 97–106.

Raj, B., Virtanen, T., Chaudhuri, S., and Singh, R. (2010) Non-negative matrix factorization based compensation of music for automatic speech recognition. *Proceedings of the 11th Annual Conference of International Speech Communication Association (INTERSPEECH)*, pp. 717–720, Makuhari, Japan.

Sawada, H., Araki, S., Mukai, R., and Makino, S. (2007) Grouping separated frequency components by estimating propagation model parameters in frequency-domain blind source separation. *IEEE Transactions on Audio, Speech, and Language Processing*, **15**(5), 1592–1604.

Sawada, H., Kameoka, H., Araki, S., and Ueda, N. (2011) New formulations and efficient algorithms for multichannel NMF. *Proceedings of IEEE Workshop on Applications of Signal Processing to Audio and Acoustics (WASPAA)*, pp. 153–156, New Paltz, NY, USA.

Sawada, H., Kameoka, H., Araki, S., and Ueda, N. (2013) Multichannel extensions of non-negative matrix factorization with complex-valued data. *IEEE Transactions on Audio, Speech, and Language Processing*, **21**(5), 971–982.

Schuijers, E., Breebaart, J., Purnhagen, H., and Engdegaard, J. (2004) Low complexity parametric stereo coding. *Proceedings of the 116th Audio Engineering Society Convention*, Berlin, Germany.

Simmer, K.U., Bitzer, J., and Marro, C. (2001) Post-filtering techniques, in *Microphone Arrays* (eds. M. Brandstein et al.), pp. 39–60. Springer, Berlin.

Smaragdis, P. and Brown, J. (2003) Non-negative matrix factorization for polyphonic music transcription. *Proceedings of IEEE Workshop on Applications of Signal Processing to Audio and Acoustics (WASPAA)*, pp. 177–180.

Sunder, K., He, J., Tan, E.L., and Gan, W.S. (2015) Natural sound rendering for headphones: Integration of signal processing techniques. *IEEE Signal Processing Magazine*, **32**(2), 100–113.

Taal, C.H., Hendriks, R.C., Heusdens R., and Jensen J. (2011) An algorithm for intelligibility prediction of time–frequency weighted noisy speech. *IEEE Transactions on Audio, Speech, and Language Processing*, **19**(7), 2125–2136.

Tashev, I. (2009) *Sound Capture and Processing: Practical Approaches.* John Wiley & Sons Inc., Hoboken, NJ.

Vainio, M., Suni, A., Järveläinen, H., Järvikivi, J., and Mattila, V.V. (2005) Developing a speech intelligibility test based on measuring speech reception thresholds in noise for English and Finnish. *Journal of the Acoustical Society of America*, **118**(3), 1742–1750.

Valimaki, V., Franck, A., Ramo, J., Gamper, H., and Savioja, L. (2015) Assisted listening using a headset: Enhancing audio perception in real, augmented, and virtual environments. *IEEE Signal Processing Magazine*, **32**(2), 92–99.

Vincent, E., Gribonval, R., and Févotte, C. (2006) Performance measurement in blind audio source separation. *IEEE Transactions on Audio, Speech, and Language Processing*, **14**(4), 1462–1469.

Virtanen, T. (2007) Monoaural sound source separation by nonnegative matrix factorization with temporal continuity and sparseness criteria. *IEEE Transactions on Audio, Speech, and Language Processing*, **15**(3), 1066–1074.

Warusfel, O. (2003) LISTEN HRTF database, http://recherche.ircam.fr/equipes/salles/listen/index.html (accessed May 29, 2017).

Weninger, F., Wöllmer, M., Geiger, J., Schuller, B., Gemmeke, J.F., Hurmalainen, A., Virtanen, T., and Rigoll, G. (2012) Non-negative matrix factorization for highly noise-robust ASR: To enhance or to recognize? *IEEE International Conference on Acoustics, Speech, and Signal Processing (ICASSP)*, pp. 4681–4684. IEEE.

Part III

Signal-Dependent Spatial Filtering

10

Time–Frequency Domain Spatial Audio Enhancement

Symeon Delikaris-Manias[1] and Pasi Pertilä[2]

[1] *Department of Signal Processing and Acoustics, Aalto University, Finland*
[2] *Department of Signal Processing, Tampere University of Technology, Finland*

10.1 Introduction

Signal enhancement from noise and interference has been a research problem of interest for several decades. The two main approaches of research in enhancement are single channel and multiple channels, with the latter gaining more popularity due to the increase in computational power and the fact that most modern devices are equipped with multiple sensors. On the other hand, the field of monophonic speech enhancement has also seen increased research efforts in recent years due to the limited requirements on microphone sensors; approaches based on deep neural network (DNN) learning are the most common, for example in Narayanan and Wang (2013), Wang *et al.* (2013), Wang and Wang (2013), Weninger *et al.* (2014), Erdogan *et al.* (2015), and Williamson *et al.* (2016). While arguably the monophonic case is the most general and challenging scenario, the availability of multiple microphones broadens the number of options to pursue for enhancement.

Multi-microphone devices enable flexible recording of sound sources in the presence of interferers, noise, and reverberation. The most common enhancement techniques for microphone arrays are based on the design of directional filters or beamforming. Directional filtering with microphone arrays is a class of methods that allow focusing in specific directions inside a sound field. Such methods can be utilized for beamforming or, in the case of multiple outputs that can cover the sound field, sound field reproduction (Van Veen and Buckley, 1988; Vilkamo and Delikaris-Manias, 2015). An illustration of a beamforming application is shown in Figure 10.1. Multiple sound and noise sources are simultaneously active in a reverberant room, and the task is to focus on one specific direction while suppressing interferers, diffuse sound, and noise. Such algorithms are ideal candidates for general scenarios where high intelligibility and perceived quality are required (Brandstein and Silverman, 1997; Merimaa, 2002; Paleologu *et al.*, 2010; Ahonen and Pulkki, 2010). Practical applications for such techniques include teleconferencing (Cohen *et al.*, 2009) or recording simultaneous instruments for music production outside of a studio environment (Braasch, 2005).

Parametric Time–Frequency Domain Spatial Audio, First Edition. Edited by Ville Pulkki,
Symeon Delikaris-Manias, and Archontis Politis.
© 2018 John Wiley & Sons Ltd. Published 2018 by John Wiley & Sons Ltd.
Companion Website: www.wiley.com/go/pulkki/parametrictime-frequency

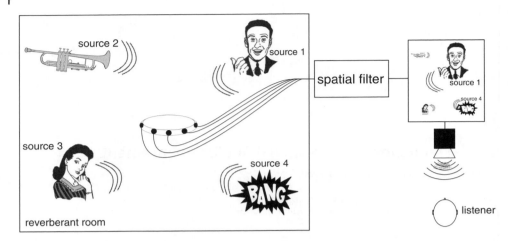

Figure 10.1 Typical scenario where multiple sound and noise sources are active in a reverberant room. The task is to enhance the direction of sound from a specific area.

There are numerous techniques for approaching the problem of signal enhancement. The selection of the appropriate enhancement technique depends mainly on the application. For example, in cases where sound quality is the priority, the enhancement algorithm can be chosen to be less aggressive in terms of filtering. In cases where signal retrieval is the priority, audible artifacts that are generated by the enhancement algorithm might be of less importance. This chapter provides a brief overview of the techniques, with a focus on post-filtering techniques that are relevant to the contents of the book.

10.2 Signal-Independent Enhancement

The most basic beamforming techniques are *signal independent* and do not assume anything about the nature of the signals or the environment. A basic block diagram illustrating these simple operations is shown in Figure 10.2. The beamformer output, y, is calculated by applying a set of user-defined weights $\mathbf{w}_{ds}$ to the microphone signals, $\mathbf{x} = [x_1, x_2, \ldots, x_q]$, with an optional time-alignment block (left), as in delay-and-sum beamforming, or a set of weights $\mathbf{w}_{ls}$ estimated from the array steering vectors and/or the array geometry (right), as in least-squares-based beamforming.

These techniques are based on performing simple operations between the microphone signals, such as summing or subtracting signals from spaced arrays (Eargle, 2004), or using conventional delay-and-sum and filter-and-sum beamformers (Brandstein and Silverman, 1997). The delay-and-sum beamformer algorithm estimates the time delays of signals received by each microphone of an array and compensates for the time difference of arrival (Bitzer and Simmer, 2001). By aligning and summing the microphone input signals, the directionality of the microphone array can be adjusted in order to create constructive interference for the desired propagating sound wave and destructive interference for sound waves originating from all other directions. Narrow directivity patterns can be obtained, but this usually requires a large spacing between

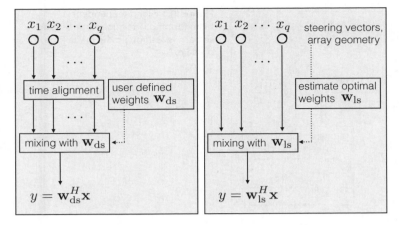

Figure 10.2 Signal-independent beamforming where microphone array input signals **x** can be either in the time or the frequency domain, depending on the weights $\mathbf{w}_{\mathrm{ds}}$ and $\mathbf{w}_{\mathrm{ls}}$.

the microphones and a large number of microphones. In a practical system, these constraints on the array size and the number of sensors result in a trade-off between spatial selectivity and noise amplification. However, some of these techniques, such as delay-and-sum, have been utilized in real-time systems due to their minimal computational requirements.

In the same class of *signal-independent* techniques, a closely spaced microphone array technique that uses least-squares minimization (Van Trees, 2001) has been proposed and can be applied to beamforming for sound reproduction (Bertet *et al.*, 2006). In this technique, the microphone signals are summed together in the same or opposite phase with different gains and frequency equalization, where the targets are directional patterns following the spherical harmonics of different orders. Such techniques, also referred to as pattern matching, have also been used in many applications where the target patterns can be head-related transfer functions (Atkins, 2011; Rasumow *et al.*, 2016), narrow beamformers for sound recording (Farina *et al.*, 2010; Cobos *et al.*, 2012), or loudspeaker panning functions utilized in sound reproduction systems (Cobos *et al.*, 2012; Zotter and Frank, 2012). The resulting response has tolerable quality only within a limited frequency window. For compact microphone arrays there is a trade-off between directionality and amplification of the self-noise of microphones in the low frequency region. At high frequencies the directional patterns are deformed above the spatial aliasing frequency. A discussion on the performance of spherical microphone arrays at low frequencies and above aliasing is provided in Chapter 2.

10.3   Signal-Dependent Enhancement

10.3.1   Adaptive Beamformers

A microphone signal contains a set of clean signals disturbed by noise and reverberation. Due to the linear superposition of acoustic waves, interference from other sources can be modeled as added components in the clean signal. *Signal-dependent* or

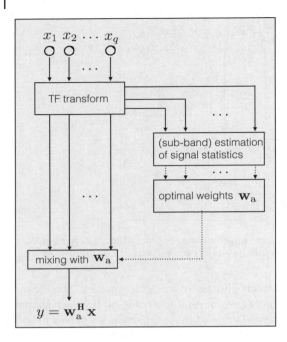

Figure 10.3 Signal-dependent beamforming. The microphone input signal **x** is assumed to be in the time domain.

adaptive beamformers, such as the linearly constrained minimum variance (LCMV) or the minimum variance distortionless response (MVDR) beamformers (Van Veen and Buckley, 1988) steer a beam pattern adaptively such that the noise and the interferers are attenuated while the desired signal(s) are passed with the user-defined gains.

The basic block diagram for these techniques is illustrated in Figure 10.3 where TF indicates a time-frequency transform. The signal statistics are estimated and used to formulate a set of optimal weights. In more detail, these techniques formulate complex-valued beamforming weights in sub-bands using constrained optimization based on the short-time estimate of the microphone signal covariance matrix. The resulting beam-former weights are such that they minimize the energy of the output of the beamformer, thus suppressing the effects of the interferers, the diffuse sounds, and the microphone self-noise. Although MVDR beamforming provides an optimal solution, in practical systems it usually does not provide sufficient noise reduction in highly diffuse/reverberant environments. Adaptive beamformer designs that constrain the microphone self-noise amplification may also result in reduced spatial resolution towards the low frequencies, which is especially evident when utilizing compact microphone arrays. As a consequence of the inability to suppress the directionally distributed interfering sound energy, the diffuse sound is degraded. This is especially true in typical acoustic conditions and at large wavelengths with respect to the array size. In such scenarios, the optimum point of the beamformers can be unsatisfactory in terms of suppressing the interferers and reverberation.

Another class of adaptive beamformers are described as part of the informed spatial filters that combine beamforming with noise reduction (Thiergart *et al.*, 2013, 2014; Thiergart and Habets, 2013, 2014). These adaptive filters use a model of the sound field based on the signal received by the microphone array. The model can consist of a finite number of plane waves, along with the diffuse sound component and the noise component. The effectiveness of this spatial filter relies on the accurate estimation of the noise statistics, the number and direction of plane waves, and the diffuse sound power. The studies present good performance in acoustic conditions matching the sound field model; however, the scalability to typical acoustic conditions and with arbitrary or spherical microphone arrays is yet to be explored.

10.3.2   Post-Filters

In adaptive beamformers there is a trade-off between directional selectivity and noise amplification which can be observed in the directivity factor (Bitzer and Simmer, 2001; Ito *et al.*, 2010). As a consequence, low directional selectivity results in spectral imbalance and undesired reproduction of diffuse sound in the beamformer output. Noise reduction, in a spectral sense, can be further improved using post-filtering/masking techniques. In traditional time–frequency (TF) masking approaches, the observed noisy magnitude spectrogram is multiplied with a real-valued mask in order to remove noise and keep the desired signal. The key concept is to apply low values in the TF regions that are dominated by noise, and pass the target signal components unchanged. The mask can therefore be applied for the reduction of noise, reverberation, and interference. Time–frequency masking is commonly applied at the output of the beamformer to adjust the spectrum to better match that of the desired source signal (Simmer *et al.*, 2001). Such adaptive filters modify the energy of the optimal beamformer in frequency bands, while the phase spectrum is inherited. However, errors in estimating the model parameters result in errors in the spectral estimates. The error can result in short sparsely spaced peaks in the time–frequency representation of the post-filter, which can become perceptually evident when the signal is transformed back to the time domain as musical noise (Esch and Vary, 2009; Takahashi *et al.*, 2010). Typical smoothing techniques, for example using a one-pole recursive filter design, and spectral floor adjustments have been applied to mitigate these artifacts. Applications of these techniques are shown in Chapter 11.

An illustrational block diagram is provided in Figure 10.4, where a post-filter/mask, G, is applied to the output of a beamformer, $\mathbf{w}_a^H \mathbf{x}$. The post-filter can be estimated from the input signals or predicted by a learning system that is trained using a database of example input features and target outputs.

10.3.3   Post-Filter Types

The post-filter/mask can be binary, real, or complex. One of the first real-valued post-filters for suppressing room reverberation was introduced by Allen *et al.* (1977). The assumption of this post-filter is that the noise in the sound scene received at each sensor of the microphone array is uncorrelated. Further work on this topic consists of an algorithm proposed by McCowan and Bourlard (2003), where the sound field was modeled with a coherence function for a spherically isotropic field to identify correlated noise. The introduction of this post-filter provided a generalization of the Zelinski post-filter (Zelinski, 1988). Extensions of these post-filters are shown in Leukimmiatis and Maragos (2006). These post-filters are only capable of reducing uncorrelated or correlated noise in the beamforming output, and rely on the output of the beamformer for the suppression of the interference. Another technique for highly correlated noise was proposed

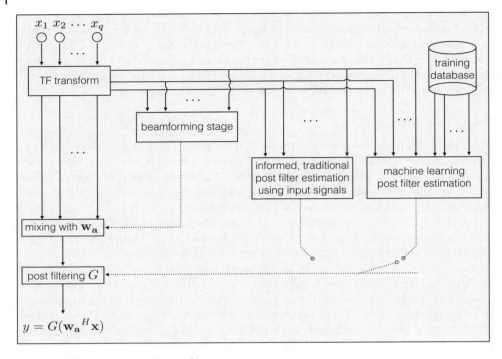

Figure 10.4 Beamforming with post-filtering.

by Fischer and Simmer (1996) based on a generalized sidelobe canceler. Unfortunately, these methods are characterized by poor performance at low frequencies when the correlation between microphone signals is high (Bitzer and Simmer, 2001). From a signal processing perspective, the optimal multi-channel filter in the minimum mean squared error (MMSE) sense is the multi-channel Wiener filter (Van Trees, 2001; Doclo *et al.*, 2015), which has been shown to be equivalent to an MVDR beamformer followed by a single-channel Wiener filter (SCWF; Simmer *et al.*, 2001).

The ideal binary mask (IBM) was introduced first, and has been considered as the goal of computational auditory scene analysis (CASA) by Wang (2005), since it essentially identifies the spectral contribution of a target source from interference and noise. The binary decision is made based on comparing a local signal-to-noise ratio (SNR) of each TF unit to a selected threshold value. However, due to the binary values, some artifacts are typically introduced and the selection of the threshold affects the enhancement quality. The ideal ratio mask (IRM) is a continuous mask with real values in the range [0, 1], in contrast to binary values. The approach is less susceptible to musical artifacts (Srinivasan *et al.*, 2006). It has been argued that the real-valued IRM may be more closely related to auditory processes than the IBM (see Hummersone *et al.*, 2014). Recently, the utilization of complex masks has been investigated for speech enhancement by Erdogan *et al.* (2015) and Williamson *et al.* (2016) using DNNs. The reason for using complex masks is to alter the phase of the enhanced signal in addition to the magnitude. The motivation for the processing of phase can be attributed to studies such as those done by Paliwal *et al.* (2011), which show that the enhanced signal phase also affects the enhancement quality.

10.3.4 Estimating Post-Filters with Machine Learning

Obtaining the IBM is a binary classification problem, whereas estimating the IRM and complex masks can be formulated as a regression task. Different machine-learning techniques exist for supervised regression of non-linear functions such as logistic regression, random forests (Breiman, 2001), support vector regression (Smola and Schölkopf, 2004), neural networks (Haykin, 1999), and Gaussian processes (Williams and Rasmussen, 1996), to name a few. All of these methods are suitable for TF-mask-based speech enhancement, but the introduction of DNNs has re-established the interest in neural networks (NNs) due to state-of-the-art results in several recognition tasks (see LeCun *et al.*, 2015). The NN consist of layers of units, except for the input layer that contains only the input values or features. In a fully connected feed-forward neural network (FFNN), inside a layer each unit receives the output of each unit in the previous layer as its inputs. The unit weights each input value separately and then transforms the sum into an output value using an activation function. The output of the last layer is then the prediction of the target value corresponding to the input. The non-linearity of activation functions allows the representation of non-linear relationships between the input and output values. Common to all supervised methods is the requirement for labeled training data, that is, a database of features and corresponding targets. A microphone-array-based enhancement algorithm that utilizes NN is demonstrated in Chapter 12.

Single-channel features for mask prediction typically include spectral magnitude information, with the possible use of a perceptually motivated filterbank and traditional speech processing features such as linear prediction coding (LPC) or mel-frequency cepstral coefficients (MFCC), while target values consist of TF masks or direct magnitude spectral values.

10.3.5 Post-Filter Design Based on Spatial Parameters

The assumption of the sparsity of the source signals is also utilized in another technique, directional audio coding (DirAC; shown in Chapter 5), which is a method to capture, process, and reproduce spatial sound over different reproduction setups. The most prominent DOA and the diffuseness of the sound field are measured as spatial parameters for each time–frequency position of sound. The DOA is estimated as the opposite direction of the intensity vector, and the diffuseness is estimated by comparing the magnitude of the intensity vector with the total energy. In the original version of DirAC, the parameters are utilized in reproduction to enhance the audio quality. A variant of DirAC has been used for beamforming (Kallinger *et al.*, 2009), where each time–frequency position of sound is amplified or attenuated depending on the spatial parameters and a specified spatial filter pattern. In practice, if the DOA of a time–frequency position is far from the desired direction, it is attenuated. Additionally, if the diffuseness is high, the attenuation is reduced since the DOA is considered to be less certain. However, in cases when the assumption of W-disjoint orthogonality is violated and two audio signals are active in the same time–frequency position, intensity-based DOA provides inaccurate data, and artifacts may occur.

A post-filter using the phase information of a microphone array was used to learn how to predict the IRM with a neural network using simulated speech signals in different noise and reverberant conditions by Pertilä and Nikunen (2014). The way the post-filter is applied to the beamformer output makes it conceptually similar to single-channel TF-masking. Multi-channel speech separation has recently been proposed by Nugraha *et al.* (2016) by combining DNN and the expectation-maximization (EM) algorithm. Neural-network-based post-filtering techniques are presented in Chapter 12.

Recently, some techniques have been proposed that assume that the signals arriving from different directions to the microphone array are sparse in the time–frequency domain, that is, one of the sources is dominant at one time–frequency position (Faller, 2008). Each time–frequency tile is then attenuated or amplified according to spatial parameters analyzed for the corresponding time–frequency position, which leads to the formation of the beam. A microphone array consisting of two cardioid capsules in opposite directions has been proposed by Faller (2007) for such a technique. Correlation measures between the cardioid capsules and Wiener filtering are used to reduce the level of coherent sound in one of the microphone signals. This produces a directive microphone, whose beamwidth can be controlled, as further explained in Chapter 13. An inherent result is that the width varies depending on the sound field. For example, with few speech sources in conditions with low levels of reverberation, a prominent narrowing of the cardioid pattern is obtained. However, with many uncorrelated sources, and in a diffuse field, the method does not change the directional pattern of the cardioid microphone at all. The method is still advantageous, as the number of microphones is low, and the setup does not require a large space.

A recent class of beamformers, which includes the technique proposed by Delikaris-Manias and Pulkki (2013), estimates the energy of the target signal using cross-spectrum estimates of two beam patterns with the constraints that their maximum sensitivity and equal phase are in the look direction. When these constraints are met, the effect of the noise is suppressed when the time-averaging interval is increased, while the energy of the source in the look direction is retained. This class of techniques omits the requirement for knowledge of the sound field, and thus provides robustness for complex acoustic scenes. The derivation of the post-filter in the cross-pattern coherence algorithm (CroPaC; Delikaris-Manias and Pulkki, 2013) relies on the calculation of the cross-spectrum of two static beam patterns. The application of CroPaC is feasible with any order of microphone input, and the directional shape of the beam can be altered by changing the formation of the directional patterns of the microphones from which the post-filter is computed.

A technique was proposed by Delikaris-Manias *et al.* (2016) to estimate a post-filter utilizing the cross-spectrum technique of two beam patterns where the selection of the beam patterns is based on minimization techniques. The first beam pattern is static, and corresponds to a spatially narrow beam pattern having equal spatial selectivity and unity gain across the frequency region of interest. The constant directional selectivity provides the grounds for spectrally balanced energy estimation in the presence of spatially spread noise. As a result of this design, this first beam pattern is characterized by high microphone self-noise gain at low frequencies, especially for compact microphone arrays (Rafaely, 2005). The second pattern is formulated adaptively in time and frequency using constrained optimization, with the constraint of suppressing the interferers while retaining the desired features of orthogonality and unity zero-phase gain towards the look direction. By these means, the proposed method inherits the noise-robust features of the prior technique in the class, while providing the adaptive suppression of the discrete interferers. Using signals originating from spatially selective but noisy beam patterns for spatial parametric estimation has previously also been shown to be effective in the context of spatial sound reproduction for loudspeaker setups and headphones by Vilkamo and Delikaris-Manias (2015) and Delikaris-Manias *et al.* (2015), respectively. The derivation of the proposed algorithm is shown in Chapter 11 in the spherical harmonic domain, and the experiments are performed using a uniform spherical microphone array such that constant beamforming performance towards all azimuthal and elevation positions is realizable. For such spherical microphone array designs, different methods for distributing microphones almost uniformly on the surface of a sphere can be considered: spherical t-designs (Rafaely, 2015), or by utilizing electron equilibrium (Saff and Kuijlaars, 1997). However, the proposed technique can also be applied to non-uniform microphone array configurations, penalised with varying directional filtering performance for different look directions.

References

Ahonen, J. and Pulkki, V. (2010) Speech intelligibility in teleconference application of directional audio coding. *40th International Audio Engineering Society Conference: Spatial Audio: Sense the Sound of Space.* Audio Engineering Society.

Allen, J.B., Berkley, D.A., and Blauert, J. (1977) Multimicrophone signal-processing technique to remove room reverberation from speech signals. *Journal of the Acoustical Society of America*, **62**, 912–915.

Atkins, J. (2011) Robust beamforming and steering of arbitrary beam patterns using spherical arrays. *IEEE Workshop on Applications of Signal Processing to Audio and Acoustics (WASPAA)*, pp. 237–240. IEEE.

Bertet, S., Daniel, J., and Moreau, S. (2006) 3D sound recording with higher order Ambisonics – objective measurements and validation of a 4th order spherical microphone. *Audio Engineering Society Convention 120.* Audio Engineering Society.

Bitzer, J. and Simmer, K.U. (2001) Superdirective microphone arrays, in *Microphone Arrays* (eds. Brandstein M. and Ward, D.), pp. 19–38. Springer, Berlin.

Braasch, J. (2005) A loudspeaker-based 3D sound projection using virtual microphone control (VIMIC). *Audio Engineering Society Convention 118.* Audio Engineering Society.

Brandstein, M.S. and Silverman, H.F. (1997) A practical methodology for speech source localization with microphone arrays. *Computer Speech Language*, **11**(2), 91–126.

Breiman, L. (2001) Random forests. *Machine Learning*, **45**(1), 5–32.

Cobos, M., Spors, S., Ahrens, J., and Lopez, J.J. (2012) On the use of small microphone arrays for wave field synthesis auralization. *45th International Conference of the AES: Applications of Time–Frequency Processing in Audio*, Helsinki, Finland.

Cohen, I., Benesty, J., and Gannot, S. (2009) *Speech Processing in Modern Communication: Challenges and Perspectives*, vol. 3. Springer Science & Business Media, New York.

Delikaris-Manias, S. and Pulkki, V. (2013) Cross-pattern coherence algorithm for spatial filtering applications utilizing microphone arrays. *IEEE Transactions on Audio, Speech, and Language Processing*, **21**(11), 2356–2367.

Delikaris-Manias, S., Vilkamo, J., and Pulkki, V. (2015) Parametric binaural rendering utilizing compact microphone arrays, *IEEE International Conference on Acoustics, Speech, and Signal Processing (ICASSP)*, pp. 629–633.

Delikaris-Manias, S., Vilkamo, J., and Pulkki, V. (2016) Signal-dependent spatial filtering based on weighted-orthogonal beamformers in the spherical harmonic domain. *IEEE/ACM Transactions on Audio, Speech, and Language Processing*, **24**(9), 1511–1523.

Doclo, S., Kellermann, W., Makino, S., and Nordholm, S.E. (2015) Multichannel signal enhancement algorithms for assisted listening devices: Exploiting spatial diversity using multiple microphones. *IEEE Signal Processing Magazine*, **32**(2), 18–30.

Eargle, J. (2004) *The Microphone Book*. Taylor & Francis US, Boca Raton, FL.

Erdogan, H., Hershey, J., Watanabe, S., and Le Roux, J. (2015) Phase-sensitive and recognition-boosted speech separation using deep recurrent neural networks. *IEEE International Conference on Acoustics, Speech, and Signal Processing (ICASSP)*, pp. 708–712. IEEE.

Esch, T. and Vary, P. (2009) Efficient musical noise suppression for speech enhancement system. *IEEE International Conference on Acoustics, Speech, and Signal Processing (ICASSP)*, pp. 4409–4412. IEEE.

Faller, C. (2007) A highly directive 2-capsule based microphone system. *Proceedings of the 123rd AES Convention*. Audio Engineering Society.

Faller, C. (2008) Modifying the directional responses of a coincident pair of microphones by postprocessing. *Journal of the Audio Engineering Society*, **56**(10), 810–822.

Farina, A., Capra, A., Chiesi, L., and Scopece, L. (2010) A spherical microphone array for synthesizing virtual directive microphones in live broadcasting and in post production. *40th International Conference of the Audio Engineering Society: Spatial Audio: Sense the Sound of Space*. Audio Engineering Society.

Fischer, S. and Simmer, K.U. (1996) Beamforming microphone arrays for speech acquisition in noisy environments. *Speech Communication*, **20**(3), 215–227.

Haykin, S. (1999) *Neural Networks: A Comprehensive Foundation*. Prentice-Hall, Englewood Cliffs, NJ.

Hummersone, C., Stokes, T., and Brookes, T. (2014) On the ideal ratio mask as the goal of computational auditory scene analysis, in *Blind Source Separation: Advances in Theory, Algorithms and Applications* (eds. Naik, G.R. and Wang, W.), chapter 12. Springer, New York.

Ito, N., Ono, N., Vincent, E., and Sagayama, S. (2010) Designing the Wiener post-filter for diffuse noise suppression using imaginary parts of inter-channel cross-spectra. *IEEE International Conference on Acoustics, Speech, and Signal Processing (ICASSP)*, pp. 2818–2821. IEEE.

Kallinger, M., Ochsenfeld, H., Del Galdo, G., Kuech, F., Mahne, D., and Thiergart, O. (2009) A spatial filtering approach for directional audio coding. *Audio Engineering Society Convention 126*, pp. 1–10. Audio Engineering Society.

LeCun, Y., Bengio, Y., and Hinton, G. (2015) Deep learning. *Nature*, **521**(7553), 436–444.

Leukimmiatis, S. and Maragos, P. (2006) Optimum post-filter estimation for noise reduction in multichannel speech processing. *14th European Signal Processing Conference*, pp. 1–5. IEEE.

McCowan, I. and Bourlard, H. (2003) Microphone array post-filter based on noise field coherence. *IEEE Transactions on Speech and Audio Processing*, **11**(6), 709–716.

Merimaa, J. (2002) Applications of a 3-D microphone array. *Audio Engineering Society Convention 112*. Audio Engineering Society.

Narayanan, A. and Wang, D. (2013) Ideal ratio mask estimation using deep neural networks for robust speech recognition. *IEEE International Conference on Acoustics, Speech, and Signal Processing (ICASSP)*, pp. 7092–7096, Vancouver, Canada.

Nugraha, A.A., Liutkus, A., and Vincent, E. (2016) Multichannel audio source separation with deep neural networks. *IEEE/ACM Transactions on Audio, Speech, and Language Processing*, **24**(9), 1652–1664.

Paleologu, C., Benesty, J., and Ciochină, S. (2010) Variable step-size adaptive filters for echo cancellation, in *Speech Processing in Modern Communication* (eds. Cohen, I., Benesty, J., and Gannot, S.), pp. 89–125. Springer, New York.

Paliwal, K., Wójcicki, K., and Shannon, B. (2011) The importance of phase in speech enhancement. *Speech Communication*, **53**(4), 465–494.

Pertilä, P. and Nikunen, J. (2014) Microphone array post-filtering using supervised machine learning for speech enhancement. *Annual Conference of the International Speech Communication Association (Interspeech)*.

Rafaely, B. (2005) Analysis and design of spherical microphone arrays. *IEEE Transactions on Speech and Audio Processing*, **13**(1), 135–143.

Rafaely, B. (2015) *Fundamentals of Spherical Array Processing*. Springer, New York.

Rasumow, E., Hansen, M., van de Par, S., Püschel, D., Mellert, V., Doclo, S., and Blau, M. (2016) Regularization approaches for synthesizing HRTF directivity patterns. *IEEE/ACM Transactions on Audio, Speech, and Language Processing*, **24**(2), 215–225.

Saff, E.B. and Kuijlaars, A.B. (1997) Distributing many points on a sphere. *The Mathematical Intelligencer*, **19**(1), 5–11.

Simmer, K.U., Bitzer, J., and Marro, C. (2001) Post-filtering techniques, in *Microphone Arrays: Signal Processing Techniques and Applications* (eds. Brandstein, M. and Ward, D.), pp. 39–60. Springer, Berlin.

Smola, A.J. and Schölkopf, B. (2004) A tutorial on support vector regression. *Statistics and Computing*, **14**(3), 199–222.

Srinivasan, S., Roman, N., and Wang, D. (2006) Binary and ratio time–frequency masks for robust speech recognition. *Speech Communication*, **48**(11), 1486–1501.

Takahashi, Y., Saruwatari, H., Shikano, K., and Kondo, K. (2010) Musical-noise analysis in methods of integrating microphone array and spectral subtraction based on higher-order statistics. *EURASIP Journal on Advances in Signal Processing*, **2010**(1), 431347.

Thiergart, O. and Habets, E.A. (2013) An informed LCMV filter based on multiple instantaneous direction-of-arrival estimates. *IEEE International Conference on Acoustics, Speech, and Signal Processing (ICASSP)*, pp. 659–663. IEEE.

Thiergart, O. and Habets, E.A.P. (2014) Extracting reverberant sound using a linearly constrained minimum variance spatial filter. *IEEE Signal Processing Letters*, **21**(5), 630–634.

Thiergart, O., Taseska, M., and Habets, E.A. (2013) An informed MMSE filter based on multiple instantaneous direction-of-arrival estimates. *Proceedings of the 21st European Signal Processing Conference (EUSIPCO)*, pp. 1–5. IEEE.

Thiergart, O., Taseska, M., and Habets, E.A.P. (2014) An informed parametric spatial filter based on instantaneous direction-of-arrival estimates. *IEEE/ACM Transactions on Audio, Speech, and Language Processing*, **22**(12), 2182–2196.

Van Trees, H.L. (2001) *Detection, Estimation, and Modulation Theory*, Part I. John Wiley & Sons, Inc., New York.

Van Veen, B.D. and Buckley, K.M. (1988) Beamforming: A versatile approach to spatial filtering. *IEEE ASSP Magazine*, **5**(2), 4–24.

Vilkamo, J. and Delikaris-Manias, S. (2015) Perceptual reproduction of spatial sound using loudspeaker-signal-domain parametrization. *IEEE/ACM Transactions on Audio, Speech, and Language Processing*, **23**(10), 1660–1669.

Wang, D. (2005) On ideal binary mask as the computational goal of auditory scene analysis, in *Speech Separation by Humans and Machines* (ed. Divenyi, P.), chapter 12. Kluwer Academic Publishers, Dordrecht.

Wang, Y., Han, K., and Wang, D. (2013) Exploring monaural features for classification-based speech segregation. *IEEE Transactions on Audio, Speech, and Language Processing*, **21**(2), 270–279.

Wang, Y. and Wang, D. (2013) Towards scaling up classification-based speech separation. *IEEE Transactions on Audio, Speech, and Language Processing*, **21**(7), 1381–1390.

Weninger, F., Eyben, F., and Schuller, B. (2014) Single-channel speech separation with memory-enhanced recurrent neural networks. *IEEE International Conference on Acoustics, Speech, and Signal Processing (ICASSP)*, pp. 3709–3713, Florence, Italy.

Williams, C.K.I. and Rasmussen, C.E. (1996) Gaussian processes for regression, in *Advances in Neural Information Processing Systems 8* (ed. Touretzky, D.S. and Hasselmo, M.E.), pp. 514–520. MIT Press, Cambridge, MA.

Williamson, D.S., Wang, Y., and Wang, D. (2016) Complex ratio masking for monaural speech separation. *IEEE/ACM Transactions on Audio, Speech, and Language Processing*, **24**(3), 483–492.

Zelinski, R. (1988) A microphone array with adaptive post-filtering for noise reduction in reverberant rooms. *International Conference on Acoustics, Speech, and Signal Processing (ICASSP)*, pp. 2578–2581. IEEE.

Zotter, F. and Frank, M. (2012) All-round ambisonic panning and decoding. *Journal of the Audio Engineering Society*, **60**(10), 807–820.

11

Cross-Spectrum-Based Post-Filter Utilizing Noisy and Robust Beamformers

Symeon Delikaris-Manias and Ville Pulkki

Department of Signal Processing and Acoustics, Aalto University, Finland

11.1   Introduction

In this chapter we review a post-filtering technique that can provide additional attenuation of spatial noise, interferers, and diffuse sound to the output of a beamformer with minimal spectral artifacts. Contrary to conventional post-filters such as the ones discussed in Chapter 10, the post-filter shown here employs the cross-spectrum of signals originating from coincident beamformers. The cross-spectrum aims to estimate the target signal energy originating from a specific look direction by using two beamformers with the same phase and unity gain towards that direction. This concept is demonstrated in Figure 11.1, where the spectrograms and beam patterns of two different beamformers are shown. The beamformers in the example are of first (dipole) and second order (quadrupole), where the order refers to the directional selectivity of the beamformers. The higher the order, the more directionally selective the beamformer is. Their polar patterns are shown on the left of Figure 11.1, and the black and white colors indicate the polarity. Both beamforming signals are generated from a spherical compact microphone array of 16 sensors and 0.15 cm radius by applying a set of complex weights. For compact microphone arrays, the higher the order of the beamformer, the more the low-frequency microphone noise is boosted. This difference is evident in the spectrograms of the beamformers shown on the right of Figure 11.1. Low-frequency noise is amplified in the case of the higher-order beamformer (quadrupole).

The computation of the cross-spectrum between these two beamformers can be illustrated as a multiplication of their directivity patterns. The cross-spectral density between two signals captured by such beamformers is maximum when their directional patterns use

- the same sensitivity, and
- equal phase

towards the same direction. This is visualized in Figure 11.2. However, the product of these beam patterns results in a mirror beam pattern with negative phase. This negative

Parametric Time–Frequency Domain Spatial Audio, First Edition. Edited by Ville Pulkki,
Symeon Delikaris-Manias, and Archontis Politis.
© 2018 John Wiley & Sons Ltd. Published 2018 by John Wiley & Sons Ltd.
Companion Website: www.wiley.com/go/pulkki/parametrictime-frequency

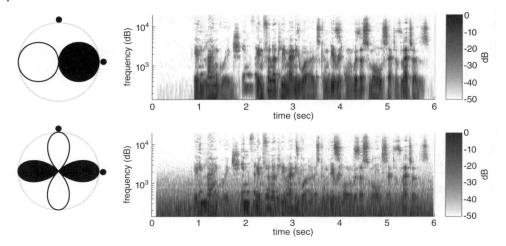

Figure 11.1 Beam patterns (left) with their corresponding spectrograms (right) of two beamforming signals. The background color indicates phase information: black for positive phase and white for negative phase.

part of the resulting beam pattern can be truncated by applying a half-wave rectification process – see Figure 11.2.

Multiplying the time–frequency representation of two beamformer signals does not produce an audio signal, but results in a real-valued parameter that estimates signal activity for a given direction. This parameter, which we will show later in this chapter is within the interval [0, 1], can be used as a soft masker. It can be applied as a single-channel spectral post-filter to the output of a single microphone or beamformer signal to provide suppression of noise, interferers, and diffuse sound.

The initial step is to calculate the two beamforming signals with different directional patterns, such as those shown in Figure 11.2. In detail, the post-filter is based on the estimation of the cross-spectral density in each time–frequency position between two coincident beamformer signals. It is computed as the time-averaged normalized cross-spectral density and is then used to modulate the output of a third signal at the same time–frequency positions. The third signal can be either a signal-dependent or signal-independent beamformer, or simply a single microphone signal. The output of this third signal is attenuated at those time–frequency positions where the two beamforming signals are uncorrelated in a spectral and directional sense. The calculation of

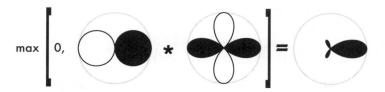

Figure 11.2 Visualization of multiplication between different two-dimensional pick-up beam patterns and the half-wave rectification process (denoted as a max operator). Black indicates positive phase and white negative phase.

this post-filter is feasible with a microphone array comprising of at least three sensors. The directional shape of the "post-filter beam" can be altered by changing the formation of the coincident beamformers.

Two variations of the algorithm are shown here based on the same principles but using different types of beamformers:

- A deterministic approach based on the cross-spectrum of two static beamformers. The selection of the two static beamformers is conducted by the user. Static beam-formers are signal-independent beamformers and their weights can be precalculated offline.
- A statistical approach based on the cross-spectrum of a static and an adaptive beam-former. In this case, one beam pattern with constant directional selectivity is defined according to the microphone array specifications. The second one is signal depen-dent, and is defined according to a constrained optimization that minimizes spatial noise.

The remainder of this chapter is organized as follows: Section 11.2 provides the nota-tion, the signal model, and background on practical beamforming techniques where an arbitrary target directivity pattern is desired. In Section 11.3, the two variations of the non-linear directional filtering algorithm are shown, including the design principles for the weight coefficients for the constant and the adaptive beamformers. Implementa-tion examples with simulated and real microphone arrays are shown in Section 11.4 in reverberant environments with single- or multiple-talker scenarios. Conclusions are presented in Section 11.5.

11.2 Notation and Signal Model

Matrices and vectors are denoted with bold-faced symbols, where upper case denotes a matrix. The entries of the vectors are expressed with the same symbols in lower-case italics with a subscript index; for example, x_q denotes the qth element of the input sig-nal vector $\mathbf{x} \in \mathbb{C}^{Q \times 1}$. The time–frequency representation of the inputs of a Q-channel microphone array signal are denoted by

$$\mathbf{x}(k, n) = \begin{bmatrix} x_1(k, n) \\ x_2(k, n) \\ \vdots \\ x_Q(k, n) \end{bmatrix}, \tag{11.1}$$

where k is the frequency index and n is the time index. The time–frequency represen-tation of the signals can be obtain with a short-time Fourier transform or a filterbank approach. Assuming a spherical coordinate system where $\Omega = (\theta, \phi)$ with elevation $\theta \in [0, \pi]$ and azimuth $\phi \in [0, 2\pi)$, we define a steering vector $\mathbf{v}(\Omega_j, k) \in \mathbb{C}^{Q \times 1}$. For a target sound source at a look direction Ω_o, the signal model for the microphone array considered in this work is

$$\mathbf{x}(k, n) = \mathbf{v}(\Omega_o, k) s_o(k, n) + \mathbf{n}(k, n) + \sum_{j=1}^{J} \mathbf{v}(\Omega_j, k) s_j(k, n), \tag{11.2}$$

where $s_0(k, n)$ is the desired sound arriving directly from the look direction Ω_0, $\mathbf{v}(\Omega_o, k)$ is the steering vector for that direction, $\mathbf{n}(k, n) \in \mathbb{C}^{Q \times 1}$ is the microphone self-noise vector, and the term $\sum_{j=1}^{J} \mathbf{v}(\Omega_j, k)\mathbf{s}_j(k, n)$ models the combined sounds from $j = 1, \dots, J$ interferers originating from J directions. In a direction-filtering scenario the target signal is commonly $\mathbf{v}(\Omega_o, k)s_o(k, n)$. By assuming infinite J, this definition covers the potential room reverberation and background ambience or a purely diffuse field. The signal model in Equation (11.2) assumes that the target source signal and each of the interferers are mutually uncorrelated, that is, $E[s_o(k, n)s_j^*(k, n)] = 0 \ \forall j \in J$.

The single-channel output beamformer signal is denoted by

$$y_b(k, n) = \mathbf{w}^H(k, n)\mathbf{x}(k, n), \tag{11.3}$$

where $\mathbf{w} \in \mathbb{C}^{Q \times 1}$ are the weights, which can be defined to be signal independent, $\mathbf{w}(k)$, or signal dependent, $\mathbf{w}(k, n)$. There are various schemes for how to calculate the complex multipliers $\mathbf{w}$, which depend entirely on the application. An overview for spherical microphone arrays is provided in Chapter 2. Noise-robust, signal-dependent beamforming techniques are discussed thoroughly in the previous chapter. In this chapter, beamforming techniques that aim to synthesize specific target pick-up patterns, which are also mentioned in the literature as virtual microphone design, are briefly discussed.

11.2.1 Virtual Microphone Design Utilizing Pressure Microphones

Virtual microphones refer to beamformer signals with specific, user-defined pick-up beam patterns. Such beamforming techniques can be implemented either in the space, the spherical/cylindrical, or in the spatial frequency domain (Josefsson and Persson, 2006; Li and Duraiswami, 2007; Farina *et al.*, 2010a; Delikaris-Manias *et al.*, 2013; Rafaely, 2015). The principles, however, are the same:

- Define a target pick-up pattern $\mathbf{t}(\Omega, k)$.
- Utilize optimization techniques to estimate the optimal weights to synthesize $\mathbf{t}(\Omega, k)$ using the microphone array steering vectors.
- Apply the weights to the microphone array signal to obtain the directional characteristics of the target pick-up pattern.

Virtual microphone design or pattern-matching techniques can be formulated as follows: a target pick-up pattern, $\mathbf{t}(\Omega, k) \in \mathbb{C}^{1 \times M}$, is defined at Ω_M points. Using a set of steering vectors $\mathbf{V}(\Omega, k) \in \mathbb{C}^{Q \times M}$ obtained with measurements or theoretical approximations, the target directivity pattern can be synthesized by minimization (Van Trees, 2004):

$$\mathbf{w}_{\text{opt}}(k) = \arg\min_{\mathbf{w}} \left\{ |\mathbf{t}(\Omega, k) - \mathbf{w}(k)\mathbf{V}(\Omega, k)|_2^2 \right\}, \tag{11.4}$$

where the $\mathbf{w}_{\text{opt}} \in \mathbb{C}^{1 \times Q}$ are the complex multipliers to be applied to the input vector $\mathbf{x}$ to synthesize $\mathbf{t}$. In the case of a regularized inverse, the inverse problem using Tikhonov regularization is defined as

$$\mathbf{w}_{\text{opt}}(k) = \arg\min_{\mathbf{w}} \left\{ |\mathbf{t}(\Omega, k) - \mathbf{w}(k)\mathbf{V}(\Omega, k)|_2^2 + |\lambda(k)\mathbf{w}(k)|_2^2 \right\}, \tag{11.5}$$

where λ is a weighting parameter (Deschamps and Cabayan, 1972; Kirkeby *et al.*, 1998).

In matrix form, the solutions to the non-regularized and regularized inverse solutions are

$$\mathbf{w}_{\text{opt}}(k) = \mathbf{t}(\Omega, k)\mathbf{V}^+(\Omega, k), \tag{11.6}$$

where $^+$ denotes the Moore–Penrose pseudo-inverse for unconstrained designs, and

$$\mathbf{w}_{\text{opt}}(k) = \mathbf{t}(\Omega, k)\mathbf{V}^{\text{H}}(\Omega, k)[\mathbf{V}(\Omega, k)\mathbf{V}^{\text{H}}(\Omega, k) + \beta \mathbf{I}_{Q \times Q}]^{-1} \tag{11.7}$$

for regularized inversion, with β denoting regularization parameters and $\mathbf{I}$ the identity matrix. For inversion using Tikhonov regularization, frequency-dependent approaches (Fazi and Nelson, 2007; Farina *et al.*, 2010b) and singular value decomposition (SVD; Gauthier *et al.*, 2011) have been proposed. Additional constraints can be implemented in the minimization problem described in Equation (11.4), such as the white noise gain (WNG) constraint (Atkins, 2011). Such constraints have a direct effect on the directivity pattern of the resulting beamformer. Typical acoustic sensors feature frequency-dependent directivity with omnidirectional response at low frequencies and narrow directivity at high frequencies. For efficient directivity synthesis, knowledge about the directivity and orientation of each sensor is desirable. Therefore, in both inverse problems of Equations (11.6) and (11.7), information about the array manifold $\mathbf{V}(\Omega, k)$ provides a much more accurate virtual microphone design. This manifold can be defined as the set of dense measured impulse responses of the array inside an anechoic environment, simulated numerically, or approximated theoretically when information about the array and the type of sensors is available.

11.3 Estimation of the Cross-Spectrum-Based Post-Filter

The post-filter is designed to estimate the energy of the target sound source arriving at the microphone array from the look direction, namely Ω_0. The primary use for the estimate is to formulate, in the time–frequency domain, that is, for each (k, n), an adaptive, real-valued attenuator G that modulates the output of a beamformer, $y_b(k, n)$, to provide the output signal

$$y_G(k, n) = G(k, n) \odot y_b(k, n)$$
$$= G(k, n) \odot (\mathbf{w}^{\text{H}}(k, n)\mathbf{x}(k, n)), \tag{11.8}$$

where G is the attenuator (post-filter) and $\odot$ denotes the Hadamard product between two vectors. The output signal is expected to consist of the spectro-temporal features close to a desired source signal s_0, as indicated in the signal model in Equation (11.2), for the desired look direction. In particular, the aim of the proposed post-filter is to overcome the trade-off in conventional beamforming between directional selectivity and microphone self-noise amplification, which results in less directionally selective beam patterns at low frequencies. This effect is especially pronounced for compact microphone arrays, and results in both spectral imbalance in the output signal and inability to suppress the interferers and the room effect satisfactorily. The adaptive gain value, G, is formulated such that the output of the beamformer, y_b, has the same amplitude spectrum as the target signal, although the phase spectrum of the beamformer output is inherited.

The cross-spectrum-based post-filter, referred to as the cross-pattern coherence (CroPaC) post-filter, can be used to enhance sounds originating from a desired direction while attenuating signals and noise from other directions. The main idea behind this approach is that a sound source is captured coherently by two coincident beamformers only when the direction of arrival (DOA) of the sound source corresponds to the look direction of the beamformers. In all other cases, the cross-spectral density between the signals is low. A post-filter that utilizes the cross-spectral density of such beamformers can potentially also obtain low values in the low-frequency region in a diffuse sound field, which addresses the main drawbacks of existing spatial-filtering techniques.

11.3.1 Post-Filter Estimation Utilizing Two Static Beamformers

The static-beamformer approach of CroPaC is based on the following principles:

- Two sets of signal-independent complex weights are utilized to provide beamformer outputs with constant directional selectivity.
- The cross-spectrum is estimated between two static beamformers with the same look direction and equal sensitivity.
- The cross-spectrum estimates the contribution of each beamformer for the look direction.

The overall block diagram of the static CroPaC is shown in Figure 11.3. The static beamforming stage synthesizes the beamformers using least-squares minimization.

The cross-spectrum between the outputs of the two static beamformers is

$$\Phi_{y_{b_1}b_2} = E[y_{b1}^* y_{b2}], \tag{11.9}$$

where $y_{b1} = \mathbf{w}_{b1}^H \mathbf{x}$ and $y_{b2} = \mathbf{w}_{b2}^H \mathbf{x}$ are the time–frequency representations of the signals from two coincident beamformers with the same unity look direction, distortionless responses, and equal phase. A noticeable difference when comparing CroPaC with other post-filters, as they are discussed in Section 11.1 and in Chapter 10, is that in those cases the microphone signals are typically scaled and aligned before the calculation of the post-filter, while in the present case the beamforming signals utilized in Equation (11.9) are design to be coincident already in the offline beamforming stage. The post-filter is then calculated by applying the real-part operator in the cross-spectrum of Equation (11.9) and applying a normalization:

$$G = \frac{\mathrm{Re}\left[\Phi_{y_{b_1}b_2}\right]}{\sum_{l=1}^{L} \Phi_{y_{b_l}b_l}}, \tag{11.10}$$

where $\Phi_{y_{b_l}b_l}$ are the auto-power spectral densities of the beamforming output signals b_l with directional patterns selected such that

$$\sum_{l=1}^{L} \Phi_{y_{b_l}b_l} = cS_0, \tag{11.11}$$

where S_0 is a signal from a microphone with omnidirectional characteristics and the noise profile of y_{b_l}, and should be satisfied for all plane waves with DOA of Ω and $c \in \mathbb{Z}^+$.

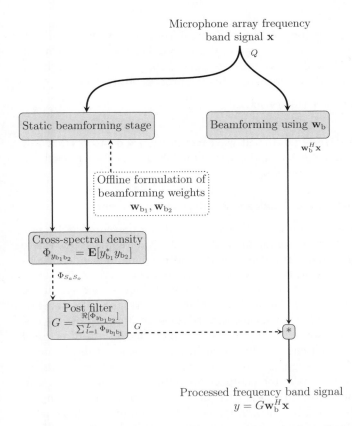

Figure 11.3 Block diagram of the static CroPaC method. Two directionally narrow beamformer weights $\mathbf{w}_{b_1}$ and $\mathbf{w}_{b_2}$ are formulated offline. The thick lines denote multiple microphone signals, the thin lines denote single-channel signals, and the dashed lines denote parametric information.

This is illustrated in Figure 11.4 for two first-order beam patterns. The sum of their energies constitutes the omnidirectional pick-up pattern.

The normalization process in Equation (11.10) ensures that with all inputs the calculated post-filter value is bounded in the interval $[-1, 1]$, and that values near unity are obtained only when the signals b_1 and b_2 are equivalent in both phase and magnitude. In this study, G is a normalized cross-pattern spectral density.

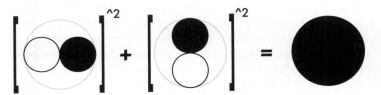

Figure 11.4 Visualization of the sum of the energies between two first-order beam patterns. The background color indicates phase information: black for positive phase and white for negative phase.

Half-Wave Rectification

Post-filter values near unity imply the presence of a sound from the look direction. Values that are below zero indicate that the sound of the analyzed time–frequency frame originates from a direction where there is a mismatch in the phase of the beamformers. Negative values can be truncated by using a rectification process, expressed as

$$G(k, n) = \frac{(1 + \beta)|G(k, n)| + (1 - \beta)G(k, n)}{2}. \tag{11.12}$$

For $\beta = 0$, Equation (11.12) corresponds to half-wave rectification and ensures that only non-negative values are used. In particular, the part of the lobe that is chosen results in a unique beamformer in the look direction.

11.3.2 Post-Filter Estimation Utilizing a Static and an Adaptive Beamformer

A statistical approach to the estimation of the CroPaC post-filter is presented in this section. The post-filtering algorithm presented in Section 11.3.1 is for applications where a static beamformer design is more suitable. However, using signal-dependent techniques, an adaptive beamformer can also be utilized in the cross-spectrum estimation to provide additional interferer attenuation. In this case, the post-filter is estimated using a static and an adaptive beamformer. The static beamformer corresponds to a narrow beam pattern having equal spatial selectivity and distortionless response across the frequency region of interest. In contrast to adaptive beamformers, a constant-directional beamformer provides a balanced energy estimation in the presence of directionally spread noise. As a result of this design, this first beamformer is prone to noise amplification at low frequencies, especially for compact microphone arrays. The second beamformer is formulated adaptively in time and frequency using constrained optimization, with the constraint of suppressing the interferers while retaining an orthogonality feature with the first beamformer and unity gain towards the look direction. By these means, the proposed method aims to exploit the noise-robust features of the conventional adaptive beamforming techniques, while providing additional suppression of interferers and spatial noise. Using signals originating from directionally selective but noisy beam patterns for parameter estimation has previously been shown also to be effective in the context of spatial sound reproduction for loudspeaker setups (Vilkamo and Delikaris-Manias, 2015) and headphones (Delikaris-Manias *et al.*, 2015), as discussed in Chapter 8.

The derivation of the proposed algorithm is provided in the spherical harmonic domain and experiments are performed using a uniform spherical microphone array. The motivation here is that such arrays provide minimal deviation in beamforming performance towards all azimuthal and elevation directions. To design such spherical microphone arrays, different methods for almost uniformly distributing microphones on the surface of a sphere can be considered: spherical t-designs (Rafaely, 2015), or by utilizing electron equilibrium (Saff and Kuijlaars, 1997). However, the proposed technique can be applied to non-uniform microphone array configurations as well, penalized with a varying filtering performance for different look directions.

The adaptive beamformer version of CroPaC is based on the following principles:

- The post-filter is formulated using short-time stochastic analysis.
- The cross-spectral estimates of two beamformers with orthogonal weights are applied to estimate the signal activity in the look direction.

- The first set of weights, $\mathbf{w}_a \in \mathbb{C}^{(L+1)^2 \times 1}$, is signal independent and provides a beam pattern with constant directional selectivity within the selected frequency bandwidth.
- The second set of weights, $\mathbf{w}_o \in \mathbb{C}^{(L+1)^2 \times 1}$, is adaptively designed. This design simultaneously provides the desired orthogonality features for suppressing the spatial and sensor noise, as well as suppression of discrete interferers at the sound scene.

The initial step for estimating the adaptive CroPaC is to estimate the microphone covariance matrix

$$\mathbf{C}_x = E[\mathbf{x}(k, n)\mathbf{x}^H(k, n)], \tag{11.13}$$

where $\mathbf{C}_x \in \mathbb{C}^{Q \times Q}$ and $E[\cdot]$ denotes the statistical expectation operator. In a typical implementation, the covariance matrix is estimated using an average over finite time frames, typically within the range of tens of milliseconds, or by a recursive scheme. The diagonal terms of the covariance matrix denote the auto-spectral densities of each microphone, Φ_{qq}, and the off-diagonal terms the cross-spectral densities between all pairs of microphones $\Phi_{qq'}$ for $q \neq q'$.

Let us assume the signal model first in the sensor domain as in Equation (11.2), with all the components assumed to be mutually incoherent. Omitting the time–frequency indices, their stochastic responses in the covariance matrix can be expressed as:

$$\mathbf{C}_x = \mathbf{C}_{dir} + \mathbf{C}_n + \mathbf{C}_R, \tag{11.14}$$

where

$$\mathbf{C}_{dir} = \mathbf{v}(\Omega_0)\mathbf{v}^H(\Omega_0)E[s_0(k, n)s_0^*], \tag{11.15}$$

$$\mathbf{C}_n = E[\mathbf{n}\mathbf{n}^H], \text{ and} \tag{11.16}$$

$$\mathbf{C}_R = \sum_{r=1}^{R} \mathbf{v}(\Omega_r)\mathbf{v}^H(\Omega_r)E[|s_r|^2], \tag{11.17}$$

where $\mathbf{C}_{dir}$, $\mathbf{C}_n$, and $\mathbf{C}_{int}$ are the covariance matrices of the target signal and the noise, and the combined matrix of interferers and ambience, respectively. The covariance can then be expressed in the spherical harmonic domain as

$$
\begin{aligned}
\mathbf{C}_{lm} &= \mathbf{W}_{SH}\mathbf{C}_x\mathbf{W}_{SH}^H, \\
&= \mathbf{W}_{SH}[\mathbf{C}_{dir} + \mathbf{C}_n + \mathbf{C}_d + \mathbf{C}_R]\mathbf{W}_{SH}^H, \\
&= \mathbf{C}_{SH_{dir}} + \mathbf{C}_{SH_n} + \mathbf{C}_{SH_d} + \mathbf{C}_{SH_R}, \\
&= \mathbf{C}_{SH_{dir}} + \mathbf{C}_{SH_{nd}} + \mathbf{C}_{SH_R},
\end{aligned} \tag{11.18}
$$

where $\mathbf{C}_{SH_{dir}}$, $\mathbf{C}_{SH_{nd}} = \mathbf{C}_{SH_n} + \mathbf{C}_{SH_d}$, and $\mathbf{C}_{SH_R}$ are the spherical harmonic covariance matrices of the target signal, of the noise and the diffuse sound signals combined, and of the interferers, respectively. Matrix $\mathbf{W}_{SH}$ transforms the microphone signal covariance matrix from the space domain to the spherical harmonic domain and is given by

$$\mathbf{W}_{SH} = \alpha_w \mathbf{B}_l^{-1}\mathbf{Y}_{lm}^H(\Omega_q), \tag{11.19}$$

where $\alpha_w = \frac{4\pi}{Q}$ for uniform distribution of the microphones on the sphere. The matrix containing the spherical harmonics $\mathbf{Y}$ and the diagonal equalization matrix $\mathbf{B}_l$ are given by

$$
\mathbf{Y}_{lm}(\Omega_q) = \begin{bmatrix}
Y_{00}(\Omega_1) & Y_{00}(\Omega_2) & \cdots & Y_{00}(\Omega_Q) \\
Y_{-11}(\Omega_1) & Y_{-11}(\Omega_2) & \cdots & Y_{-11}(\Omega_Q) \\
Y_{10}(\Omega_1) & Y_{10}(\Omega_2) & \cdots & Y_{10}(\Omega_Q) \\
Y_{11}(\Omega_1) & Y_{11}(\Omega_2) & \cdots & Y_{11}(\Omega_Q) \\
\vdots & \vdots & \ddots & \vdots \\
Y_{LL}(\Omega_1) & Y_{LL}(\Omega_2) & \cdots & Y_{LL}(\Omega_Q)
\end{bmatrix}^{\mathrm{T}}
\tag{11.20}
$$

and

$$
\mathbf{B}_l = \begin{bmatrix}
b_0 & & & & & \\
& b_1 & & & & \\
& & b_1 & & & \\
& & & b_1 & & \\
& & & & \ddots & \\
& & & & & b_L
\end{bmatrix},
\tag{11.21}
$$

respectively. Direct inversion of matrix $\mathbf{B}_l$ might cause a boost of the low-frequency energy. Different approaches for the design of matrix $\mathbf{B}_l$ are discussed in Chapter 2.

The cross-spectrum of the two beamformers, required for the estimation of the post-filter, can be formulated as

$$
\Phi_{S_a S_o} = \mathbf{w}_a^{\mathrm{H}} \mathbf{C}_{lm} \mathbf{w}_o,
\tag{11.22}
$$

where $\mathbf{C}_{lm} \in \mathbb{C}^{(L+1)^2 \times (L+1)^2}$ is the covariance matrix of the spherical harmonic signals, which can be estimated from the microphone array signal covariance matrix as

$$
\mathbf{C}_{lm} = \mathbf{W}_{\mathrm{SH}} \mathbf{C}_x \mathbf{W}_{\mathrm{SH}}^{\mathrm{H}}.
\tag{11.23}
$$

The covariance matrix estimate of the spherical harmonic signals contains the energies and the inter-channel coherences of the spherical harmonic signals.

The overall block diagram of the proposed technique is shown in Figure 11.5. In the following subsections, the choice and design of the two weight vectors is shown, followed by the estimation of the post-filter G. The post-filter G is calculated in the spherical harmonic domain using the spherical harmonic signals, whereas the beam-former that is modulated with G is formulated in the sensor domain. The performance of the sensor and spherical-harmonic-domain beamformers has been studied by Rafaely (2015), and is equivalent for spherical microphone arrays with an ideal uniform sampling scheme and under the assumption of a spherical isotropic field (Sun *et al.*, 2010).

Design of the Static Analysis Beam Patterns

The analysis weights $\mathbf{w}_a$ are formulated such that they correspond to a beamformer with narrow and constant directivity at all frequency bands and unity zero-phase response towards the look direction.

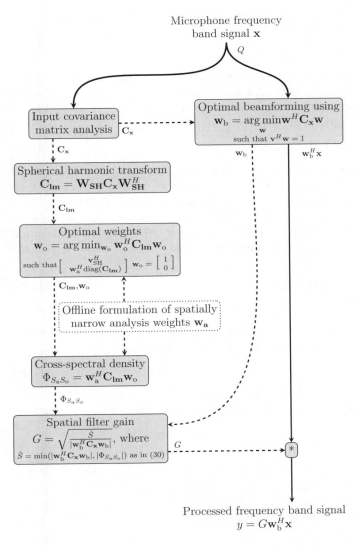

Figure 11.5 Block diagram of the adaptive CroPaC method. Narrow beam pattern weights $\mathbf{w}_a$ are formulated offline and vector $\mathbf{v}$ is the steering vector towards the look direction. The thick lines denote multiple microphone signals, the thin lines denote single-channel signals, and the dashed lines denote parametric information. Source: Delikaris-Manias 2016. Reproduced with permission of IEEE.

Unconstrained narrow beamforming designs result in constant high directional selectivity for a given look direction, suppressing other directions according to their directional response. An option of fulfilling the requirements for narrow static analysis beam patterns are regular beam patterns that are rotationally symmetric around the look direction $\Omega_0 = (\theta_0, \phi_0)$ (Duraiswami *et al.*, 2010; Rafaely, 2015). Although the constant directional selectivity of the beam pattern in all frequencies may result in noise amplification at low frequencies, it is utilized only for the estimation of a post-filter parameter and not for generating an audio signal.

Design of the Robust Synthesis Beam Patterns

The second set of beam pattern weights $\mathbf{w}_o$ is designed to be weighted-orthogonal with respect to $\mathbf{w}_a$ and constrained by unity zero-phase gain towards Ω_0, while optimized to minimize the beamformer output energy. This design is detailed and motivated in the following. The covariance matrix of the target signal is defined as

$$\mathbf{C}_{SH_{dir}} = \mathbf{v}_{SH}(\Omega_0)\mathbf{v}_{SH}^H(\Omega_0)S, \tag{11.24}$$

where $\mathbf{v}_{SH}$ is the steering vector in the spherical harmonic domain and $S = E[s_0 s_0^*]$ is the energy of the sound arriving from the look direction. The expected value of the diffuse field covariance matrix $\mathbf{C}_{SH_{nd}}$ is a diagonal matrix, as shown by Jarrett and Habets (2012).

In the proposed method, $\Phi_{S_a S_o}$ is employed for the estimation of S. In order to avoid a systematic bias in the estimate, we examine the case of only a single sound source, without spatial or microphone noise, or interferers with $\mathbf{C}_{SH_R} = \mathbf{C}_{SH_{nd}} = 0$. In this case, Equations (11.22) and (11.24) give

$$\hat{\Phi}_{S_a S_o} = \underbrace{\mathbf{w}_a^H \mathbf{v}_{SH}(\Omega_0)}_{=1} \mathbf{v}_{SH}^H(\Omega_0)\mathbf{w}_o S. \tag{11.25}$$

For this scenario, it is required that $\Phi_{S_a S_o} = S$; thus, the first constraint is distortionless response, which can be achieved as unity gain in the look direction,

$$\mathbf{v}_{SH}^H(\Omega_0)\mathbf{w}_o = 1. \tag{11.26}$$

The second constraint is derived with the assumption that the constrained optimization in the design of weights $\mathbf{w}_o$ will suppress the energy of the diffuse noise and the microphone noise:

$$\mathbf{w}_a^H \mathbf{C}_{SH_{nd}} \mathbf{w}_o = 0. \tag{11.27}$$

The diagonal matrix $\mathbf{C}_{SH_{nd}}$, which includes the combined information on diffuse sound and noise, needs to be approximated. In the present study, we approximate it by the diagonal of the spherical harmonic signals covariance matrix $\mathbf{C}_{lm}$. The expected value of this estimator is exact for the cases where there are no sources or discrete interferers.

By utilizing these constraints in the beam pattern optimization, we arrive at weights that suppress the interferers directionally, while maintaining unity gain in the look direction Ω_0, and providing the weighted orthogonality features accounting for the diffuse and microphone noise at the cross-spectral estimate. The minimization problem is formulated as

$$\hat{\mathbf{w}}_o = \arg \min_{\mathbf{w}_o} \mathbf{w}_o^H \mathbf{C}_{lm} \mathbf{w}_o$$

$$\text{subject to } \mathbf{v}_{SH}^H(\Omega_0)\mathbf{w}_o = 1 \tag{11.28}$$

$$\text{and } \mathbf{w}_a^H \text{diag}(\mathbf{C}_{lm})\mathbf{w}_o = 0.$$

The above formulation resembles a linearly constrained minimization problem (Van Veen and Buckley, 1988); however, it has the property of having the weight vector $\mathbf{w}_a$

as part of the constraints. With the method of Lagrange multipliers, the solution to the constrained minimization problem in Equation (11.28) is

$$\hat{\mathbf{w}}_{\mathrm{o}} = \mathbf{C}_{\mathrm{lm}}^{-1}\mathbf{A}^{\mathrm{H}}\left[\mathbf{A}\mathbf{C}_{\mathrm{lm}}^{-1}\mathbf{A}^{\mathrm{H}}\right]^{-1}\mathbf{b}, \tag{11.29}$$

where

$$\mathbf{A} = [\mathbf{v}_{\mathrm{SH}}(\Omega_0) \quad \mathrm{diag}(\mathbf{C}_{\mathrm{lm}})\mathbf{w}_{\mathrm{a}}]^{\mathrm{H}} \tag{11.30}$$

and

$$\mathbf{b} = [1 \quad 0]^{\mathrm{T}}. \tag{11.31}$$

The proof is provided in Delikaris-Manias *et al.* (2016).

Calculation of the Post-Filter

To ensure a distortionless response at the output of the proposed algorithm, the estimated cross-spectrum $\Phi_{S_a S_o}$ is only utilized when it is smaller than the output energy of the beamformer; the output energy $\hat{S}$ is determined as

$$\hat{S} = \min\left(\left|\mathbf{w}_{\mathrm{b}}^{\mathrm{H}}\mathbf{C}_{\mathbf{x}}\mathbf{w}_{\mathrm{b}}\right|, \left|\Phi_{S_a S_o}\right|\right). \tag{11.32}$$

The post-filter $G \in [0, 1]$ in Equation (11.8) that adaptively equalizes the energy of the robust beamformer determined by the weight vector $\mathbf{w}_{\mathrm{b}}$ to obtain the energy according to the estimate $\hat{S}$ is given by

$$G = \sqrt{\frac{\hat{S}}{\left|\mathbf{w}_{\mathrm{b}}^{\mathrm{H}}\mathbf{C}_{\mathbf{x}}\mathbf{w}_{\mathrm{b}}\right|}}. \tag{11.33}$$

11.3.3 Smoothing Techniques

Although a post-filtering process can provide higher noise and interferer suppression than conventional beamforming techniques, it may suffer from artifacts. The value of G is calculated according to the cross-spectral densities between microphone signals for each time–frequency frame. In a real recording scenario, the levels of sound sources with different directions of arrival may fluctuate rapidly and result in rapid changes in the calculated values of G. By modulating the beamformer output signal $\mathbf{w}^{\mathrm{H}}\mathbf{x}$, as in Equation (11.8), with the post-filter G, audible artifacts can be generated. The main cause is the relatively fast fluctuation of the post-filter estimates, which introduces a high variance in the G values in the interval $[0, 1]$ at each time–frequency frame. This is evident especially in low-SNR conditions, and is caused by the modulation of the input signal with the parameter G. This phenomenon can be explained as an inaccurate parameter estimation in the post-filter calculation, and results in randomly spaced spectral peaks in the filtering function. In the time domain, these translate to very short sinusoids with frequencies varying from one time frame to another. The resulting auditory event is commonly referred to as a musical noise or bubbling effect. Similar effects have been reported in Wiener-based post-filters, as discussed in Section 5.9.2, and intensity-based spatial filtering techniques (Kallinger *et al.*, 2009).

Smoothing techniques are employed to mitigate such artifacts. These can be either static or adaptive. In previous post-filters, such as the ones suggested by McCowan and

Bourlard (2003) or Delikaris-Manias and Pulkki (2013), the auto- and cross-spectral densities are smoothed with a recursive filter. There are two parameters that can be adjusted to reduce annoying artifacts: a further averaging/smoothing coefficient of the post-filter, and the spectral floor. Averaging can be employed in the time axis to reduce the variance of the post-filter function. An exponential or logarithmic frequency-dependent averaging coefficient controls how much smoothing is performed across the frequency axis. The spectral floor defined as the lower allowed limit for the post-filter is adjusted to higher values with the cost of decreasing the algorithm's performance. The values for both parameters depend on whether the target signal is present at the frame (k, n) as well as on the background noise level.

Static Smoothing

This type of averaging or smoothing is essentially a single-pole recursive filter, defined as

$$G(k, n) = \alpha(k)G(k, n) - (1 - \alpha(k))G(k, n - 1), \tag{11.34}$$

where $G(k, n)$ are the smoothed gain coefficients for frequency k and time i, and $\alpha(k)$ are the smoothing coefficients for each frequency k. The values of static smoothing remain the same for each type frame and can be defined so that more/less averaging can be applied for specific frequency ranges.

Adaptive Smoothing

This type of smoothing depends on the signal activity and the background noise at each time–frequency block. The signal activity detection can be performed, for example, with a voice activity detector or the minimum statistics approach. The minimum statistics approach estimates the instantaneous SNR, under the assumptions that the speech and the noise are statistically independent and that the power of the speech signal decays to the power of the noise level (Martin, 2001). The SNR is estimated by averaging the instantaneous SNR over each sensor of the microphone array. The time–frequency-dependent adaptive smoothing coefficient α is then given by

$$\alpha(k, n) = \frac{\text{SNR}(k, n)}{1 + \text{SNR}(k, n)}, \tag{11.35}$$

where SNR is the SNR estimated for each time–frequency frame. Upper and lower bounds are applied so that α is limited in $[\alpha_{\min}, \alpha_{\max}]$. The post-filter is then smoothed according to the recursive formula in Equation (11.34) to reduce the estimator variance. For a high estimate of SNR, the averaging coefficient decreases to unity, meaning that less averaging is performed. When the SNR is very low, α is near unity and more averaging is performed. Sudden changes in the post-filtering function occur only at signal onsets or offsets. The values of the smoothing parameters on those changes are crucial since they are related to the amount of audible musical noise.

Spectral Floor

In a reverberant environment with multiple instantaneous talkers, the fluctuations of $G(k, n)$ may vary significantly, especially in the presence of background noise. In spite of the time-averaging process, these fluctuations may still produce audible musical noise.

Therefore, a lower bound λ is imposed on G to prevent the resulting values from reaching below a certain level:

$$G(k,n) = \begin{cases} G(k,n) & \text{if } G(k,n) \geq \lambda, \\ \lambda & \text{if } G(k,n) < \lambda. \end{cases} \tag{11.36}$$

The spectral floor λ of the derived parameter G can be adjusted according to the application, and it is a trade-off between the effectiveness of the post-filtering method and the preservation of the quality of the output signal. The effect of λ on the annoyance caused by the artifacts has been studied with listening tests by Delikaris-Manias and Pulkki (2013). The parameters under test were different levels of spectral floor under different SNR conditions. The motivation for the listening tests was to tune the CroPaC post-filter for minimal artifacts. For the specific array, the results showed that a value of 0.2 produced imperceptible artifacts when compared to reference signals for SNR values down to 10 dB.[1]

11.4 Implementation Examples

The performance of the post-filter proposed in this chapter is shown, first in ideal conditions and afterwards in simulated and real rooms.

11.4.1 Ideal Conditions

In this part, no microphone array is employed and hence there is no effect of the encoding process such as noise amplification and/or spatial aliasing. The two beamformer outputs were generated directly in the spherical harmonic domain for order $L = 3$, under different sound field scenarios: one target source with one and multiple interferers, both defined as white noise sources, and diffuse sound. The diffuse sound is generated as a set of 10 000 white Gaussian noise sources with zero mean at random azimuthal and elevation positions with a direct-to-diffuse ratio of 5 dB.

Static CroPaC Results
For the case of the static CroPaC, both weights $\mathbf{w}_1$ and $\mathbf{w}_2$ are applied to the spherical harmonic matrix $\mathbf{Y}_{lm}$ to select the desired harmonic. The spherical harmonics are defined at a uniform grid Ω_s to visualize their directivity. In this case the weights are simply applied to select a specific spherical harmonic:

$$b_1(\Omega_s) = \mathbf{w}_1^H \mathbf{Y}(\Omega_s), \tag{11.37}$$
$$b_2(\Omega_s) = \mathbf{w}_2^H \mathbf{Y}(\Omega_s), \tag{11.38}$$

where $\mathbf{w}_1$ selects the spherical harmonic with $l = 1, m = 1$ and $\mathbf{w}_2$ the $l = 2, m = 2$. The cross-spectrum of the two beamformers is the element-wise multiplication

$$b_{cs}(\Omega_s) = \left[\mathbf{w}_1^H \mathbf{Y}(\Omega_s)\right] \odot \left[\mathbf{w}_2^H \mathbf{Y}(\Omega_s)\right]. \tag{11.39}$$

The resulting beam patterns and their product are shown in Figure 11.6.

---

[1] Sound samples are available at http://research.spa.aalto.fi/projects/cropac/soundExamples/index.html.

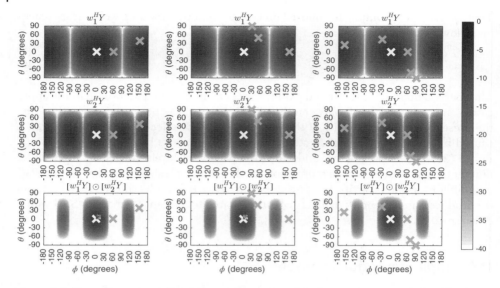

Figure 11.6 Three-dimensional directivity pattern, projected on the two-dimensional plane, for the first static beamformer (top row), for the second static beamformer (middle row), and for their product (bottom row) in a diffuse field with the presence of different numbers of interferers. The white x denotes the position of the target sound source, and the gray x the positions of the interferers.

Adaptive CroPaC Results

For the case of the adaptive CroPaC the static beamformer weights, $\mathbf{w}_a$, are synthesized theoretically as a rotationally symmetric beamformer as described by Rafaely (2015). The adaptive beamformer weights are synthesized according to Equation (11.29):

$$b_a(\Omega_s) = \mathbf{w}_a^H \mathbf{Y}(\Omega_s) \tag{11.40}$$

for the static beam pattern, and

$$b_o(\Omega_s) = \mathbf{w}_o^H \mathbf{Y}(\Omega_s) \tag{11.41}$$

for the adaptive beam pattern. The cross-spectrum of the two beamformers is the element-wise multiplication

$$b_{cs}(\Omega_s) = \left[\mathbf{w}_a^H \mathbf{Y}(\Omega_s)\right] \odot \left[\mathbf{w}_o^H \mathbf{Y}(\Omega_s)\right], \tag{11.42}$$

with $\mathbf{w}_a^H \mathbf{w}_o = 0$. The resulting beam patterns are shown in Figure 11.7 for scenarios with one, two, three, and four interferers in a diffuse field. Each column in Figure 11.7 represent different sound scenarios with a target source, marked with a black x, and the interferers, each marked with a gray x. The static beam pattern remains constant for all the cases of different numbers of interferets (top row). The adaptive beamformer places nulls in the positions of the interferers, as in the conventional adaptive beamformers (middle row). The proposed post-filter is the result of the cross-spectral density of the two beam patterns. In this example, the directional selectivity is estimated by the element-wise multiplication of the patterns $b_{cs}(\Omega_s)$, as in Equation (11.39), and the resulting beam pattern is shown in Figure 11.7 (bottom row).

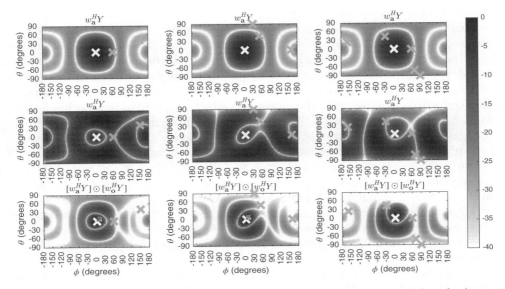

Figure 11.7 Three-dimensional directivity pattern, projected on the two-dimensional plane, for the static beamformer (top row), for the adaptive beamformer (middle row), and for their product (bottom row) in a diffuse field with the presence of different numbers of interferers. The white x denotes the position of the target sound source and the gray x the positions of the interferers.

11.4.2 Prototype Microphone Arrays

The CroPaC post-filter has been shown to be effective in simulated scenarios. To study its performance in real acoustic conditions, various compact prototype microphone arrays have been built. These are shown in Figure 11.8, and consist of open and rigid cylindrical arrays, but also mobile-phone-like arrays made out of styrofoam or wooden bodies. In this section, results from some of these arrays are discussed.

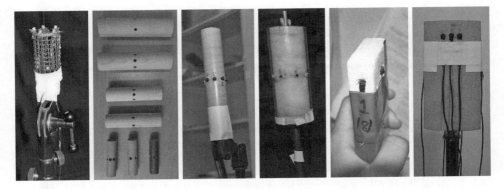

Figure 11.8 Prototype microphone arrays built and tested with the CroPaC post-filtering scheme.

Uniform Cylindrical Microphone Array

An eight-microphone, rigid-body, cylindrical array of radius 1.3 cm and height 16 cm was employed with DPA 4060 sensors placed equidistantly in the horizontal plane every 45°. The microphones were mounted on the perimeter at the half-height of the rigid cylinder. Although only five sensors were required in theory in a unified circular array to deliver microphone components of the second order, the additional sensors provided an increased aliasing frequency and higher SNR when compared to an array having the same radius with fewer sensors. The array is shown in Figure 11.8 (third array from the left). The array was positioned in the center of a room with a measured reverberation time of $RT_{60} = 500$ ms, mounted on a tripod. The sound field scenario consisted of two loudspeakers placed at 0° and 90° in the azimuthal plane, 1.5 m away from the microphone array, transmitting speech signals simultaneously. The background noise in the room was mainly from a computer and air-conditioning noise. Two objective measures are employed to evaluate the performance of CroPaC and are compared to those of the McCowan post-filter: the frequency-weighted segmental SNR enhancement (segSNRE), and the mel-frequency cepstrum coefficients (MFCC) distance. Both instrumental measure results indicated better performance when compared to previous post-filters (Delikaris-Manias and Pulkki, 2013). Spectrograms of the microphone input, the output of an MVDR, and the output of the CroPaC post-filter are shown in Figure 11.9.

Arbitrary Microphone Array on a Mobile-Like Device

A seven-channel microphone array was fitted on a rigid rectangular object imitating a mobile device of size $5.5 \times 2 \times 11$ cm. The microphone array prototype is shown in Figure 11.8 (last array on the right). Three DPA 4060 microphones were fitted in the front of the array and four in the back. The array manifold was measured in an anechoic chamber, with the array placed on a turntable and a loudspeaker at 2 m distance in the same

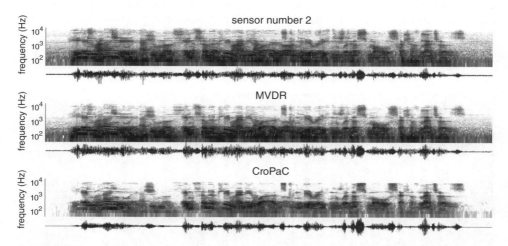

Figure 11.9 Waveforms and spectrograms for the noisy microphone output, the MVDR output, and the CroPaC output. The MVDR provides limited attenuation due to the compact size of the array. The CroPaC post-filter provides additional attenuation, especially in the low-frequency region.

plane. The turntable performed 5° rotations and one measurement was obtained per rotation with the swept-sine technique (Farina, 2000), resulting in a total of 72 measurements. The virtual microphone design technique discussed in Section 11.2.1 was then used to generate the signals of four directional microphones: two dipoles with their positive phase at 0° and 90°, and two quadrupoles with their positive phase at at 0° and 45°.

The performance of the post-filter was tested with a recording scenario having four sound sources at various levels of background noise. The recording scenarios were generated in a listening room with $RT_{60} = 300$ ms. The sound sources were at 0°, 90°, 180°, and 270°. The background noise was generated with four additional loudspeakers in the corners of the room. The performance of the array was tested by steering the directional microphones towards each sound source and calculating the CroPaC output. The smoothing coefficient α was calculated according to the instantaneous SNR from Equation (11.35), and G was smoothed according to Equation (11.35). All signals were transformed into the time–frequency domain with a three-band multi-resolution STFT and cutoff frequencies of 0.8, 3, and 16 kHz. The window sizes of each band were of 1024, 256, and 64 samples, and the hopsizes 512, 128, and 32 samples. The sampling frequency was set at 44.1 kHz. The results indicated an SNR improvement over MVDR and a lower MFCC distance due to musical noise produced in the output, as shown in Delikaris-Manias and Pulkki (2014).

Spherical Microphone Array

The adaptive version of CroPaC was tested in a multi-talker scenario in a room with dimensions of $5.4 \times 6.3 \times 2.7$ m. A microphone array was placed approximately in the center of the room. The microphone array was a compact rigid spherical array of radius 1.5 cm consisting of 16 microphones in a nearly uniform arrangement. The distribution of the microphones on the sphere was performed based on three-dimensional sphere-covering solutions (Conway and Sloane, 2013). A spherical microphone array room simulator was used to calculate the impulse responses (Jarrett *et al.*, 2012) based on the image source method (Allen and Berkley, 1979). The acquired impulse responses were used to simulate acoustical scenarios with multiple interferers and different reverberation times. Conventional distance measures such as SNRE, MFCC, log-area ratio and cepstrum distance were used to compare the CroPaC output to a static beamformer, an adaptive beamformer, and a multichannel Wiener filter. In all cases the CroPaC outperformed the previous algorithms, as shown in Delikaris-Manias *et al.* (2016).

11.5 Conclusions and Further Remarks

Capturing a target signal of interest in a reverberant environment with a number of competing sound sources is a challenging task. While typical microphone array signal processing techniques, such as adaptive beamforming, provide a flexible framework for such tasks, they impose constraints on the geometrical properties of the array, and typically require large arrays with many sensors. In this chapter we have shown the formulation and examples of a technique for directional filtering using compact microphone arrays that can achieve superior reduction of spatial noise. The technique operates as a post-filter at the output of a beamformer, adaptively in the time–frequency domain. In contrast to previous post-filtering techniques which only suppress diffuse sound, the proposed technique applies additional interferer attenuation. Its formulation is based on the idea of estimating a soft masker by utilizing the cross-spectral density of the output of two beamformers: either two static beamformers or a static and an adaptive one. Both beamformers are designed with unity gain in the look direction. By taking the cross-spectrum, the orthogonality of the weights results in a reduction of spatial and microphone noise, while the unity gain in the look direction captures the energy of the target sound signals. The post-filter is then calculated and applied as a time–frequency soft masker to the output of another beamformer which provides the initial attenuation. Sound examples for the static and adaptive version of the algorithm and an acoustic camera implementation, as in McCormack *et al.* can be found at

- http://research.spa.aalto.fi/projects/cropac/soundExamples/index.html
- http://research.spa.aalto.fi/publications/papers/SHspatialfilter/
- http://research.spa.aalto.fi/publications/papers/acousticCamera/

11.6 Source Code

Example Matlab code is provided in Listing 11.1. The code requires two audio files (which can be downloaded from http://research.spa.aalto.fi/projects/cropac/soundExamples/index.html), generates reverberant noisy microphone input signals, and processes them with the static CroPaC post-filtering algorithm.

Listing 11.1: CroPaC Post-Filtering Example

```
1   function [s,x] = demo_cropac()
2   %------------------------------------
3   % Cross Pattern Coherence algorithm for spatial filtering applications
4   %------------------------------------
5   % Symeon Delikaris-Manias and Ville Pulkki
6   % Copyright Aalto University, Dept Signal Processing and Acoustics 2017
7   %------------------------------------
8
9   % anechoic signals: download at:
10  % http://research.spa.aalto.fi/projects/cropac/...
11  % soundExamples/soundsamples/2voices_anechoic.wav
12  [sig,fs] = audioread('2voices_anechoic.wav');
13  sigLength = (size(sig,1));
14  sig1 = sig(:,1);
15  sig1=sig1./max(abs(sig1))*2; % normalize
16  sig2 = sig(:,2);
17  sig2=sig2./(max(abs(sig2))); % normalize
18
19  % noise signals
20  [b1,a1]=butter(1,(40/fs)*2);
21  g1=2^6;
22  [b2,a2]=butter(2,(42/fs)*2);
23  g2=g1^2;
24  n=(rand(length(sig1),7)-0.5);
25  no(:,1)=n(:,1);
```

```
26  no(:,2)=filter(b1,a1,n(:,2))*g1;
27  no(:,3)=filter(b1,a1,n(:,3))*g1;
28  no(:,4)=filter(b2,a2,n(:,4))*g2;
29  no(:,5)=filter(b2,a2,n(:,5))*g2;
30  no=0.017*no;
31
32  % microphone input signals
33  interferer_angle = 90;
34  s(:,1)=sig1+sig2+no(:,1);                            % omnidirectional
35  s(:,2)=sig1+sig2*cosd(interferer_angle)+no(:,2);     % dipole at 0^o
36  s(:,3)=sig1+sig2*sind(interferer_angle)+no(:,3);     % dipole at 90^o
37  s(:,4)=sig1+sig2*cosd(2*interferer_angle)+no(:,4);   % quadrupole at 0^o
38  s(:,5)=sig1+sig2*sind(2*interferer_angle)+no(:,5);   % quadrupole at 90^o
39
40  % Add reverberation
41  for i=1:2
42      rev=[zeros(fs/100,1); rand(fs/4,1)-0.5 / sqrt(10)...
43          * max(s(:,1))]/200;
44      revsig=conv(rev, sig1+sig2);
45      s(:,1)=s(:,1) + revsig(1:length(s(:,1)));
46      s(:,2)=s(:,2) + revsig(1:length(s(:,1)))*cosd((i*36));
47      s(:,3)=s(:,3) + revsig(1:length(s(:,1)))*sind((i*36));
48      s(:,4)=s(:,4) + revsig(1:length(s(:,1)))*cosd(2*(i*36));
49      s(:,5)=s(:,5) + revsig(1:length(s(:,1)))*sind(2*(i*36));
50  end
51
52  % TF transform
53  winsize=1024;
54  hopsize=512;
55  for m=1:size(s,2)
56      [S(:,:,m),time,frequency] = stft(s(:,m),winsize,hopsize,fs);
57  end
58
59  % Calculate cross-pattern coherence post filter (dipole*quadrupole)
60  G=2*real(conj(S(:,:,2)).*S(:,:,4))./...
61      ( abs(S(:,:,2).^2)+abs(S(:,:,3).^2)...
62      +abs(S(:,:,4).^2)+abs(S(:,:,5).^2) );
63
64  % spectral floor
65  Gmin=0.1;
66  G=max(G,Gmin);
67
68  % smoothing
69  clear Gav
70  a=ones(size(G,1),1)*0.2;
71  Gav=zeros(size(G,1),size(G,2)); Gav(:,1)=G(:,1);
72  for t=2:size(G,2)
73      Gav(:,t)=a.*Gav(:,t-1)+(1-a).*G(:,t);
74  end
75
76  % enhanced output signal
77  X = max(0,(S(:,:,1))).*Gav;
```

```
78  x = istft(X,winsize,hopsize);
79  soundsc(s(:,1),fs);          % omni signal
80  pause(ceil(sigLength/fs))
81  soundsc(real(x),fs); % test signal
82
83  % plots
84  figure(2)
85  subplot(211);surf(time, frequency(1:end/2),...
86      20*log10(abs(S(1:end/2,:,1))));
87  title('omni microphone')
88  subplot(212);surf(time, frequency(1:end/2),...
89      20*log10(abs(X(1:end/2,:))));
90  title('CroPaC output')
91  for n= 1:2
92      subplot(2,1,n)
93      shading flat
94      set(gca,'YScale','log','YLim',[frequency(2) 20000],'XLim',[0,17])
95      ylabel('frequency (Hz)')
96      xlabel('time (seconds)')
97      view(0,90)
98      caxis([-40 40])
99  end
100
101 end
102
103 % forward TF function
104 function [X,time,frequency] = stft(x,winsize,hopsize,fs)
105
106 %   [X,Tw] = stft(x,wintype,winsize,hopsize)
107 %  INPUT:
108 %   x = input signal
109 %   winzise = size of the window for the TF analysis
110 %   hopsize = size of the overlap between two windows
111 %
112 %  OUTPUT:
113 %   X frequency-time signal from the stft analysis of input x
114
115 if size(x,1)<size(x,2)
116     x=x';
117 end
118 if nargin < 2;
119     winsize = 512;
120     hopsize = 256;
121     fs      = 44100;
122 end
123 if nargin == 4;
124     fs      = 44100;
125 end
126 x      = [zeros(hopsize,1);x;zeros(winsize,1)];
127 w      = sqrt(hanning(winsize,'periodic')');
128 Nwin   = floor((length(x)-winsize)/hopsize)+1;
129 X      = zeros(winsize,Nwin);
```

```
130  index = 0;
131  for i=1:Nwin
132      x_temp = x(index+1:index+winsize,1);
133      x_w    = x_temp' .* w;
134      X(:,i) = fft(x_w);
135      index  = index + hopsize;
136  end
137  % save frequency and time vectors
138  time      = (0:size(X,2)-1)*hopsize/(fs);
139  frequency = (0:size(X,1)-1)*(fs)/winsize;
140
141  end
142
143  % inverse TF function
144  function x = istft(X, winsize,hopsize)
145  %[x] = istft(X,winsize,hopsize)
146  %  INPUT:
147  %   X = input signal in stft domain: (frequency)x(time)
148  %   winzise = size of the window
149  %   hopsize = size of the overlap between two windows
150  %
151  %  OUTPUT:
152  %   x time domain signal from the istft analysis of input X
153
154  Nwin  = size(X,2);
155  x     = zeros(Nwin*hopsize+winsize,1);
156  w     = sqrt(hanning(winsize,'periodic'));
157  index = 0;
158  for ii = 1:Nwin
159      temp                       = ifft(X(:,ii));
160      x(index+1:index+winsize) = x(index+1:index+winsize) +temp.*w;
161      index                      = index + hopsize;
162  end
163  x = x(hopsize+1:end-winsize-hopsize/4);
164
165  end
```

References

Allen, J.B. and Berkley, D.A. (1979) Image method for efficiently simulating small-room acoustics. *Journal of the Acoustical Society of America*, **65**(4), 943–950.

Atkins, J. (2011) Robust beamforming and steering of arbitrary beam patterns using spherical arrays. *IEEE Workshop on Applications of Signal Processing to Audio and Acoustics (WASPAA)*, pp. 237–240. IEEE.

Conway, J.H. and Sloane, N.J.A. (2013) *Sphere Packings, Lattices and Groups*, Grundlehren der mathematischen Wissenschaften vol. 290. Springer Science & Business Media, New York.

Delikaris-Manias, S. and Pulkki, V. (2013) Cross-pattern coherence algorithm for spatial filtering applications utilizing microphone arrays. *IEEE Transactions on Audio, Speech, and Language Processing*, **21**(11), 2356–2367.

Delikaris-Manias, S. and Pulkki, V. (2014) Parametric spatial filter utilizing dual beamformer and snr-based smoothing. *55th International Conference of the Audio Engineering Society: Spatial Audio*. Audio Engineering Society.

Delikaris-Manias, S., Valagiannopoulos, C.A., and Pulkki, V. (2013) Optimal directional pattern design utilizing arbitrary microphone arrays: A continuous-wave approach. *Audio Engineering Society Convention 134*. Audio Engineering Society.

Delikaris-Manias, S., Vilkamo, J., and Pulkki, V. (2015) Parametric binaural rendering utilizing compact microphone arrays. *IEEE International Conference on Acoustics, Speech, and Signal Processing (ICASSP)*, pp. 629–633.

Delikaris-Manias, S., Vilkamo, J., and Pulkki, V. (2016) Signal-dependent spatial filtering based on weighted-orthogonal beamformers in the spherical harmonic domain. *IEEE/ACM Transactions on Audio, Speech, and Language Processing*, **24**(9), 1507–1519.

Deschamps, G.A. and Cabayan, H.S. (1972) Antenna synthesis and solution of inverse problems by regularization methods. *IEEE Transactions on Antennas and Propagation*, **20**(3), 268–274.

Duraiswami, R., Gumerov, N.A., and Donovan, A.E. (2010) Spherical sound scene analysis. *2nd International Symposium on Ambisonics and Spherical Acoustics*, Paris, France.

Farina, A. (2000) Simultaneous measurement of impulse response and distortion with a swept-sine technique. *Audio Engineering Society Convention 108*. Audio Engineering Society.

Farina, A., Capra, A., Chiesi, L., and Scopece, L. (2010a) A spherical microphone array for synthesizing virtual directive microphones in live broadcasting and in post production. *40th International Audio Engineering Society Conference: Spatial Audio: Sense the Sound of Space*. Audio Engineering Society.

Farina, A., Capra, A., Chiesi, L., and Scopece, L. (2010b) A spherical microphone array for synthesizing virtual directive microphones in live broadcasting and in post production. *40th International Conference of the Audio Engineering Society: Spatial Audio: Sense the Sound of Space*. Audio Engineering Society.

Fazi, F.M. and Nelson, P.A. (2007) The ill-conditioning problem in sound field reconstruction. *Audio Engineering Society Convention 123*. Audio Engineering Society.

Gauthier, P.A., Camier, C., Pasco, Y., Berry, A., Chambatte, E., Lapointe, R., and Delalay, M.A. (2011) Beamforming regularization matrix and inverse problems applied to sound field measurement and extrapolation using microphone array. *Journal of Sound and Vibration*, **330**(24), 5852–5877.

Jarrett, D., Habets, E., Thomas, M., and Naylor, P. (2012) Rigid sphere room impulse response simulation: Algorithm and applications. *Journal of the Acoustical Society of America*, **132**(3), 1462–1472.

Jarrett, D.P. and Habets, E.A.P. (2012) On the noise reduction performance of a spherical harmonic domain tradeoff beamformer. *IEEE Signal Processing Letters*, **19**(11), 773–776.

Josefsson, L. and Persson, P. (2006) *Conformal Array Antenna Theory and Design*. John Wiley & Sons, Chichester.

Kallinger, M., Del Galdo, G., Kuech, F., Mahne, D., and Schultz-Amling, R. (2009) Spatial filtering using directional audio coding parameters. *IEEE International Conference on Acoustics, Speech, and Signal Processing*, pp. 217–220.

Kirkeby, O., Nelson, P.A., Hamada, H., and Orduna-Bustamante, F. (1998) Fast deconvolution of multichannel systems using regularization. *IEEE Transactions on Speech and Audio Processing*, **6**(2), 189–194.

Li, Z. and Duraiswami, R. (2007) Flexible and optimal design of spherical microphone arrays for beamforming. *IEEE Transactions on Audio, Speech, and Language Processing*, **15**(2), 702–714.

Martin, R. (2001) Noise power spectral density estimation based on optimal smoothing and minimum statistics. *IEEE Transactions on Speech and Audio Processing*, **9**(5), 504–512.

McCormack, L., Delkaris-Manias, S., and Pulkki, V. "Parametric Acoustic Camera for Real-Time Sound Capture, Analysis and Tracking," Proceedings of the 20th International Conference of Digital Audio Effects (DAFx-17), Edinburgh, UK, September 5–9, 2017.

McCowan, I. and Bourlard, H. (2003) Microphone array post-filter based on noise field coherence. *IEEE Transactions on Speech and Audio Processing*, **11**(6), 709–716.

Rafaely, B. (2015) *Fundamentals of Spherical Array Processing*. Springer, New York.

Saff, E.B. and Kuijlaars, A.B. (1997) Distributing many points on a sphere. *The Mathematical Intelligencer*, **19**(1), 5–11.

Sun, H., Yan, S., and Svensson, U.P. (2010) Space domain optimal beamforming for spherical microphone arrays. *IEEE International Conference on Acoustics, Speech, and Signal Processing (ICASSP)*, pp. 117–120.

Teutsch, H. (2007) *Modal Array Signal Processing: Principles and Applications of Acoustic Wavefield Decomposition*. Springer, Berlin.

Van Trees, H.L. (2004) *Detection, Estimation, and Modulation Theory*. John Wiley & Sons, Chichester.

Van Veen, B.D. and Buckley, K.M. (1988) Beamforming: A versatile approach to spatial filtering. *IEEE ASSP Magazine*, **5**(2), 4–24.

Vilkamo, J. and Delikaris-Manias, S. (2015) Perceptual reproduction of spatial sound using loudspeaker-signal-domain parametrization. *IEEE/ACM Transactions on Audio, Speech, and Language Processing*, **23**(10), 1660–1669.

12

Microphone-Array-Based Speech Enhancement Using Neural Networks

Pasi Pertilä

Department of Signal Processing, Tampere University of Technology, Finland

12.1 Introduction

As discussed Chapter 10, the noise reduction capacity of beamforming can in practice be rather modest, and the use of post-filtering is often called for to further reduce the noise and interference in the beamformer's output by using time–frequency (TF) masking. The Wiener filter is theoretically an optimal method (in the mean squared error sense) for noise suppression, but it requires the noise power spectrum (or that of the target signal) to be available during operation. This is problematic in typical real-world scenarios, where only the noisy target signal is observed and no explicit noise (or target) signal is available. A traditional speech enhancement approach is to update the estimates of the noise parameters during silence periods of speech. In environments where the noise statistics do not change significantly until the next update is available, this approach can achieve good noise suppression. Different variants of this technique have been developed in the past (see, for example, Diethorn, 2004). However, relying on a voice activity detection scheme inherently increases the system's complexity and decreases its robustness. Furthermore, real-world noise is often dynamic, which violates the assumption of noise stationarity. The errors made in the parameter estimates required by the approach ultimately lead to unwanted distortions in the enhanced signal and to the loss of noise suppression capacity.

This chapter investigates the use of artificial neural networks (ANNs) in learning to predict TF masks from the noisy input data. During operation, this approach only requires the noisy input values, and does not necessitate the estimation of signal or noise parameters. Instead, the approach requires a database of inputs and corresponding target values, so that the neural network can be trained to learn the input–output mapping.

Artificial neural networks are inspired by the operation of biological neural networks, where individual neurons receive inputs from other connected neurons. Based on the total amount of input stimulus to a neuron, it can be triggered to send a nerve impulse to other connected neurons. The human brain is estimated to have approximately

Parametric Time–Frequency Domain Spatial Audio, First Edition. Edited by Ville Pulkki, Symeon Delikaris-Manias, and Archontis Politis.
© 2018 John Wiley & Sons Ltd. Published 2018 by John Wiley & Sons Ltd.
Companion Website: www.wiley.com/go/pulkki/parametrictime-frequency

100 billion neurons, and individual neurons in the human nervous system can receive inputs from 1 up to 100 000 other neurons (Bear *et al.*, 2006). The artificial neural network consists of interconnected units that have weights associated with their inputs. The output of a unit is the weighted sum of the inputs passed through an activation function, which can be linear or non-linear. If the activation function is set as non-linear it allows the network itself to be non-linear, which is suitable for dealing with tasks that are inherently non-linear, such as the generation mechanism of speech signals (Haykin, 1999). The artificial neural network can be defined as a massive parallel system made of individual simple units that have the natural capacity to store and retrieve experiential knowledge. It is thought to represent the brain in two ways: it can acquire knowledge of an environment through a learning process, and the weights of the network are used to store the acquired knowledge (Haykin, 1999).

The physical speech production mechanics starts by the creation of airflow in the lungs, which is then transformed by the glottis located in the larynx. The periodic vibration of the vocal cords results in voiced speech, which consists of harmonic frequencies. A turbulent airflow results in noise-like unvoiced speech. According to the source-filter model, the vocal tract then acts as a tube with resonances and anti-resonances. The sound produced depends on the different positions of the articulators: vocal cords, velum, tongue, teeth, lips, and jaw, and the manner of articulation (O'Shaughnessy, 2000). The effective bandwidth of speech spans up to approximately 7 kHz, since speech production is most effective at these frequencies (O'Shaughnessy, 2000). This chapter focuses on TF mask prediction for speech enhancement in dynamic noise environments using artificial neural networks. The network therefore has to learn how to separate naturally occurring spectral patterns of speech from other everyday sounds. The task is non-trivial due to the complexity of speech production, and because each individual speaker has a unique voice.

This chapter is organized as follows. In Section 12.2, the enhancement framework of microphone array signals using beamforming with post-filtering (TF masking) is reviewed. Then, an overview of the supervised learning framework used for the TF mask-based speech enhancement is presented. Several successful recognition and classification systems employ features obtained from the magnitude spectrum of a noisy input signal. In the case of multichannel or binaural input signals, interchannel and inter-aural features also become available. These cues are related to the spatial properties of the incoming wavefront, and are described together with the common magnitude-based features. The section then ends with a summary of different TF mask designs. Artificial neural networks are reviewed and their training is discussed in Section 12.3, with a specific focus on supervised learning using feed-forward neural networks. Common methods for network weight learning and performance generalization are discussed. The differences between deep neural networks (DNNs) and traditional neural networks are briefly summarized. To make the discussion more concrete, an example case of TF mask-based supervised denoising is presented using simulated monophonic sawtooth signals in Section 12.4. The effect of using different magnitude-based features to learn mask prediction is investigated in conjunction with using different neural network structures. In Section 12.5, the effectiveness of feed-forward neural networks for a real-world enhancement application is explored using recordings from everyday noisy environments, where a microphone array is used to capture the signals. Both spectral and spatial features are explored for TF mask prediction. The generalization capability of the trained neural networks is evaluated by investigating the mask prediction error. Estimated instrumental intelligibility and signal-to-noise ratio (SNR) scores are evaluated to measure how well the predicted masks improve speech quality, using networks trained on different input features. Section 12.6 presents the conclusions; it is followed by Section 12.7, which contains the source code for the mask prediction using simulated data, and for the extraction of spatial features from a microphone array input.

12.2 Time–Frequency Masks for Speech Enhancement Using Supervised Learning

12.2.1 Beamforming with Post-Filtering

The jth microphone signal of an array with M microphones is modeled as a delayed source signal embedded in additive noise. The sound decay of a distant source is approximately equal for the array microphones, and is therefore omitted. By using the short-time Fourier transform (STFT), the model for the spectrum of the jth microphone signal is

$$m_j(k, t) = s(k, t) \cdot a_j(k, t) + n_j(k, t), \tag{12.1}$$

where $s(k, t)$ denotes the source signal, $n_j(k, t)$ is independent and identically distributed noise between the microphones, $a_j(k, t)$ denotes the jth index of a steering vector, the frequency index is denoted by $k = [0, \dots, N - 1]$, N is the length of the discrete Fourier transform (DFT), and t is time frame index. The steering vector, defined as

$$a_j(k, t) \equiv \exp(-J\omega_k \tau_j), \tag{12.2}$$

where $J = \sqrt{-1}$, $\omega_k = 2\pi k / N$, delays the source signal according to the sound propagation time τ_j between the source position $\mathbf{r} \in \mathbb{R}^3$ and the microphone $\mathbf{p}_j \in \mathbb{R}^3$, where $\tau_j = F_s \cdot c^{-1} \|\mathbf{p}_j - \mathbf{r}\|$. The unit of delay is samples with sampling frequency F_s; $\| \cdot \|$ denotes the ℓ_2-norm, and c is the speed of sound in air.

A beamformer combined with a single-channel mask is often referred to as beamforming with post-filtering. The general filter-and-sum beamformer with post-filtering can be written as

$$u(k, t) = \underbrace{\left[\sum_{j=0}^{M-1} m_j(k, t) \cdot w_j(k, t) \right]}_{u_{\mathrm{BF}}(k,t)} \cdot H(k, t), \tag{12.3}$$

where $u_{\mathrm{BF}}(k, t)$ denotes the beamformer output signal, the enhanced signal is denoted by $u(k, t)$, and the post-filter (TF mask) is denoted by $H(k, t)$. In the case of delay-and-sum beamforming, the weights are defined as $w_j(k, t) \equiv a_j^*(k, t) = \exp(J\omega_k \tau_j)$, and $(\cdot)^*$ denotes the complex conjugate taken to undo the effects of the propagation delay in order to time-align the microphone signals before summation. The relative weight with

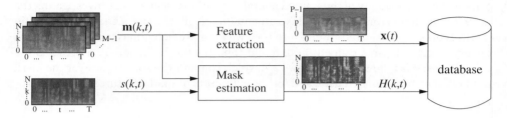

Figure 12.1 An illustration of building a database for supervised mask learning. The input feature vector $\mathbf{x}(t)$ at time t is extracted from the noisy microphone array signal spectrograms $\mathbf{m}(k, t)$, $k = 0, \ldots, N - 1$ at time frame t. The corresponding target mask values $H(k, t)$ are obtained using microphone array signals and a clean reference signal $s(k, t)$.

respect to an arbitrary reference microphone (here, microphone index 0) is sufficient to perform delay-and-sum beamforming:

$$w_j(k, t) = \frac{a_j^*(k, t)}{a_0^*(k, t)} = \frac{\exp(j\omega_k\tau_j)}{\exp(j\omega_k\tau_0)} = \exp(j\omega_k(\tau_j - \tau_0)) = \exp(j\omega_k\Delta\tau_{j,0}), \qquad (12.4)$$

where the wavefront's time difference of arrival (TDOA) in a pair of microphones $\{j, i\}$ is defined as $\Delta\tau_{j,i} \equiv \tau_j - \tau_i$. Therefore, only the relative TDOA values with respect to a reference microphone need to be available in order to time-align the source emitted wavefronts.

12.2.2 Overview of Mask Prediction

This section presents an overview of the supervised learning framework for mask prediction using ANNs. During the learning phase the ANN first predicts the target values from the input features and then changes its internal state (network weights) based on the error between the predicted output and actual target. The input data is typically processed several times over during the training phase. To facilitate this process a database of input and target values is built. The ANN weights are kept unchanged in the testing phase.

Figure 12.1 illustrates extracting feature vectors $\mathbf{x}(t) \equiv [x_0(t), x_1(t), \ldots, x_{P-1}(t)]^T$ from the noisy microphone array signals $\mathbf{m}(k, t) \equiv [m_0(k, t), m_1(k, t), \ldots, m_{M-1}(k, t)]^T$, where P denotes the feature vector length. The clean input signal $s(k, t)$ together with the noisy input signals $\mathbf{m}(k, t)$ are used to produce the mask values $H(k, t)$.

Once the database is constructed, the ANN is trained to predict a mask value $\hat{H}(k, t)$ from the input feature vector $\mathbf{x}(t)$ – see Figure 12.2. The prediction error is used to correct the weights of the ANN using a training algorithm. Finally, after the training

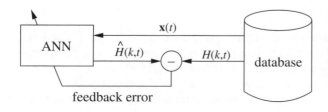

Figure 12.2 An illustration of training an artificial neural network (ANN) to predict target mask values $H(k, t)$ from features $\mathbf{x}(t)$. The weights of the ANN are adjusted based on the feedback error.

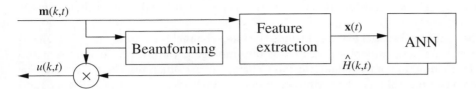

Figure 12.3 An illustration of ANN-based mask prediction for beamformed signals. The features obtained from the input signal are used to predict the mask using the trained ANN. The predicted mask is then applied to the beamformed signal as a post-filter.

phase, the features are inserted into the input of the ANN and propagated through the network using the learned weights. The ANN output then represents the prediction of the corresponding mask. Figure 12.3 presents the use of mask prediction in conjunction with beamforming.

12.2.3 Features for Mask Learning

Magnitude-based features have been used to predict the mask values from the input signal in monophonic denoising and de-reverberation. The most simple type of features includes only the log-magnitude spectrum of the noisy input signal using STFT (Han *et al.*, 2015).

Instead of processing the features in full STFT resolution, more compact features can be extracted to reduce the size of the input vectors and the amount of data. Features such as linear prediction coefficients can compactly model the spectral envelope of a signal, which in the case of speech corresponds to the resonance frequencies of the vocal tract filter (see, for example, O'Shaughnessy, 2000). The mel-frequency cepstral coefficients (MFCCs) sum the log-energies of equally spaced and triangularly weighted mel frequency scale filterbanks, and decorrelate the values using a discrete cosine transform (Davis and Mermelstein, 1980). The non-linear mel scale results from human pitch perception, and therefore the mel filterbank retains perceptually more relevant spectral content than a filterbank with, for example, linear band division. Similarly, the logarithm is taken to produce perceptually more meaningful magnitude values in contrast to linear magnitude. The mel-frequency cepstral coefficients are a widely used feature in several audio-based recognition and classification systems. Similarly to linear prediction analysis, the mel-frequency cepstral coefficients can be used to estimate the spectral envelope of a speech signal. Other hearing-motivated features such as gammatone-frequency cepstral coefficients proposed by Shao *et al.* (2009) and relative spectral transforms with perceptual linear prediction (RASTA-PLP) proposed by Hermansky *et al.* (1992) have been used to obtain features for learning binary masks for speech enhancement by Wang *et al.* (2013). Other features used for mask learning include the amplitude modulation spectrogram used by Kim *et al.* (2009).

The results obtained with deep learning in automatic speech recognition, a task related to speech enhancement, have shown a preference for more primitive features such as direct log-mel filterbank outputs over mel-frequency cepstral coefficients (Deng *et al.*, 2013). Moreover, recent work demonstrates that directly using the time-domain multi-microphone input can be used to learn the acoustic model of an automatic speech recognition system (Sainath *et al.*, 2015). While deep learning methods operating on the time domain have significant potential, we focus here on using the established acoustic features.

The traditional spatial features include the inter-channel time difference (ICTD) and inter-channel level difference (ICLD) of the impinging wavefront. To represent the ICTD as a function of frequency, the inter-channel phase difference (ICPD) is used. For a stereo or binaural recording with two input channels $m_L(k, t)$ (left microphone) and $m_R(k, t)$ (right microphone), the ICPD can be defined as

$$\text{ICPD}_{L,R}(k, t) = \angle m_L(k, t) - \angle m_R(k, t), \tag{12.5}$$

and the ICLD in dB is

$$\text{ICLD}_{L,R}(k, t) = 20 \log_{10}(|m_L(k, t)|) - 20 \log_{10}(|m_R(k, t)|). \tag{12.6}$$

With binaural signals, these two cues are also known as inter-aural time and level differences (ITD and ILD, correspondingly), which are important in the process of how humans locate sounds (Blauert, 1996). For a microphone array with M microphones the number of unique pairs is $M(M - 1)/2$. The number of pairwise features thus increases quadratically as microphones are added to the array.

Using the notation of complex variables ($z = a \exp(\jmath\varphi)$), the jth microphone signal's phase component can be written using the propagation signal model of Equation (12.1) as $\varphi_j(k, t) = \varphi_s(k, t) - \omega_k \tau_j$, where $\varphi_s(k, t)$ is the phase of the source signal. Using this definition, the propagation delay can be written as $\tau_j = -\omega_k^{-1}(\varphi_j(k, t) - \varphi_s(k, t))$. Substituting this into the definition of TDOA results in $\Delta\tau_{j,i} \equiv \tau_j - \tau_i = -\omega_k^{-1} \cdot (\varphi_j(k, t) - \varphi_i(k, t))$. Therefore, the difference between the frequency-dependent phase values can be written using the TDOA values that depend on the frequency in a linear manner:

$$\varphi_i(k, t) - \varphi_j(k, t) = \omega_k \cdot (\tau_j - \tau_i) = \omega_k \Delta\tau_{j,i}. \tag{12.7}$$

Recall that the TDOA $\Delta\tau_{j,i}$ is used to time-align signals in the beamformer, and is assumed to be available. If a frequency component is dominated by an interference source in some direction other than the target direction, the instantaneous phase difference measurement $\angle m_i(k, t) - \angle m_j(k, t)$ will not agree with the phase difference predicted by the TDOA value $\Delta\tau_{j,i}$. To quantify this, a phase-based feature vector that takes every microphone pair $\{j, i\}$ into account can be written as

$$f(k, t) = \frac{2}{M(M - 1)} \sum_{\forall\{j,i\}} \cos((\angle m_i(k, t) - \angle m_j(k, t)) - \omega_k \Delta\tau_{j,i}), \tag{12.8}$$

where $f(k, t)$ is a scalar value in the range $[-1, +1]$. The cosine function wraps the phase values to account for spatial aliasing ($\pm 2\pi$ ambiguity). The maximum value of 1 indicates perfect agreement between measured phase difference and expected phase difference for a TF point. Note that the dimension of the feature $f(k, t)$ is constant with respect to the number of microphones. A related form of Equation (12.8) that uses the direction of arrival vector to obtain the time differences was presented by Pertilä and Nikunen (2014), where the feature was used to predict the Wiener filter. A Matlab function is provided to obtain the phase-based features $f(k, t)$ from a microphone array signal with an arbitrary number of microphones in Section 12.7.

The phase difference error distribution between a pair of sensors and that of the corresponding cosine function inside Equation (12.8) can be represented as a function of the SNR of the pair of received signals (Mustiere *et al.*, 2016). A method to estimate the

SNR at each frequency bin using only the phase difference measurement was presented by the same authors. The SNR estimate was in turn used to obtain the Wiener filter (see, for example, Diethorn, 2004). This provides a theoretical basis and insight into the use of phase differences in TF mask prediction.

Temporal Information

Observed speech signals result from the complex movement of the speech production organs. As in any dynamic system, the knowledge of current and past system states is useful for estimating the future state. Similarly, past observations can be used to aid in predicting the current state. The features discussed so far represent only a snapshot in time, but could be aggregated to convey temporal information. One technique to include temporal information into the input feature is to utilize features from past time frames. Utilizing future time frames is also feasible, but this will introduce a delay to the system in order for it to be causal. Feature vectors from temporally adjacent frames could be concatenated with the current feature vector to make a new input feature vector. Alternatively, the difference between the current feature vector and an adjacent one could be considered as an addition to a feature vector, and such features are often referred as delta features. An immediate drawback of this approach is that concatenating feature vectors increases the input vector's length, and therefore increases the complexity of the ANN, which leads to an increase in the required amount of training data, memory requirements, and computational cost.

12.2.4 Target Mask Design

There are several possibilities to form a target TF mask $H(k, t)$ to be estimated. Here, we review masks extracted using a monophonic noisy input signal $m(k, t) = s(k, t) + n(k, t)$, modeled as the sum of the target signal $s(k, t)$ and noise $n(k, t)$, since, the multichannel Wiener filter can be represented by an MVDR beamformer and a single-channel post-filter, refer to chapter 10. The TF masking operation with a real-valued mask is simply the multiplication of the noisy signal and the mask, $u(k, t) = H(k, t) \cdot m(k, t)$ – see Equation (12.3). In the case of a continuous real-valued mask, the optimal (frequency-domain) Wiener filter is defined as (Diethorn, 2004)

$$H_W(k, t) = \frac{\phi_{ss}(k, t)}{\phi_{ss}(k, t) + \phi_{nn}(k, t)}, \tag{12.9}$$

where the power-spectral densities are defined for the target signal and noise as $\phi_{ss}(k, t) = \mathbb{E}[|s(k, t)|^2]$, $\phi_{nn}(k, t) = \mathbb{E}[|n(k, t)|^2]$, correspondingly, where $\mathbb{E}[\cdot]$ denotes the expectation operator. In practice, the power-spectral densities $(\phi_{ss}(k, t), \phi_{nn}(k, t))$ need to be estimated. A simplified method is to use the squared values of the magnitude spectra,

$$H_W(k, t) = \frac{|s(k, t)|^2}{|s(k, t)|^2 + |n(k, t)|^2}. \tag{12.10}$$

The Wiener filter is bounded in the range [0, 1]. A more general approach is the parameterized Wiener filter (Diethorn, 2004), written using the squared magnitude spectra as

$$H_G(k, t) = \left[1 - \left(\frac{|n(k, t)|}{|m(k, t)|}\right)^\gamma\right]^\beta, \tag{12.11}$$

which can be used to represent other enhancement techniques such as magnitude subtraction ($\gamma = 1, \beta = 1$), power subtraction ($\gamma = 2, \beta = 1/2$), and Wiener filtering ($\gamma = 2, \beta = 1$). Another feasible form of TF mask is the magnitude ratio mask,

$$H_R(k, t) = \frac{|s(k, t)|}{|m(k, t)|}, \tag{12.12}$$

where the value range is $[0, \infty]$. The masking operation can also be written using the magnitude and phase spectra explicitly as

$$u(k, t) = H(k, t) \cdot |m(k, t)| \exp(J\angle m(k, t)). \tag{12.13}$$

The above formulation of the magnitude-based real-valued masking reveals that it does not modify the phase spectrum, since the noisy signal's phase spectrum is retained in the signal reconstruction. Estimating the phase spectrum of the noise component and utilizing it in the signal reconstruction has been shown to enhance the sound quality, as demonstrated by Paliwal *et al.* (2011). Furthermore, utilization of a ground truth noise component for phase modifications in conjunction with traditional magnitude-based enhancement showed significant quality improvements (Paliwal *et al.*, 2011).

To include the phase modification into the masking operation, the use of complex masks has been introduced by Erdogan *et al.* (2015). The complex ratio mask can be defined simply by

$$H_{CR}(k, t) = \frac{s(k, t)}{m(k, t)}. \tag{12.14}$$

The change in the enhanced signal's phase can be seen by examining the complex mask multiplication, where the mask's phase is added to the noisy signal's phase,

$$u(k, t) = |H_{CR}(k, t)| \cdot |m(k, t)| \exp(J(\angle m(k, t) + \angle H_{CR}(k, t))). \tag{12.15}$$

Erdogan *et al.* (2015) proposed taking only the real part of the complex ratio mask in order to avoid learning complex mask values, but to still include some of the phase information. A representation of the complex ratio mask using the real and imaginary parts was derived by Williamson *et al.* (2016):

$$H_{CR}(k, t) = \Re\{H_{CR}(k, t)\} + J \cdot \Im\{H_{CR}(k, t)\}, \tag{12.16}$$

where the real and imaginary part coefficients are real-valued. This allows us to obtain a real-valued target for the neural network. The complex mask can then be reconstructed in the enhancement stage from the network output.

It is noted that the ratio masks with an unrestricted value range can potentially cause large unwanted target values (both in mask estimation and prediction), causing distortions in the output. In practice, to avoid boosting the signal, limiting the mask values using clipping or soft truncation has been applied (Williamson *et al.*, 2016).

12.3 Artificial Neural Networks

The input $\mathbf{x}$ of an artificial neural network can be expressed in the form of a vector of real values, $\mathbf{x} \equiv \mathbf{y}^{(1)} = [y_1^{(1)}, y_2^{(1)}, \ldots, y_{D_1}^{(1)}]^T$, where D_1 is the number of input values. A feedforward neural network consists of L layers; layers $l = 2, \ldots, L$ contain artificial neurons (units), and D_l is the number of units in layer l. The input layer $l = 1$ contains only the input values $\mathbf{y}^{(1)}$ and does not contain units. Inside each unit $j = 1, \ldots, D_l$ of each layer l the input values $y_i^{(l-1)}$, $i = 1, \ldots, D_{l-1}$, are multiplied with the corresponding weight value $w_{j,i}^{(l)}$. The resulting sum is

$$z_j^{(l)} = \sum_{i=1}^{D_{l-1}} w_{j,i}^{(l)} y_i^{(l-1)} + b_j^{(l)}, \tag{12.17}$$

where $b_j^{(l)}$ is the bias weight of unit j at layer l. Each unit's net input $z_j^{(l)}$ is then transformed with an activation function $h(\cdot)$ to obtain the output activation of the unit, $y_j^{(l)} = h(z_j^{(l)})$. The function $h(\cdot)$ is typically a non-linear and differentiable function such as the logistic function or hyperbolic tangent function that limits the output value range. If only linear activation functions are used, the output is a linear combination of the input values representable with a single hidden layer. In contrast, a non-linear activation function allows the network to represent more complicated non-linear relationships between the input and output.

To obtain the output of a neural network, a forward pass is performed by propagating the input vector $\mathbf{y}^{(1)}$ through the network. A recursive matrix equation can be given to obtain the output activation values of layer l:

$$\mathbf{y}^{(l)} = h^{(l)}(\mathbf{W}^{(l)}\mathbf{y}^{(l-1)} + \mathbf{b}^{(l)}), \tag{12.18}$$

where $\mathbf{y}^{(l)}, \mathbf{b}^{(l)} \in \mathbb{R}^{D_l}$, and $\mathbf{W}^{(l)} \in \mathbb{R}^{D_l \times D_{l-1}}$. This type of network is called a feed-forward neural network since the layer inputs are always from the previous layer and not from the current layer or any subsequent layer. The last layer's output $\mathbf{y}^{(L)} \in \mathbb{R}^{D_L}$ represents the network output, where D_L is the number of elements in the output vector.

Figure 12.4 illustrates the architecture of a feed-forward neural network with one hidden layer. The hidden layer's jth unit is depicted in detail with every operation involved in transforming the input vector into output values of the network. The activation function is depicted as the hyperbolic tangent $h^{(2)}(z_j^{(2)}) \equiv \tanh(z_j^{(2)})$, which essentially saturates large input values, and $y_j^{(2)} \in [-1, 1]$.

12.3.1 Learning the Weights

The objective is to learn a function mapping from input to output using a set of training value pairs $(\mathbf{x}_t, \mathbf{t}_t)$, where $\mathbf{t}_t \in \mathbb{R}^{D_L}$ is the desired target vector, $\mathbf{x}_t \in \mathbb{R}^{D_1}$ is the input feature vector, and $t = 1, \ldots, T$, where T is the number of input vectors.

A measure of how well the network can map the input sample t to the desired target is obtained by examining the difference in the predicted output values and the actual target. A cost function that depends on the network weights $\mathbf{W} \equiv \{\mathbf{W}^{(l)}\}_{l=2}^{L}$, biases $\mathbf{B} \equiv \{\mathbf{b}^{(l)}\}_{l=2}^{L}$, and the training data $\{\mathbf{x}_t, \mathbf{t}_t\}_{t=1}^{T}$ is denoted as $E(\mathbf{W}, \mathbf{B}; \mathbf{x}_t, \mathbf{t}_t)$. A common cost function for regression is the sum of squared errors, $E(\mathbf{W}, \mathbf{B}; \mathbf{x}_t, \mathbf{t}_t) = \frac{1}{2}\|\mathbf{t}_t - \mathbf{y}_t^{(L)}\|^2$, and given T training samples the cost function is written as

$$E(\mathbf{W}, \mathbf{B}) = \frac{1}{2T}\sum_t E(\mathbf{W}, \mathbf{B}; \mathbf{x}_t, \mathbf{t}_t) = \frac{1}{2T}\sum_t \|\mathbf{t}_t - \mathbf{y}_t^{(L)}\|^2. \tag{12.19}$$

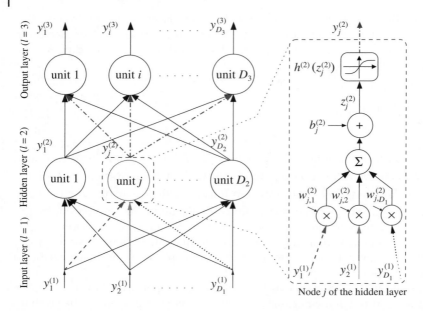

Figure 12.4 Illustration of a forward pass of a feed-forward neural network (FFNN) with one hidden layer. The contents of the jth hidden unit are depicted on the right: each element of the input vector $\mathbf{y}^{(1)} = [y_1^{(1)}, y_2^{(1)}, \dots, y_{D_1}^{(1)}]^{\mathrm{T}}$ is multiplied with the corresponding weight of the hidden unit $w_{j,1}^{(2)}, w_{j,2}^{(2)}, \dots,$ $w_{j,D_1}^{(2)}$ and summed together with the bias value $b_j^{(2)}$. The output of this linear operation is denoted as $z_j^{(2)}$ [see Equation (12.17)], which is then transformed with an activation function to obtain the output value of the unit, $y_j^{(2)} = h^{(2)}(z_j^{(2)})$ [see Equation (12.18)].

Naturally, a small value of the cost function in Equation (12.19) is desired as it means the network has a good capability to predict the outputs given the input values. The free parameters of the network are the weights and biases, and a training algorithm is needed to minimize the cost function by changing these parameters.

Back-Propagation Algorithm
The cost function can be minimized by the method of gradient descent, where the function parameters are changed towards the direction of the negative gradient. The gradients of the cost function with respect to the weights and biases in each layer, $(\partial/\partial \mathbf{W}^{(l)})E(\mathbf{W}, \mathbf{B}; \mathbf{x}_t, \mathbf{t}_t)$ and $(\partial/\partial \mathbf{b}^{(l)})E(\mathbf{W}, \mathbf{B}; \mathbf{x}_t, \mathbf{t}_t)$, need to be evaluated. The back-propagation algorithm is an efficient technique for finding the gradient of cost with respect to the weights of each layer by applying the chain rule of differentiation in order to perform weight updates iteratively (Haykin, 1999). The back-propagation algorithm is summarized as follows – see Nielsen (2015) for a detailed derivation:

1. Randomly initialize weights $\mathbf{W}$ and biases $\mathbf{b}$.
2. Initialize the network input, $\mathbf{y}_t^{(1)} \leftarrow \mathbf{x}_t$.
3. Obtain the output value $\mathbf{y}_t^{(L)}$ using the forward pass in Equation (12.18).
4. For the output layer L, calculate the error term $\delta_t^{(L)} = -(\mathbf{t}_t - \mathbf{y}_t^{(L)}) \odot h'^{(L)}(\mathbf{z}_t^{(L)})$, where $\odot$ denotes the element-wise product, and $h'(\cdot)$ denotes the derivative of the activation function $h(\cdot)$.

5. For layers $l = L - 1, \ldots, 2$, compute the error term $\delta_t^{(l)} = \mathbf{W}^{(l+1)\mathrm{T}} \delta_t^{(l+1)} h'^{(l)}(\mathbf{z}_t^{(l)})$.
6. Compute the gradients of weights and biases: $\Delta \mathbf{W}_t^{(l)} = \delta_t^{(l)} (\mathbf{y}_t^{(l-1)})^{\mathrm{T}}$, $\Delta \mathbf{b}_t^{(l)} = \delta_t^{(l)}$.
7. Repeat steps 2–6 for $t = 1, \ldots, T$, and apply averaged gradients to change the weights:
$\mathbf{W}^{(l)} = \mathbf{W}^{(l)} - \frac{\eta}{T} \sum_t \Delta \mathbf{W}_t^{(l)}$, $\mathbf{b}^{(l)} = \mathbf{b}^{(l)} - \frac{\eta}{T} \sum_t \Delta \mathbf{b}_t^{(l)}$.

Here, η is the learning rate, a so-called hyper-parameter. Other hyper-parameters include the number of hidden units per layer, and the number of hidden layers, which are not learned in the training phase, in contrast to the model parameters, that is, the weights and biases. The learning rate η controls the learning speed and stability of the convergence. Steps 2–7 are referred to as a single epoch, and network training typically requires multiple epochs.

In step 7, the gradients are averaged over the whole sample set T, and then applied. This is referred to as "full batch mode" or "offline mode," and results in the most accurate gradient, but is inherently insufficient due to the large amount of computation per single update. Alternatively, since the error is obtained for each input and target value pair t, the gradient can be changed after each sample has been processed, which is called "stochastic mode" or "online mode." In "mini-batch mode," the gradient is averaged over a small set of samples before updating the weights. The size of the mini-batch can vary, but it should contain diverse examples of the training data. Several variants of applying the gradient to the weights have been developed to speed up convergence and/or increase the optimization stability, such as adding weight momentum, or using local learning rates for weights based on gradient sign changes (LeCun *et al.*, 2012).

12.3.2 Generalization

Overfitting, Underfitting, and Regularization
In *overfitting*, the network essentially starts to memorize the input data examples. As a result, the error on the training set decreases in each epoch, but such a network will not generalize to unseen data. Conversely, in *underfitting* a network does not have the capacity to learn sufficiently from the data to model it. This problem is commonly referred to as the bias–variance dilemma, due to the breakdown of the output error as a sum of a squared bias and variance (see, for example, Haykin, 1999). Such problems can prevent the network from performing well on unseen data, that is, it may not be able to generalize.

In the weight-decay method, a penalty (also known as ℓ_2 regularization) is added to the cost function of Equation (12.19):

$$E(\mathbf{W}, \mathbf{B}) = \frac{1}{2T} \sum_t E(\mathbf{W}, \mathbf{B}; \mathbf{x}_t, \mathbf{t}_t) + \frac{\lambda}{2} \sum_{l=2}^{L} \sum_{j=1}^{D_l} \sum_{i=1}^{D_{l-1}} [W_{ji}^{(l)}]^2, \tag{12.20}$$

where λ is the regularization parameter, and $W_{ji}^{(l)}$ is element (j, i) of weight matrix $\mathbf{W}^{(l)}$. This causes the weights to be divided into ones that have a large influence on the network, and ones with very little influence (excess weights). Without restricting the network complexity, the excess weights can lead to poor network generalization. The weight regularization encourages such weights to take values near zero, which improves generalization (Haykin, 1999). The derivative of the added penalty term in the cost function with respect to the weight matrix is just the weight matrix multiplied by λ. Therefore, gradient descent will push the weight values towards zero with a strength depending on the parameter λ. Selecting the model hyper-parameters, such as the number of layers and amount of regularization, can be performed using cross-validation, for example – this is discussed below.

Example Signal

To create an example signal, $T = 15$ samples were drawn from a simple target function $t(x) = \frac{1}{2}\sin(2\pi x)$ with equal spacing of features in the range $x \in [0, 1]$. Target values were then corrupted with additive Gaussian noise. These value pairs (x_t, t_t) were then used to train four different neural networks. The networks used a non-linear (tanh) activation function. Training was stopped when the mean square error (MSE) stopped decreasing. Another set of test examples (without noise) was created to span the same range with more samples to see how the network estimated the target values for these new data points.

The first network contained three hidden layers and 50 units per layer. Figure 12.5(a) depicts the noisy training data (circles) and the corresponding predicted training values (crosses). Note that the network perfectly fits the training values, indicated by near-zero MSE. The continuous line shows the predicted values for the test points. Note that the predicted values go through the training points, but clearly model the noise of the training data instead of the underlying function depicted by the test data. Although the training MSE is almost zero, the testing MSE is high, which is typical in overfitting.

Figure 12.5(b) considers a network with a single hidden unit. The model approximates the training data with a straight line, and does not learn the non-linear target function. The MSE values of the training and the testing data are large, which is typical in underfitting.

Figure 12.5(c) depicts the outputs of a network with two hidden units. Now, the model is just expressive enough to learn the underlying function, but is not complex enough to learn the noise of the training data. Both the training and the testing MSE values are small, which suggests that the network can model the function well.

Figure 12.5(d) displays the most complex network's result again, but this time with a weight decay value of $\lambda = 0.001$. The MSE values of the training and testing data are small, which indicates that the network modeled the function well. If the regularization parameter λ were too large, the network would underfit. Conversely, if λ were too small, the network would start to overfit, similarly to the unregularized network, where effectively $\lambda = 0$.

Cross-Validation

In cross-validation, the data is divided into two randomly drawn sets, training and validation, where only the training set is used to update the weights. The validation set error gives information about the generalization capability of the network and can be used to evaluate the network performance. Different strides of cross-validation exist, where the simple holdout technique uses only the two sets. The first set is used to train the network, and the other set is used to evaluate the network's performance. In K-fold cross-validation the data is divided into K sets, trained using $K - 1$ sets, and tested on the Kth set. The process is performed K times by rotating the sets to eventually exploit all the available data in training and in testing. The network's performance is estimated as the average performance of the test sets. In leave-one-out cross-validation, the extreme

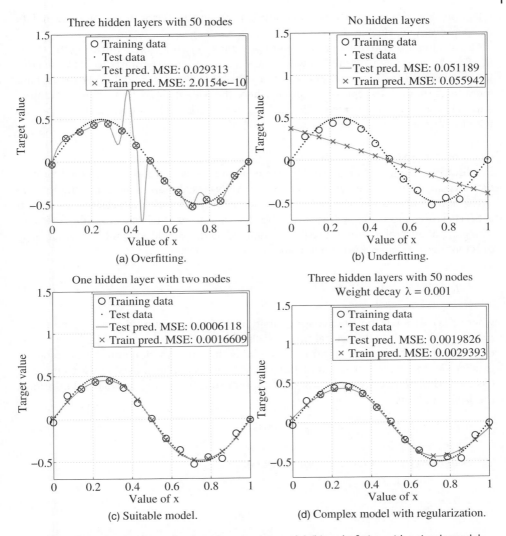

Figure 12.5 Examples of: (a) overfitting with a complex model, (b) underfitting with a simple model, (c) successful fitting with a suitable model, and (d) successful fitting with a complex model with regularization.

case of K-fold cross-validation, all but one example is used for training, while the performance is evaluated on the remaining data point.

Cross-validation can also be used for model selection to decide the network parameters such as shape (width, depth) or amount of regularization. In this approach, the dataset is divided into three parts: training, validation, and testing. For each parameter combination, a network is trained using the training set. Then, the validation set is used to select the model that has lowest MSE. The testing set is then used to estimate the selected network's performance, since the error of the validation set was used as a selection criterion and might be too optimistic about the performance on unseen data (see, for example, Haykin, 1999).

Cross-validation can further be used to prevent a network from overfitting the training data by stopping the training when the validation set error starts to increase after it has decreased for a number of epochs. The point where the validation error starts to increase is thought to represent a point where the weights start to overfit the training data. This approach is referred to as "early stopping." The performance is then reported with the test set.

Ensemble Methods

In addition, having a set of separately trained neural networks and combining their predictions can reduce the estimation error. For ANN-based regression, even simple averaging of the outputs of independently trained neural networks reduces the prediction MSE (Perrone and Cooper, 1993), which is referred to as the basic ensemble method. The general ensemble method (GEM) is defined as the weighted combination of N_{nets} different neural networks (trained using the training set),

$$f_{\text{GEM}}(\mathbf{x}) \equiv \sum_{i=0}^{N_{\text{nets}}-1} \alpha_i f_i(\mathbf{x}), \tag{12.21}$$

where $f_i(\mathbf{x})$ denotes the ith network's mapping of the input $\mathbf{x}$ to the output, where $i = 0, \dots, N_{\text{nets}} - 1$. In the special case of scalar output, the error covariance is defined as $C_{i,j} \equiv \mathbb{E}[(f_i(\mathbf{x}) - f(\mathbf{x}))(f_j(\mathbf{x}) - f(\mathbf{x}))]$, where $f(\mathbf{x})$ is the true function and the expectation is taken over a separate cross-validation set, which has not been used in the training. It has been shown that the optimal weights α_i that sum to one ($\sum_i \alpha_i = 1$) and minimize the MSE function

$$\text{MSE}[f_{\text{GEM}}(\mathbf{x})] \equiv \sum_{i,j} \alpha_i \alpha_j C_{i,j} \tag{12.22}$$

are, as derived by Perrone and Cooper (1993),

$$\alpha_i = \frac{\sum_j C_{i,j}^{-1}}{\sum_k \sum_j C_{k,j}^{-1}}. \tag{12.23}$$

Pre-Processing

Apart from developing optimization techniques, the data can be pre-processed to suit training with neural networks. Shifting and scaling of the input variables to have zero mean and unit standard deviation can speed up learning. For the ith index of a feature vector at time t, $x_t^{(i)}$, the mean is $\mu_i = \frac{1}{T} \sum_{t=1}^{T} x_t^{(i)}$, the standard deviation is $\sigma_i = \sqrt{\frac{1}{T} \sum_{t=1}^{T} (x_t^{(i)} - \mu_i)^2}$, and the normalization is written as

$$\tilde{x}_t^{(i)} = \frac{x_t^{(i)} - \mu_i}{\sigma_i}. \tag{12.24}$$

Similarly, decorrelating the input variables can be useful, since correlated features can slow down the convergence of weight learning (LeCun *et al.*, 2012).

12.3.3   Deep Neural Networks

Recently, deep neural networks (DNNs) have provided state-of-the-art results in different types of classification tasks – see LeCun *et al.* (2015) for a list of examples. The performance gain of DNNs can be attributed to (i) circumventing the use of hand-engineered input features, such as the mel-frequency cepstral coefficients or linear prediction coefficients, by using the raw input signal instead, and (ii) using a large number of stacked layers (typically 5–20).

Successful training of DNNs requires large amounts of data to learn from, and consequently it requires increased computing capacity. The development and utilization of graphics processing units (GPUs) has allowed multiple-times speedups in the training times of neural networks. In addition, using the rectified linear unit (ReLu) $h(z) = \max(z, 0)$ as the activation function of the hidden units results in faster learning in networks with multiple layers in deep supervised training compared to traditional activation functions such as the sigmoid or hyperbolic tangent (LeCun *et al.*, 2015).

While DNNs are an effective learning machine, they are slow to train and overfitting is an issue. A regularization technique by Srivastava *et al.* (2014) referred to as "dropout" has been highly effective in addressing these issues. In the dropout procedure, a randomly chosen set of units is turned off during the feed-forward pass, for example 50% of the units in each layer. They don't therefore affect the error, and the resulting "thinned" network has to model the training batch. For the next batch, different units are turned off. This introduces separate robust "networks" inside the DNN that each model the data in order not to co-adapt too much. During test time, all units are turned on and scaled down to obtain a combined prediction of the target function, which shares a resemblance to the model ensemble technique.

12.4   Mask Learning: A Simulated Example

For illustrative purposes, this section investigates the TF mask learning for enhancement of a simple sawtooth signal. A network is given examples of short sawtooth waves with a randomly chosen fundamental frequency together with the desired mask, which, if multiplied with the noisy signal's spectrum, would produce a spectrum of the clean signal. The network then learns to predict the mask for a general sawtooth wave. Different neural network structures are experimented with to illustrate the effect of the hyper-parameters such as network depth, width, and amount of regularization. The experiments are restricted to monophonic signals instead of multiple input channels for simplicity, and to focus on the problem of TF mask learning with neural networks.

The monophonic sawtooth signal is obtained as

$$s_{\text{saw}}(l) = l \cdot f_0 - \left\lfloor \frac{1}{2} + l \cdot f_0 \right\rfloor, \tag{12.25}$$

where l is time (s), and f_0 is the fundamental frequency (Hz), that is, the lowest harmonic frequency. The sawtooth signal spectrum contains harmonics $k = 1, 2, 3, \ldots$, with amplitude $\frac{1}{k}$, that is, $1, \frac{1}{2}, \frac{1}{3}, \ldots$ The sawtooth signal is then corrupted with Gaussian noise $n(l)$,

$$m(l) = s_{\text{saw}}(l) + n(l). \tag{12.26}$$

Here, the length of a single sawtooth sequence was 64 ms. The sawtooth was linearly attenuated to zero amplitude after a short onset. A silence period was added after each sawtooth signal with a random duration between 0 ms and 160 ms. Such sequences were then concatenated – see Figure 12.7(b) for an illustration.

12.4.1 Feature Extraction

As illustrated in Figure 12.1, features $\mathbf{x}(t)$ and corresponding target values $H(k, t)$ are required to construct a database for supervised learning. The features $\mathbf{x}(t)$ were extracted from the noisy microphone input $m(k, t)$. However, the first practical issue related to network design is that several alternatives for features exist. The lower half of the log-magnitude spectrum corresponding to the real part of the spectrum was used as an input feature, denoted by $\mathbf{x}^{(0)} \in \mathbb{R}^{N/2+1}$, where N is the length of the DFT. This feature does not contain any hand-engineering, that is, it is not designed to capture specific knowledge that is thought to be relevant for the particular problem. Other features could be used instead of the magnitude spectrum, or in conjunction with it. Here, features based on analyzing the periodicity of the signal were utilized. Since the sawtooth signal contains only harmonic frequencies, its autocorrelation function (ACF) can be written using the inverse Fourier transform as

$$\text{ACF}(\tau, t) = \sum_{k=0}^{N-1} m(k, t) m^*(k, t) \exp(\jmath \tau \omega_k), \tag{12.27}$$

which will contain peaks at delay values τ determined by the fundamental frequency f_0. Therefore, the autocorrelation vector of sample delay values $\tau \in [-N/2, N/2]$ was selected as an input feature $\mathbf{x}^{(1)} \in \mathbb{R}^{N+1}$. A third input feature was the peak location and amplitude of the log-magnitude spectrum, which represent a direct estimate of the fundamental frequency and its amplitude, $\mathbf{x}^{(2)} \equiv [\hat{f}_0, \hat{A}_{\max}]^T \in \mathbb{R}^2$. The fourth feature vector considered consisted of the linear prediction (LP) coefficients, which have the inherent capability to model the resonance frequencies. The linear prediction model order was selected to be dependent on the sampling frequency F_s, and the resulting feature is denoted by $\mathbf{x}^{(3)} \in \mathbb{R}^{\lceil F_s/1000+2 \rceil}$.

12.4.2 Target Mask Design

The neural network's training target was set as the frequency-domain Wiener filter. However, the sawtooth signal in Equation (12.10) contains aliasing in the spectral domain if the fundamental frequency is not an integer divisor of the sampling rate (see, for example, Välimaki, 2005). Having aliasing in the spectral estimates that are used to construct the Wiener filter would negatively affect the resulting enhancement, since undesired spectral components would not be attenuated optimally. Therefore, in addition to the time-domain signal in Equation (12.25), a corresponding synthesized target signal that contains the harmonic frequencies with the desired amplitudes was created by a adding scaled sinusoidal components to avoid aliasing:

$$\tilde{s}_{\text{saw}}(l) = \sum_{q=1}^{Q} \frac{1}{q} \sin(l \cdot 2\pi f_0 \cdot q), \tag{12.28}$$

where $Q = \lfloor F_s/(2f_0) \rfloor$ is the number of harmonics under the Nyquist frequency. The STFT was applied to $\tilde{s}_{\text{saw}}(l)$ to obtain the magnitude spectrum of the clean signal, $|s(k, t)|$, and similarly the STFT was applied to the time-domain noise signal $n(l)$ to obtain $|n(k, t)|$. The Wiener filter $H_{\text{W}}(k, t)$ from Equation (12.10) was formed using the two magnitude spectra. The neural network's target vector was then obtained directly from the Wiener filter: $\mathbf{t}_t \leftarrow [H_{\text{W}}(0, t), H_{\text{W}}(1, t), \dots, H_{\text{W}}(N/2, t)]^{\text{T}}$.

12.4.3 Neural Network Training

The effect of different values of network hyper-parameters, the network width, depth, amount of regularization, and the use of the ensemble method were experimented with to see how the mask learning was affected when using different types of input features. The Matlab programming language was used together with its Neural Network Toolbox, and the accompanying code is listed in Section 12.7.1. The number of hidden layers L_h, the network width W, and ℓ_2-regularization parameter η_β were varied in the ranges $L_h = 1, 2, 3, 4$, $W = 16, 32, 64, 128, 256, 1024$, and $\eta_\beta = 0, 0.1, 0.2, 0.3, 0.4, 0.5$. Similarly to the regularization parameter λ defined in Equation (12.20), the Neural Network Toolbox's regularization parameter η_β scales the penalty of the weight size, but in addition uses the value $1 - \eta_\beta$ to multiply the fitting error. The mask estimate $\hat{H}(k, t)$ for each parameter combination was obtained using the basic ensemble method, that is, by averaging the masks of five randomly initialized and trained neural networks.

Processing

The data was simulated with a 16 kHz sample rate, and processed in frames of 16 ms length (256 samples), with 8 ms overlap (128 samples). The STFT used a square-root Hann window. The enhanced sawtooth signal $u(t)$ was reconstructed by applying the predicted Wiener filter $\hat{H}(k, t)$ to the STFT of the input signal $m(k, t)$ and taking the inverse STFT. The signal frame obtained was weighted again with the square-root Hann window and reconstructed by summing with overlap.

Training

A database of 2000 sawtooth sequences was created with randomly drawn fundamental frequencies in the range [100, 7900] Hz, rounded to nearest integer. The resulting training signal was approximately 288 s long and consisted of different tones. The signal (and added noise) were used to obtain a database of 36 000 feature and target vectors. This database was further divided into training (70%), validation (15%), and testing (15%) sets, which were utilized during the training. The training set was used to learn the weights. The function `nnet.m` creates the test signal, extracts log-magnitude features, and performs training – see Section 12.7.1. As an example, using Matlab (v.2015b) with an NVIDIA Tesla GPU M2090, the command `nnet(512,2,0.3)` took 6 min to converge after 1519 epochs. Without the GPU (using an Intel Xeon 2.60 GHz CPU), the training took 30 times longer.

Testing

To determine the performance of the network, 50 sawtooth signals were created with fundamental frequency drawn from the subset of signals with non-aliased spectra with $f_0 \in [250, 320, 400, 500, 800, 1000, 1600, 2000, 4000]$ Hz. The test signal length was 7 s,

Table 12.1 The effect of different features for network performance with the simulated data.

Features				MSE		
$x^{(0)}$: log-mag.	$x^{(1)}$: ACF	$x^{(2)}$: $[\hat{f}_0, \hat{A}_{max}]$	$x^{(3)}$: LP coeff.	Ensemble	Mean	Network parameters
✓				0.0217	0.0225	$L_h = 2, \eta_\beta = 0.3, W = 512$
	✓			0.0206	0.0211	$L_h = 1, \eta_\beta = 0, W = 512$
		✓		0.0297	0.0299	$L_h = 3, \eta_\beta = 0, W = 256$
			✓	0.0280	0.0282	$L_h = 2, \eta_\beta = 0.1, W = 256$
✓	✓			0.0194	0.0200	$L_h = 1, \eta_\beta = 0.1, W = 512$

corresponding to 900 frames. The error between the predicted mask and ideal mask is reported using the test set samples $MSE = \frac{1}{T}\frac{1}{N/2+1} \sum_{k=0}^{(N/2)} \sum_{t=0}^{(T-1)}(\hat{H}(k,t) - H(k,t))^2$.

12.4.4 Results

Table 12.1 presents the best achieved mask prediction MSE for each of the proposed features with corresponding neural network hyper-parameters. The column "Mean" reports the average of the MSE values over the five different mask predictions, while "Ensemble" reports the MSE using the basic ensemble of network predictions.

For all features, the use of ensemble resulted in lower MSE in contrast to the average MSE of individual networks, and all subsequently discussed MSE values refer to this method. The lowest MSE (0.0206) was obtained using the autocorrelation values with a single hidden layer, 512 units in the hidden layer, without weight decay. The log-magnitude of the STFT performed nearly as well (MSE 0.0217). By concatenating these two features, an even lower MSE value was obtained (MSE 0.0194). These best features contained 129 log-magnitude coefficients from the real part of the spectrum concatenated with the 257 autocorrelation function values. The total length of the best-performing feature vector was $P = 386$.

Figure 12.6 illustrates the MSE for the best-performing features with different network parameter values. Clearly, a single hidden layer was sufficient for this problem, and the prediction capability got worse when more hidden layers were considered. Using regularization increased each network's performance. A sufficient number of units per layer was the most important hyper-parameter setting, and the best performance was obtained using between 256 and 1024 units for every number of hidden layers.

The left panel of Figure 12.7(a) illustrates the first 40 input feature vectors used by the best-performing network – stacked log-magnitude spectrum and autocorrelation values. The middle panel depicts the corresponding Wiener filter obtained from the reference signal and noise, Equation (12.10). The right panel depicts the ensemble network output, that is, the average of the predicted mask over different networks. Figure 12.7(b) depicts the corresponding time-domain signals. The noise reduction capability is evident in the gap between the sawtooth signals.

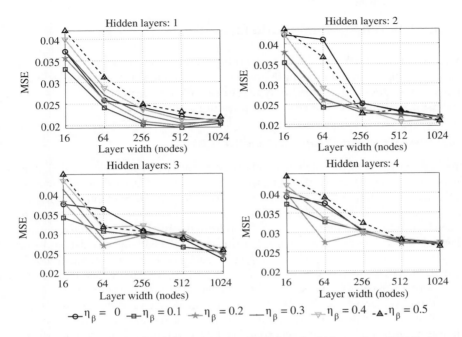

Figure 12.6 The effect of different hyper-parameters (network width, number of hidden layers, and ℓ_2-regularization parameter (η_β, refer to text) values for the networks with the best-performing features for the simulated sawtooth data.

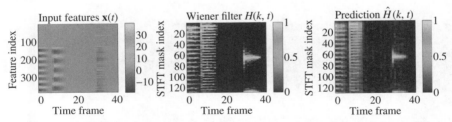

(a) Left: The neural network input features $\mathbf{x}(t)$ (stacked log-magnitude values of the input signal) and the autocorrelation values. Middle: Optimal Wiener filter. Right: Predicted ensemble mask values from the input features with the best-performing network.

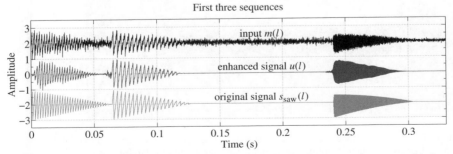

(b) Illustration of the time-domain signals. Note that offset values of $+2, -2$ are added to visualize the original input and reference signals.

Figure 12.7 Example of the first three sawtooth sequences. Panel (a) displays the features and targets, and panel (b) illustrates the corresponding time-domain signals.

12.5   Mask Learning: A Real-World Example

This section evaluates different features for learning the TF mask using the third CHiME challenge data. The accuracy of the mask prediction is evaluated and used to select network hyper-parameters for each feature type. The signal enhancement obtained is then evaluated using two objective metrics for each feature type.

12.5.1   Brief Description of the Third CHiME Challenge Data

A complete description of the challenge data is presented in Barker *et al.* (2015). The recordings contain sentences read from the tablet display in four real environments: bus (BUS), street (STR), café (CAF), and in a pedestrian area (PED). Six microphones are embedded in the tablet's edges, and a separate close-talk microphone captures the utterances – see Figure 12.8. The sentences were taken from the *Wall Street Journal* database (WSJ0) with limited word (5k) vocabulary. The third CHiME challenge dataset has been divided into training, development, and evaluation datasets. Each set contains recordings from all the environments with four different speakers (two male, two female). In total, 12 different speakers are present in the database. The training set contains 1600 sentences, the development set has 1640 sentences, and evaluation dataset has 1320 sentences. The database also contains simulated data for each set and environment, but that data was not utilized here. The background is time-varying and contains competing noise sources as well as diffuse sounds.

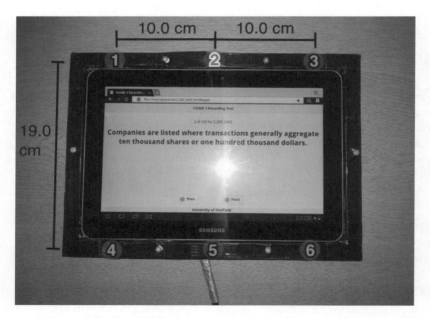

Figure 12.8 The six-microphone array mounted to a tablet computer's surface used to capture signals is depicted. The top-middle microphone faces backwards, and is referred to here as the "noise-capturing microphone." Source: Courtesy of Jon Barker, University of Sheffield. Reproduced with permission of Jon Barker.

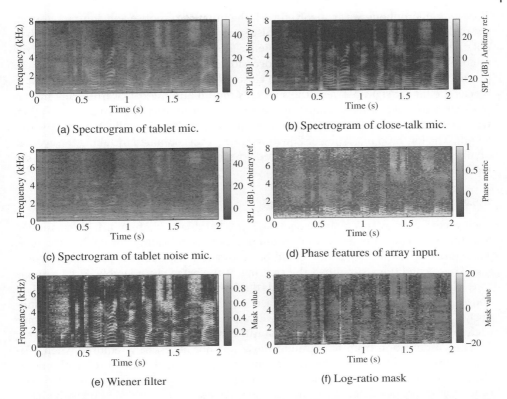

Figure 12.9 Short part of a signal of a WSJ corpus sentence uttered by a female reader in a pedestrian area, used to illustrate magnitude spectrograms in (a), (b), and (c), and phase-based features in (d). Different target masks are illustrated in (e) and (f).

Far-field microphone arrays do not have a similar possibility of using a backward-facing microphone to capture noise, which makes the third CHiME challenge data suitable for evaluating the different contributions of features and input signals to the mask learning capability of a neural network.

Figure 12.9 illustrates a recorded speech utterance read by a female in a pedestrian area with background noise. The data is captured by holding a six-microphone tablet array, where five of the microphones face upwards from the display and one microphone faces backwards.

The spectrogram of the microphone signal from one speaker is depicted in Figure 12.9(a), and Figure 12.9(b) depicts the corresponding close-talk microphone signal. The spectrogram of the outwards-facing microphone is depicted in Figure 12.9(c), and it clearly contains more background noise than the forward-facing microphone. The phase-based spatial feature $f(k, t)$ of Equation (12.8) is depicted in Figure 12.9(d). Figure 12.9(e) illustrates the Wiener filter, which was obtained using the close-talk microphone as the clean signal, while the noise signal was estimated in two steps, as described by Barker *et al.* (2015). In the first step, an STFT domain impulse response was estimated between the close-talk microphone and a tablet microphone. In the second step, the close-talk microphone was convolved with the impulse response and subtracted from the tablet microphone to estimate the noise spectrum. The Wiener filter used was then obtained as the average of individual Wiener filters corresponding to individual array microphones. Environmental noise and occasional turbulence caused by respiration are present in the signals captured with the close-talk microphone. The tablet's structure blocks the line of sight between the speaker's mouth and the backward-facing microphone. However, speech leaks into this microphone due to reverberation and diffraction, which causes low-frequency sounds to "pass through" obstacles smaller than their wavelength. The log-ratio mask [log of Equation (12.12)] was obtained by taking the average of all log-ratios between the array microphones and the close-talk microphone – see Figure 12.9(f).

12.5.2 Data Processing and Beamforming

Each recording contains a single sentence in a varying background. It was assumed here that the person holding the tablet did not move significantly during the utterance. Therefore, the speaker's position could be considered static for each separate recording. These conditions simplified the steering of the array considerably. First, a single time-alignment of the microphone signals during the recording was sufficient for delay-and-sum beamforming. This is in contrast to tracking a moving speaker and continuously adjusting the steering vector of the delay-and-sum beamforming. Secondly, since each recording was trimmed to contain only the uttered sentence, automatic processing for speech detection from background noise was not necessarily required when estimating the steering vector. The time-alignment values for the forward-facing microphones were calculated using the relative time delays $\Delta\tau_{j,0}$ – see Equation (12.4). The time delays were estimated using the generalized cross-correlation method with phase-transform weighting (GCC-PHAT; see Knapp and Carter, 1976). Parabolic peak interpolation (Smith, 2011) was utilized to obtain sub-sample steering delays. See Section 12.7 for a Matlab implementation of the time-aligment function `StaticSourceTDOA`.

The array input data was provided at 16 kHz and was processed in overlapping windows of 512 samples (64 ms) with 50% overlap using the square root of the Hann window (32 ms) during the analysis and synthesis stages. The input signals were normalized with their root-mean-square energy to reduce the effects of possible amplification gain mismatches.

Using the full STFT resolution of 256 bins would have led to large feature and target vectors. In addition, the adjacent spectral bin magnitude values of the input features were highly correlated, which is known to slow down the learning in shallow networks (LeCun *et al.*, 2012). Therefore, the magnitude spectrum was represented using 30 equally spaced mel frequency bands between 0 kHz and 8 kHz. This reduced the dimensions of the input and output layers of the networks, while perceptually relevant parts of the spectrum were preserved – see Section 12.2.3. The predicted mel frequency scale TF mask was then inverse-transferred to STFT resolution before multiplication with the spectrogram of the signal to be enhanced.

12.5.3 Description of Network Structure, Features, and Targets

The structure of the network that was applied is similar to the example of Figure 12.4. The hyperbolic tangent was used as the activation function of the single hidden layer:

$h^{(2)}(z) \equiv \tanh(z)$. The output layer's activation functions were linear, to allow a wide output value range due to the use of the log-ratio mask: $h^{(3)}(z) \equiv z$. The width of the output layer corresponded to the target mask size of 30 mel bands ($D_3 = 30$), while the input size was dependent on the feature vector size used ($D_1 = P$). The hidden layer size was selected using cross-validation, as described below.

The 1600 training sentences in the training data set were randomly divided into three subsets: training (85%, 1360 sentences, 3.5M feature vectors), validation (7.5%, 120 sentences, 0.3M feature vectors), and testing (7.5%, 120 sentences, 0.3M feature vectors). Each subset was balanced to have the same ratio of non-overlapping sentences from each of the four environments (BUS, PED, STR, CAF).

Target Description

The target value was set as the log-ratio mask between the close-talk microphone and the average of the forward-facing microphones' magnitude spectrograms. The log-ratio mask was chosen since it is easy to obtain in practice, in contrast to the Wiener filter, which requires an estimate of the noise power spectral density. Mask values outside of the range $[-40, 40]$ dB were truncated. As motivated above, the STFT resolution mask was converted to 30 mel frequency scale values. Normalization was performed for the target mask values in Equation (12.24), where mean and standard deviation values were obtained using the training subset. Therefore, during the test phase the network outputs were inverse-normalized correspondingly, after which they were inverse-transferred back to the STFT resolution from the mel frequency scale.

Input Feature Descriptions

Similarly to the targets, the feature values were also normalized with Equation (12.24) using the training subset. Table 12.2 describes the different combinations of features that were used to train the neural networks. Two basic spectral features were considered: 30 mel-frequency cepstral coefficients (MFCCs) and 19 linear prediction (LP) coefficients (with pre-emphasis).

The most basic feature consisted only of a single forward-facing microphone signal's mel-frequency cepstral coefficients and is abbreviated as "M." The addition of the backward-facing "noise" microphone's mel-frequency cepstral coefficient features is abbreviated as feature combination "MN." The linear prediction coefficients did not

Table 12.2 List of feature descriptions for mask prediction with feature dimension (P) and the number of microphones required (M).

Abbrev.	Feature description	P	M
M	MFCCs of tablet microphone 1	30	1
BL	MFCCs and LP coeff. of DSB output	49	5
MN	MFCCs of tablet forward-facing microphone and noise microphone	60	2
BLP	MFCCs, LP coeff. of DSB output, and phase features [Eq. (12.8)]	79	5
BLN	MFCCs, LP coeffs. of DSB output, MFCCs of noise microphone	79	6
BLPN	MFCCs, LP coeffs. of DSB output, MFCCs of noise microphone, phase features (12.8)	109	6

provide an improvement in the network's capability to learn the mask in terms of MSE reduction for these two cases, and was omitted. This is most likely due to the low SNR of the input signal, which is known to be detrimental to linear prediction analysis using autocorrelation (Sambur and Jayant, 1976).

The first multichannel feature consisted of mel-frequency cepstral coefficients of the beamformer's output concatenated with the delay-and-sum beamforming (DSB) output's linear prediction coefficients, and is abbreviated as "BL." The concatenation of the noise microphone's mel-frequency cepstral coefficients to these features is denoted "BLN." The magnitude spectrum features of the delay-and-sum beamforming (mel-frequency cepstral coefficients and linear prediction coefficients) were concatenated with the phase-based features of Equation (12.8) using 30 mel frequency bands, and the resulting feature is denoted by "BLP." Finally, "BLPN" denotes the additional use of the noise microphone's mel-frequency cepstral coefficient features.

12.5.4   Mask Prediction Results and Discussion

Different values of weight regularization $\lambda = (2 \times 10^{-2}, 4 \times 10^{-3}, 8 \times 10^{-4}, 1.6 \times 10^{-6}, 3.2 \times 10^{-7}, 6.4 \times 10^{-8})$ and different hidden layer widths $(30, 150, 300, 600)$ were used to select a network that best fitted the data with the cross-validation approach. Each feature combination listed in Table 12.2 was used to train networks using all hyper-parameter combinations (width and regularization) with the training subset. For each feature combination, using the ensemble method with four networks, the network configuration that resulted in the smallest MSE in the validation set during the first 50 training epochs was selected, and the performance is reported with the development and evaluation datasets.

The Keras deep learning library for the Python programming language was used with GPU acceleration by Chollet (2015). The AdaMax stochastic optimization algorithm by Kingma and Ba (2015) was used to learn the weights in 1000 sample mini-batches. Note that only the training data, in contrast to the development and evaluation datasets, were utilized during the training phase in order to report results that comply with the third CHiME challenge rules. Due to the use of mini-batch processing, GPU acceleration, and a state-of-the-art optimization algorithm, the largest network with 600 hidden units took 13 min to train the 50 epochs using the 3.5M features of the training subset with the longest features (BLPN, input dimension $P = 109$).

Figure 12.10 illustrates the mel scale log-ratio mask using the same data as in Fig. 12.9. The left panel depicts the target mask, and the right panel depicts the predicted mask.

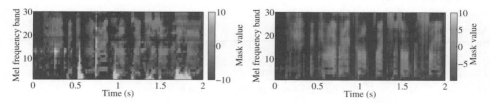

(a) Target log-ratio mask in mel frequency scale.   (b) Predicted log-ratio mask in mel frequency scale.

Figure 12.10 Example of a mask and the predicted mask with array spectral and spatial features (BLPN) for the signal in Figure 12.9. The mask values are truncated between −10 dB and 10 dB for illustrative purposes.

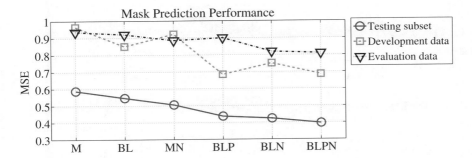

Figure 12.11 Average training MSE on different sets of data (lower is better). The test set refers to the randomly drawn 7.5% training data subset (120 sentences), whereas development and evaluation contain 1640 and 1320 recorded sentences, respectively. Refer to Table 12.2 for a description of the feature combinations. Note that the target mask values are logarithms of the ratio masks with zero mean shift and standard deviation scaling. Therefore, values are not comparable to the simulation section's MSE values of the Wiener filters.

Figure 12.11 reports the MSE values for the test subset of the training data, the development, and the evaluation data. In the test subset, the MSE of the mask predicted with a neural network using only features of a single microphone signal (M) had the highest MSE value of 0.59. Using the delay-and-sum beamforming (BL) lowered the MSE to 0.55. The lowest MSE of 0.40 was obtained by using the delay-and-sum beamforming output with spatial features and the backward-facing microphone (BLPN). The MSE values for the development and evaluation datasets followed a similar trend, but exhibited an increase in their MSE values. In addition, using the phase features in conjunction with the delay-and-sum beamforming and the noise microphone (BLPN) resulted in the lowest MSE in all datasets.

While diverse amounts of regularization were preferred by networks with different features, all networks performed best with 600 hidden units. However, the absolute number of hidden units did not have a large impact; for example, for the BLPN using 30 hidden units instead of 600 increased the MSE only by 1.8%.

Discussion

The systematic increase in the error when applied to the development and evaluation data in comparison to the test set can be partly explained by changes in the data distribution. Each division of data, that is, training, development, and evaluation, contains four different speakers in four different environment instances. The training, validation, and testing subsets of the training data contain different sentences from the same speakers in the exact same environments, which are used to learn the weights and set hyperparameters. However, the development and evaluation sets contain different speakers in different realizations of the environments. Therefore, the development and evaluation sets measure the generalization performance of the mask prediction in a more strict manner than the test subset.

To conclude, data from a microphone array mounted on a tablet computer's surface was captured and divided into three sets of data (training, development, and evaluation). Training data was used to build a database of different input features (including both magnitude and spatial) and target mask values. The effects of different input

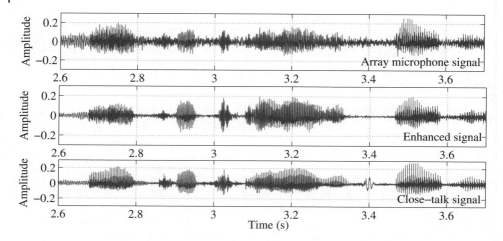

Figure 12.12 Time-domain signal excerpts of the raw input signal of a microphone (top), the delay-and-sum beamforming signal enhanced with TF masking and BLPN features (middle), and the close-talk microphone signal (bottom).

features and their combinations were investigated in the performance of neural-network-based mask prediction. Specific hyper-parameter values for each trained network (layer width and amount of regularization) were selected using cross-validation. The best mask prediction performance for all three sets of data was obtained by using the microphone array instead of a single microphone. The neural network with the lowest mask prediction error used the magnitude and the phase-based features of the forward-facing microphones along with the noise-capturing microphone's magnitude spectrum features.

12.5.5 Speech Enhancement Results

Figure 12.12 illustrates the time-domain speech signal of the raw input (top panel), enhanced speech (middle panel), and close-talk microphone signal (bottom panel).

Two different types of objective measurements were made to evaluate the quality of the enhancement. The first measure used was the frequency-weighted segmental SNR (fwSegSNR), which reports the average magnitude difference in dB between the enhanced signal $u(k, t)$ and the close-talk signal in short overlapping time steps using critical bands of hearing (Hu and Loizou, 2008).[1] The second metric was the short-time objective speech intelligibility measure[2] (STOI; Taal *et al.*, 2011), which is designed to take non-linearities into account – these are typically introduced by time–frequency processing algorithms. The STOI algorithm's output is a scalar value in the range [0, 1] that is expected to have a monotonic relation with the average intelligibility of the degraded signal.

---

[1] Twenty-five mel bands were used, and the SNR values were limited between −10 dB and 35 dB (Hu and Loizou, 2008).
[2] http://ceestaal.nl/stoi.zip.

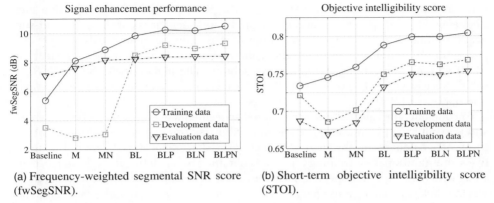

(a) Frequency-weighted segmental SNR score (fwSegSNR).

(b) Short-term objective intelligibility score (STOI).

Figure 12.13 Average speech enhancement objective scores for different methods using the three sets of data. All sentences of the training ($N = 1600$), development ($N = 1640$), and evaluation ($N = 1320$) datasets are used. Refer to Table 12.2 for a description of the feature combinations for the post-filtering.

Figure 12.13 reports the objective scores. The baseline method consisted of the basic delay-and-sum beamforming without post-filtering. The delay-and-sum beamforming is known to have modest noise suppression capabilities, but it does not introduce artifacts. The predicted masks were applied to the delay-and-sum beamforming output when multiple microphones were used (refer to Figure 12.3). The mask was applied to the spectrum of a single forward-facing microphone in cases where the mask was trained using it (M and MN).

For the 1600 sentences of the training set, all the mask predictions improved the fwSegSNR and the STOI scores over the baseline. However, since most of the training dataset was also used to train the neural networks, the results are not generalizable and do not represent the performance for unseen signals. Instead, the 1640 sentences of the development dataset and the 1320 sentences of the evaluation dataset measure the performance for signals that are not used in training and originate from different speakers and environments.

In the development dataset, the baseline fwSegSNR was lowest (3.5 dB) in comparison to the other sets. For the evaluation dataset, the baseline fwSegSNR was the highest of the sets (7.1 dB), but the intelligibility was lowest at 0.69.

Using only a single forward-facing microphone for the mask prediction with and without the backward-facing microphone (M/MN) reduced the intelligibility in contrast to the baseline in both the development and the evaluation datasets. The fwSegSNR also decreased in the development dataset, but was higher than the baseline in the evaluation dataset. The results, except for the slight increase of fwSegSNR for the evaluation dataset, indicate that the simple delay-and-sum beamforming without post-filtering performed better than single-channel TF mask enhancement in terms of fwSegSNR and intelligibility improvement.

The rest of the feature combinations used all of the five forward-facing microphones of the array. These feature combinations produced higher fwSegSNR and STOI values compared to the baseline. The STOI and fwSegSNR scores of the mask using mel-frequency cepstral coefficients and linear prediction coefficients of the delay-and-sum beamforming output (BL) were further improved by adding the backward-facing micro-phone (BLN). A similar or slightly higher gain was obtained by adding the phase features (BLP) without considering the noise microphone. The highest fwSegSNR and STOI values were obtained by combining all the spectral and spatial features with the noise microphone (BLPN).

To conclude, the microphone array-based speech enhancement using beamforming and post-filtering (TF masking) provided lower mask training error and higher objective scores compared to the single-channel TF masking. The largest gain in the objective scores (and smallest training error) was obtained by using the phase-based features along with the magnitude spectrum features of the delay-and-sum beamforming output and the backward-facing noise microphone. It is noted, however, that the delay-and-sum beamforming and the phase-based feature extraction require a method to estimate the steering delays related to the spatial position of the speaker. This adds complexity to the delay-and-sum beamforming-based enhancement over methods that use a single microphone.

12.6 Conclusions

This chapter has presented a general framework for time–frequency masking-based audio enhancement using neural networks. An introduction to neural networks was presented, including the basic back-propagation algorithm for weight learning, and methods for improving network generalization. The chapter explored the audio features used for monophonic, stereo, and multichannel input signals. Different types of target masks were summarized.

A simple monophonic time–frequency domain mask estimation example was used to explore neural network training with different types of network structures and audio features. Speech enhancement was investigated using a tablet-based microphone array, where delay-and-sum beamforming was used as a baseline method. Neural networks with different features and numbers of microphones were used for mask prediction. Mask prediction performance was found to benefit from using both the magnitude spectrum and the phase-based features in terms of the prediction error and the objective enhancement scores.

12.7 Source Code

12.7.1 Matlab Code for Neural-Network-Based Sawtooth Denoising Example

The code in Listing 12.1 first generates noisy sawtooth signals and uses the short-time Fourier transform to extract magnitude-spectrogram-based input features and corresponding target vectors, which are modeled after the Wiener filter. Then a feed-forward neural network is trained to predict the TF mask, and the error of the prediction is evaluated. Finally, the mask is applied to the noisy signal in the STFT domain and the enhanced signal is then reconstructed back into the time domain.

Listing 12.1: Sawtooth Denoising Example

```
1   % Function simulates harmonic signals in SNR range between ...
        [0,10] dB.
2   % A feed-forward neural network (FFNN) is trained to predict the
3   % freq.domain Wiener filter from log-magnitude spectrum.
4   % A test signal is then used to test the network's capability to
5   % predict the mask and clean the signal.
6   %
7   % INPUT
8   % width:  the width of each layer (except output)
9   % depth:  number of hidden layers. (the output layer is added ...
                on top)
10  % l2regu: amount of L2-norm regulalization penalty for weights.
11  %
12  % OUTPUT
13  % x_org:  the original noisy sawtooth test signal
14  % x_enh:  the enhanced sawtooth test signal
15  % net:    trained network
16  %
17  % (C) Pasi Pertila, pasi.pertila@tut.fi, 26.10.2016
18
19  function [x_org, x_enh, net] = nnet(width,depth,l2regu)
20  %
21  nsamples = [2000;50];% number of sawtooth signals: [ntrain;ntest]
22  len      = .064;     % Length of a single sawtooth (s)
23  fs       = 16e3;     % sampling rate (Hz)
24  winlen   = 256 ;     % processing window length (samples)
25  win      = [0;sqrt(hann(winlen-1))]; % square-root windowing
26  t        = linspace(0,len,fs*len)'; % time vector.
27  vdataset = {'train','test'};
28  for di = 1:length(vdataset)
29      rng(di);
30      vec_s=[];vec_n=[];x_org=[];vec_s2=[];
31      % Generare "nsamples" sawtooth waves
32      for  sample_count = 1:nsamples(di)
33          % choosen a fundamental frequency of a harmonic tone ...
                from a set
34          if  strcmp( vdataset{di}, 'train') % training signal
35              F_0 = round(rand()*(fs/2-200))+100;
36          else % Test signal
37              vF0 = [250,320,400,500,800,1000,1600,2000,4000];
38              F_0 = vF0(ceil(rand()*length(vF0)));
39          end
40          % generate a sawtooth signal (sawtooth has aliasing in ...
                spectrum)
41          sig = 2*( t*F_0 - floor(1/2  + t*F_0 ));
42          sig = release_and_smooth(sig,20);
```

```
43          % generate a signal that has all k harmonics with ...
                geometric ratio 1/k
44          % e.g. the spectral content of sawtooth signal without ...
                aliasing.
45          k = repmat([F_0:F_0:fs/2]',1,fs*len);
46          T = repmat(linspace(0,len,fs*len),size(k,1),1);
47          sig2 = 0 - 2/pi*sum(sin(2*pi* k .* T - pi)./repmat( ...
                [1:size(k,1)]',1,fs*len),1)';
48          sig2 = release_and_smooth(sig2,20);
49          % add silence to signal.
50          sil  = round(rand()*10)*winlen;
51          sig  = [sig;zeros(sil,1)];
52          sig2 = [sig2;zeros(sil,1)];
53          % add WGN to signal in order to obtain target SNR level
54          x = awgn(sig,rand()*10,'measured');
55          n = x - sig;
56          % concatenate signals over all samples into vectors
57          vec_s  = [vec_s; sig];
58          vec_s2 = [vec_s2; sig2];
59          vec_n  = [vec_n; n];
60          x_org  = [x_org; x];
61      end
62      % Short-Time Fourier Transforms of noisy signal, noise, and ...
            unalised signal
63      X = spectrogram(x_org,win,winlen-(winlen/2),winlen);
64      N = spectrogram(vec_n,win,winlen-(winlen/2),winlen);
65      S = spectrogram(vec_s2,win,winlen-(winlen/2),winlen);
66      % Produce the Wiener mask using unaliased spectrum and noi
67      W =  abs(S).^2 ./ (  abs(S).^2 + abs(N).^2 );
68      if  strcmp( vdataset{di}, 'train')
69          x_tr = log(abs(X));
70          t_tr = W;
71      else
72          x_test = log(abs(X));
73          t_test = W;
74      end
75  end
76  %%
77  % FFNN Initialization and training:
78  for ii=1:depth; L{ii}='tansig'; end
79  L{depth+1}                    = 'purelin';
80  net = newff(x_tr,t_tr,repmat(width,1,depth),L);
81  net.trainFcn                  = 'traingdx' ; %'trainscg';
82  net.divideFcn                 = 'divideblock';
83  net.performParam.regularization = l2regu;
84  net.trainParam.showWindow     = 0;
85  net.trainParam.epochs         = 2000;
86  net.trainParam.showCommandLine = 1;
```

```
87   % Training
88   net = train(net,x_tr , t_tr,'useGPU','yes','showResources','yes' );
89   %% FFNN prediction the Wiener mask for whole data, and clip in ...
         range [0,1]
90   t_hat = min(1,max(0,sim(net,x_test)));
91   MSE   = mean((t_hat(:) - t_test(:)).^2);
92   disp(['MSE: ' num2str(MSE)])
93   % Multiply the noisy spectrogram with the predicted Wiener filter
94   X_enh = X .*  t_hat ;
95   % reconstruct enhanced signal back into time-domain
96   x_enh = iSTFT( X_enh,winlen,(winlen/2),win);
97   % Plot results.
98   figure;subplot(1,3,1);imagesc(x_test(:,1:41));colorbar;
99   xlabel('Frame index');ylabel('feature index');title('Input ...
         features');
100  subplot(1,3,2);imagesc(t_test(:,1:41));colorbar
101  xlabel('Frame index');ylabel('STFT mask index');title('Wiener ...
         filter')
102  subplot(1,3,3);imagesc(t_hat(:,1:41));colorbar;
103  xlabel('Frame index');ylabel('STFT mask index');title ...
         ('Prediction');
104
105  save res.mat
106
107  % Inverse STFT
108  function x = iSTFT(X,wl,winskip,win)
109  si = 1;
110  for frame = 1: size(X,2)
111      x(si+wl-1,1)   = 0;
112      x(si:si+wl-1) = x(si:si+wl-1)+win.* ifft(X(:,frame),wl, ...
             'symmetric');
113      si = si + winskip;
114  end
115
116  % Time-domain smoothing function
117  function x = release_and_smooth(x,t)
118  x = x .* linspace(1,0,length(x))'; % release (linear attenuation)
119  x(1:t) = x(1:t)  .* exp(-logspace(1,-2,t))'; % smooth the onset
```

12.7.2   Matlab Code for Phase Feature Extraction

The code in Listing 12.2 estimates the phase-based feature vector from the microphone array signal x_input. The method first estimates the wavefront's time difference of arrival (TDOA) between the microphones for the duration of the input signal. The TDOA analysis uses parabolic peak interpolation to obtain sub-sample delay values. Then, instantaneous phase difference estimates are compared to the static phase differences resulting from the TDOA analysis. The function returns the mean cosine value of the resulting phase difference values see Equation (12.8). The code is intended to be used for static and continuously active sources.

Listing 12.2: Phase Feature Extraction Example

```
1  % Phase based features for a microphone array. The source is
2  % assumed to be static during the lenght of the input signal.
3  % INPUT:
4  % x          = microphone array input signal, matrix: ...
                  nsamples x nchans
5  % fs         = sampling rate. scalar., e.g. 16000 (hz)
6  % winlen_ms  = Processing window lenght (milliseconds) e.g. ...
                  40 (ms)
7  % OUTPUT:
8  % phaseFeats = frames x [NFFT/2+1] matrix of phase features in
9  %              range [-1,1]. Values near +1 show agreement between
10 %              the steering vector and phases of the signals.
11 %
12 % (C) Pasi Pertila, pasi.pertila@tut.fi, 21.10.2016
13
14 function phaseFeats = PhaseFeatures(x_input,fs,winlen_ms)
15 % init variables
16 M          = size(x_input,2); % number of channels
17 winlen     = 2^nextpow2(winlen_ms/1000*16000); % length of window
18 overlap    = winlen/2; % amount of overlap between processing ...
                  windows
19 NFFT       = winlen; % FFT resolution
20 vOmega     = linspace(0,pi,NFFT/2+1)'; % create the angular ...
                  frequency vector.
21 mtxOmega   = repmat(vOmega,[1,M*(M-1)/2]);
22 win_fun    = repmat(window('hann', winlen),[1 M]); % windowing ...
                  function.
23 vPair      = nchoosek(1:M,2);          % number of microphone pairs.
24 sindex     = 1;eindex = sindex + winlen - 1;frame = 0;
25
26 % Theoretical phase difference based on TDOA estimates
27 tdoa           = StaticSourceTDOA(x_input,M, vPair)
28 ThereticalDiff = repmat(tdoa,[NFFT/2+1 1]) .* mtxOmega ;
29
30 % Phase feature evaluation
31 while eindex < size(x_input,1)
32     frame = frame + 1;
33     % Take the FFT of the current frame of the input signal
34     X = fft( win_fun .* x_input(sindex:eindex,:), NFFT );
35     X = X(1:NFFT/2+1,:); % remove imaginary part
36     % measured angle difference between microphone pairs for ...
           current frame
37     MeasDiff = angle(X(:,vPair(:,2))) - angle(X(:,vPair(:,1)));
```

```
38        % The Phase features are within range [-1,+1] where -1
39        % indicate poor fit, and +1 a good fit.
40        phaseFeats(:,frame) = mean(cos(ThereticalDiff - MeasDiff),2);
41        sindex = sindex + overlap;
42        eindex = sindex + winlen - 1;
43    end
44
45    %%%%%%%%%%%%%%%%%%%%%%%%%%%%%%%%%%%%%%%%%%%%%%%%%%%%%%%%%%%%%%%%%%%%
46
47    function tdoa = StaticSourceTDOA(x_input,M, vPair)
48    % Time-difference of arrival (TDOA) for a static continously
49    % active source using parabolic peak interpolation for ...
         subsample delay estimation.
50    % (C) Pasi Pertila, pasi.pertila@tut.fi, 20.4.2016
51    n          = size(x_input,1);
52    win_signal = repmat(window('hann', n),[1 M]);
53    n          = n + mod(n,2); % even lenght FFT
54    X_sig      = fft(x_input .* win_signal,n);
55    % Cross-spectrum between microphone pairs over the whole signal
56    CS         = X_sig(:,vPair(:,1)) .* conj(X_sig(:,vPair(:,2)));
57    % Generalized cross-correlation with phase transform
58    gcc_phat   = fftshift( ifft( CS ./ (eps+abs( CS )),n,...
         'symmetric'),1 ) ;
59    % obtain the TDOA for each microphone pair
60    for cP = 1:size(vPair,1)
61        % index of maximum correlation value (time delay)
62        [~,mx_ind] = max( gcc_phat(:,cP) ) ;
63        % subsample interpolation of the delay value (parabolic peak
64        % interpolation)
65        subsample_tdoa = qint([gcc_phat(mx_ind-1,cP), ...
66            gcc_phat(mx_ind,cP),gcc_phat(mx_ind+1,cP)]);
67        % Convert maximum peak location into pairwise delay values
68        tdoa(cP) = +(mx_ind - (n/2 + 1)) + subsample_tdoa;
69    end
70
71    %%%%%%%%%%%%%%%%%%%%%%%%%%%%%%%%%%%%%%%%%%%%%%%%%%%%%%%%%%%%%%%%%%%%
72
73    function [p,y,a] = qint(iy)
74    % References:
75    % [2]   Smith, J.O. "Matlab for Parabolic Peak Interpolation", in
76    %       Spectral Audio Signal Processing,
77    %       http://ccrma.stanford.edu/~jos/sasp/\
78    %             Matlab_Parabolic_Peak_Interpolation.html,
79    %       online book, 2011 edition, accessed 2016-04-20.
80    % QINT - quadratic interpolation of three adjacent samples
81    p = (iy(3) - iy(1))/(2*(2*iy(2) - iy(3) - iy(1)));
82    y = iy(2) - 0.25*(iy(1)-iy(3))*p;
83    a = 0.5*(iy(1) - 2*iy(2) + iy(3));
```

References

Barker, J., Marxer, R., Vincent, E., and Watanabe, S. (2015) The third CHiME speech separation and recognition challenge: Dataset, task and baselines. *IEEE Automatic Speech Recognition and Understanding Workshop (ASRU)*. IEEE.

Bear, M., Connors, B., and Paradiso, M. (2006) *Neuroscience: Exploring the Brain*. Lippincott Williams & Wilkins, Philadelphia, PA.

Blauert, J. (1996) *Spatial Hearing: The Psychophysics of Human Sound Localization*. MIT Press, Cambridge, MA.

Chollet, F. (2015) Keras. https://github.com/fchollet/keras (accessed May 31, 2017).

Davis, S. and Mermelstein, P. (1980) Comparison of parametric representations for monosyllabic word recognition in continuously spoken sentences. *IEEE Transactions on Acoustics, Speech, and Signal Processing*, **28**(4), 357–366.

Deng, L., Li, J., Huang, J.T., Yao, K., Yu, D., Seide, F., Seltzer, M., Zweig, G., He, X., Williams, J., Gong, Y., and Acero, A. (2013) Recent advances in deep learning for speech research at Microsoft. *IEEE International Conference on Acoustics, Speech, and Signal Processing (ICASSP)*, pp. 8604–8608. IEEE.

Diethorn, E. (2004) Subband noise reduction methods for speech enhancement, in *Audio Signal Processing for Next-Generation Multimedia Communication Systems* (ed. Huang, Y. and Benesty, J.), chapter 4, pp. 91–115. Kluwer Academic Publishers, Dordrecht.

Erdogan, H., Hershey, J., Watanabe, S., and Le Roux, J. (2015) Phase-sensitive and recognition-boosted speech separation using deep recurrent neural networks. *IEEE International Conference on Acoustics, Speech, and Signal Processing (ICASSP)*, pp. 708–712. IEEE.

Han, K., Wang, Y., Wang, D., Woods, W.S., Merks, I., and Zhang, T. (2015) Learning spectral mapping for speech dereverberation and denoising. *IEEE/ACM Transactions on Audio, Speech, and Language Processing*, **23**(6), 982–992.

Haykin, S. (1999) *Neural Networks: A Comprehensive Foundation*. Prentice-Hall, Englewood Cliffs, NJ.

Hermansky, H., Morgan, N., Bayya, A., and Kohn, P. (1992) RASTA-PLP speech analysis technique. *IEEE International Conference on Acoustics, Speech, and Signal Processing (ICASSP)*, vol. 1, pp. 121–124.

Hu, Y. and Loizou, P.C. (2008) Evaluation of objective quality measures for speech enhancement. *IEEE Transactions on Audio, Speech, and Language Processing*, **16**(1), 229–238.

Kim, G., Lu, Y., Hu, Y., and Loizou, P.C. (2009) An algorithm that improves speech intelligibility in noise for normal-hearing listeners. *Journal of the Acoustical Society of America*, **126**(3), 1486–1494.

Kingma, D. and Ba, J. (2015) Adam: A method for stochastic optimization. *Third International Conference on Learning Representations (ICLR)*.

Knapp, C. and Carter, G. (1976) The generalized correlation method for estimation of time delay. *IEEE Transactions on Acoustics, Speech, and Signal Processing*, **24**(4), 320–327.

LeCun, Y., Bengio, Y., and Hinton, G. (2015) Deep learning. *Nature*, **521**(7553), 436–444.

LeCun, Y., Bottou, L., Orr, G.B., and Müller, K.R. (2012) Efficient backprop, in *Neural Networks: Tricks of the Trade* (eds. Montavon, G., Orr, G.B., and Müller, K.R.), 2nd edn., pp. 9–48. Springer, Berlin.

Mustiere, F., Nakagawa, R., Wojcicki, K., Merks, I., and Zhang, T. (2016) Dual-microphone phase-difference-based SNR estimation with applications to speech enhancement. *International Workshop on Acoustic Signal Enhancement (IWAENC)*.

Nielsen, M.A. (2015) *Neural Networks and Deep Learning*. Determination Press.

O'Shaughnessy, D. (2000) *Speech Analysis*. Wiley-IEEE Press, Hoboken, NJ.

Paliwal, K., Wójcicki, K., and Shannon, B. (2011) The importance of phase in speech enhancement. *Speech Communication*, **53**(4), 465–494.

Perrone, M. and Cooper, L. (1993) When networks disagree: Ensemble methods for hydrid neural networks, in *Neural Networks for Speech and Image Processing* (ed. Mammone, R.). Chapman & Hall, London.

Pertilä, P. and Nikunen, J. (2014) Microphone array post-filtering using supervised machine learning for speech enhancement. *Annual Conference of the International Speech Communication Association (Interspeech)*.

Sainath, T.N., Weiss, R.J., Wilson, K.W., Narayanan, A., Bacchiani, M., and Senior, A. (2015) Speaker location and microphone spacing invariant acoustic modeling from raw multichannel waveforms. *Proceedings of the IEEE Automatic Speech Recognition and Understanding Workshop (ASRU)*. IEEE.

Sambur, M. and Jayant, N. (1976) LPC analysis/synthesis from speech inputs containing quantizing noise or additive white noise. *IEEE Transactions on Acoustics, Speech, and Signal Processing*, **24**(6), 488–494.

Shao, Y., Jin, Z., Wang, D., and Srinivasan, S. (2009) An auditory-based feature for robust speech recognition. *IEEE International Conference on Acoustics, Speech, and Signal Processing*, pp. 4625–4628. IEEE.

Smith, J.O. (2011) *Spectral Audio Signal Processing*. http://ccrma.stanford.edu/ jos/sasp/ (accessed May 31, 2017).

Srivastava, N., Hinton, G.E., Krizhevsky, A., Sutskever, I., and Salakhutdinov, R. (2014) Dropout: a simple way to prevent neural networks from overfitting. *Journal of Machine Learning Research*, **15**(1), 1929–1958.

Taal, C.H., Hendriks, R.C., Heusdens, R., and Jensen, J. (2011) An algorithm for intelligibility prediction of time–frequency weighted noisy speech. *IEEE Transactions on Audio, Speech, and Language Processing*, **19**(7), 2125–2136.

Välimaki, V. (2005) Discrete-time synthesis of the sawtooth waveform with reduced aliasing. *IEEE Signal Processing Letters*, **12**(3), 214–217.

Wang, Y., Han, K., and Wang, D. (2013) Exploring monaural features for classification-based speech segregation. *IEEE Transactions on Audio, Speech, and Language Processing*, **21**(2), 270–279.

Williamson, D.S., Wang, Y., and Wang, D. (2016) Complex ratio masking for monaural speech separation. *IEEE/ACM Transactions on Audio, Speech, and Language Processing*, **24**(3), 483–492.

Part IV

Applications

13

Upmixing and Beamforming in Professional Audio

Christof Faller

Illusonic GmbH, Switzerland and École Polytechnique Fédérale de Lausanne (EPFL), Switzerland

13.1 Introduction

As the earlier chapters in this book show, a wide variety of techniques has been proposed in the field of parametric time–frequency domain spatial audio. A few commercial products already exist that utilize such techniques. This chapter presents three examples, also discussing considerations of different features and quality requirements that are important in professional products. The products that are described are:

- TSL Soundfield UPM-1/X-1: A stereo to 5.1 upmix processor widely used in broadcasting.
- Schoeps SuperCMIT 2U: A professional beamformer, that is, a shotgun microphone enhanced with adaptive beamforming.
- Sennheiser Esfera: A 5.1 surround microphone using two cardioid microphone elements. The algorithm can be viewed as an upmixer optimized for the specific microphone setup.

The first product, the Soundfield UPM-1 (and its successor, the Soundfield X-1), is described with a high level of detail on the time–frequency processing. The level of detail is reduced in the descriptions of the other products, since the principles are similar. The descriptions are for illustrative purposes only. The actual products may work differently or may have been updated with modified algorithms.

13.2 Stereo-to-Multichannel Upmix Processor

13.2.1 Product Description

In stereo-to-multichannel upmixing, a stereophonic signal is rendered to a surround setup with typically five or seven loudspeakers in the horizontal plane around the listener. Conventional time-domain matrix-based methods, as described in Dressler (2000), for example, are still the most widely used upmix algorithms, where coloration

Parametric Time–Frequency Domain Spatial Audio, First Edition. Edited by Ville Pulkki,
Symeon Delikaris-Manias, and Archontis Politis.
© 2018 John Wiley & Sons Ltd. Published 2018 by John Wiley & Sons Ltd.
Companion Website: www.wiley.com/go/pulkki/parametrictime-frequency

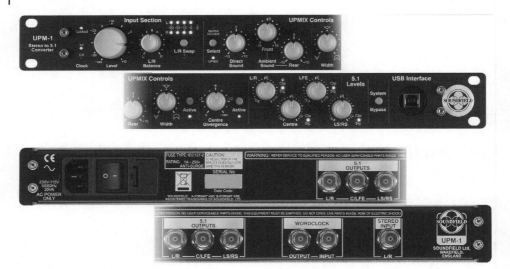

Figure 13.1 Soundfield UPM-1 front and back panels. Source: Courtesy of TSL Professional Products Ltd. Reproduced with permission of Chris Exelby.

artifacts are unavoidable due to the relatively high coherence between loudspeaker signals. Time–frequency domain stereo-to-surround upmixing was proposed more than a decade ago (Avendano and Jot, 2002; Faller, 2006; Pulkki, 2006), and provides advantages in both spatial and timbral audio quality. This section describes the signal processing methods utilized in the TSL Soundfield UPM-1 and X-1 processors used in the audio industry for this task, mainly in broadcasting.

The front and back panels of the TSL Soundfield UPM-1 upmix processor are shown in Figure 13.1. The TSL Soundfield X-1 is a network controllable version of the upmix processor without the front panel controls and with more functions. It features a digital AES stereo input and six audio output channels (three AES stereo pairs).

When the device was introduced to the market, its delay was equal to the duration of one video frame (PAL: 40 ms; NTSC: 33 ms). This made it simple for audio engineers to compensate for the audio delay by delaying the video by one frame. Later, a low-delay upmix update was provided with an end-to-end delay of 6 ms. This delay is low enough that video and audio including the additional upmixing delay are sufficiently synchronized. Currently (2017), broadcasters almost exclusively use only the low-delay version, because it is convenient when the upmix delay is so low that it can be ignored.

The Soundfield UPM-1 and X-1 are mostly used in broadcasting to convert stereo audio feeds to 5.1 surround. In addition to channel gains and left–right input balance, the following upmix parameters are available:

- Center divergence: Determines the reproduction method of center-analyzed sound; the level of sound upmixed to the C channel compared to the level of sound applied to a center-amplitude-panned virtual source created with the L and R channels. Divergence is a commonly used function in surround mixing.
- Sound stage width: By default, the front stereo image is upmixed to front left (L), right (R), and center (C) channels. When desired, the stage width can also be made wider

than the stereo image. In its extreme setting, stage width renders the stereo image to the surrounding sector right surround (RS)–R–C–L–left surround (LS).

- Direct and ambient gain: Put simply, upmix decomposes the stereo input into direct and ambient sound. Ideally, the direct part contains only sound arriving directly from the source, which is termed the dry part of the sound. Correspondingly, the ambient sound ideally contains the reflections and reverberation, known as the wet part of the sound. These signals are then rendered to multichannel loudspeaker signals. Gain can be applied to both the stereo-extracted dry and ambient sounds. These gains aim to control how dry or wet the upmix should be overall.
- Matrix decoding: Upmix can be configured to decode matrix surround encoded (Lt, Rt) two-channel audio. Negative direct sound left–right amplitude ratios imply surround positioning of direct sound.

13.2.2 Considerations for Professional Audio and Broadcast

Single-Channel Audio Quality

Time–frequency upmixing gives a vast improvement in the spatial audio quality when compared to time-domain matrix upmix. However, often the resulting loudspeaker signals are processed with time-dependent filtering and decorrelation operations. The modifications mean that when an individual loudspeaker signal is perceptually compared to the original sound tracks, a drop in audio quality is perceived. Such issues are referred to as deficiencies in single-channel audio quality. While in laboratories and in well-calibrated multichannel listening conditions, time–frequency upmixed sound material produces an impressive spatial effect in the best listening area, unacceptable single-channel quality issues may become audible in scenarios when a listener is near one of the loudspeakers, causing the single-channel sound to dominate the listening.

In the professional domain, the requirement for single-channel audio quality is higher than in the consumer domain. Special care has to be taken when applying time–frequency domain processing in upmixing to maintain high single-channel audio quality. It can be improved by the following means:

- The time–frequency processing used in upmixing changes its effect with time, which is a major source of artifacts. The artifacts can be made less audible when the speed of change is also made signal dependent. Optimally, the process should react quickly to transients in the input, and slowly when the input signal is more stable. Delicate hand-tuning of the parameters is required to reduce the speed of the time–frequency processing, in order to obtain enhancement while avoiding audible artifacts. If the speed is too slow, spatial cues do not react quickly enough and the perceived auditory spatial image is unstable; conversely; if the speed is too high, temporal signal fluctuations impair the single-channel quality.
- There is a trade-off between the degree of signal separation in the output channels and the audibility of artifacts. It general, the lower the coherence obtained for the output channels, the higher the chance of producing audible deficiencies in the output. For example, when a panning law is used that also distributes the sound at a low level to all loudspeakers, instead of panning only to the loudspeaker pair, the single-channel audio quality is improved. However, applying coherent sound to all loudspeakers with a comparable level will produce timbral and spatial artifacts.

Delay

Another desired feature for products being used in a real-time scenario such as broadcasting is a low delay between the input and output of a device. When the delay is low enough, the product can be added to the signal chain without taking special care of synchronization with other signal routes, as typically the video route has to be synchronized with the audio route. We have used time–frequency processing with delays of as low as 6 ms. To achieve such a low delay, the buffer size has to be set as low as 2 ms. The remaining 4 ms delay is caused by the short-time Fourier transform (STFT) overlap-add (= algorithmic delay), and the DSP processing time. Fundamentally, using a low delay reduces the frequency resolution, as dictated by the time–frequency uncertainty principle. Besides this limitation, other means are used to enable high quality:

- We use a relatively long time constant in temporal smoothing of the time–frequency processing parameters.
- We use a larger FFT size than would otherwise be needed, and pad the windowed input signal with zeros at the beginning and end. This reduces temporal wraparound artifacts.

13.2.3 Signal Processing

A more detailed description of the upmixing used here can be found in Faller (2006) and Faller (2007b). A high-level representation of the upmix algorithm is shown in Figure 13.2. The main processing steps are:

1. Upmixing is carried out in a time–frequency representation (subbands) as indicated by the use of the filterbank (FB) and inverse filterbank (IFB) in Figure 13.2. The matrix decoding process is applied independently to the subbands at each frequency.
2. Each subband pair of the input stereo signal is decomposed into a direct sound subband and two ambient sound subbands. The direct sound is estimated together with the direction at which it appears in the stereo image.
3. Direct sound components are mixed to the upmix output signal subbands using a pair-wise amplitude panning law (Pulkki, 1997). The center divergence parameter

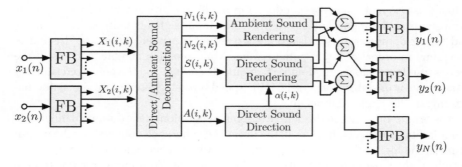

Figure 13.2 The upmix algorithm operates in a subband domain. The input stereo signal is decomposed into direct and ambient sound. Direct sound is rendered as a function of the estimated direction, and ambient sound is rendered with the goal of generating high-quality surround ambience. Source: Faller 2007b. Reproduced with permission of the Audio Engineering Society.

determines how much center signal is actually rendered to the center channel, and how much of it is rendered into the L and R channels.
4. As explained in more detail below, ambient sound is rendered in such a way that it will be perceived as surrounding the listener.

The following four subsections describe these four main processing steps in detail.

Time–Frequency Processing

A discrete STFT is used as the time–frequency representation. The STFT is implemented with a fast Fourier transform (FFT) and a sine analysis and synthesis window with 50 percent overlap.

Not every spectral coefficient of the STFT spectra is processed individually, but the spectral coefficients are grouped such that the subbands represented by the combined spectral coefficients mimic the spectral resolution of the periphery of the human auditory system. Such subbands are often termed "critical bands" (Zwicker and Fastl, 1999). This type of processing is described in detail in Faller and Baumgarte (2003) and Faller (2004). If $x(n)$ is a time-domain signal, $X(i, k)$ is its corresponding time–frequency representation, where i is the subband (critical band) index and k is the time index.

The advantages of processing the audio signal in critical bands and not with the uniform spectral resolution of the STFT are:

- Lower computational complexity because of a lower number of subbands that need to be analyzed individually.
- Fewer artifacts due to spectral smoothing that can be applied between critical bands. The computed parameters change less abruptly between the STFT frequencies, corresponding to shorter corresponding time-domain filters, reducing temporal artifacts.

Obtaining Direct and Ambient Sound

The stereo signal is decomposed into direct and ambient sound. This is achieved by assuming a signal model and solving it in a least squares sense.

It is assumed that at each time k and subband i the L and R stereo signals, $X_1(i, k)$ and $X_2(i, k)$, can be written as

$$X_1(i, k) = S(i, k) + N_1(i, k),$$
$$X_2(i, k) = A(i, k)S(i, k) + N_2(i, k), \tag{13.1}$$

where $S(i, k)$ represents direct sound, $A(i, k)$ represents the corresponding left/right amplitude ratio of the direct sound, and $N_1(i, k)$ and $N_2(i, k)$ represent the left and right ambient sound. For brevity of notation, the subband and time indices are often ignored in the following.

At each time k for each subband i the signals S, N_1, N_2, and the factor A are estimated independently.

Least Squares Estimation of S, N_1, and N_2

Given the stereo subband signals X_1 and X_2, the goal is to compute estimates of S, A, N_1, and N_2. A short-time estimate of the power of X_1 is denoted by $P_{X_1}(i, k) = \mathrm{E}\{X_1^2(i, k)\}$. For the other signals, the same convention is used – that is, P_{X_2}, P_S, and $P_N = P_{N_1} = P_{N_2}$

are the corresponding short-time power estimates. The power of N_1 and N_2 is assumed to be the same; that is, it is assumed that the amount of ambient sound is the same in the L and R channels.

For a detailed derivation of the results presented in the following, refer to Faller (2006). For each i and k, the signal S is estimated as

$$\hat{S} = w_1 X_1 + w_2 X_2$$
$$= w_1(S + N_1) + w_2(AS + N_2), \tag{13.2}$$

where w_1 and w_2 are real-valued weights. The optimal weights in a least mean squares sense are:

$$w_1 = \frac{P_S P_N}{(A^2 + 1)P_S P_N + P_N^2},$$

$$w_2 = \frac{A P_S P_N}{(A^2 + 1)P_S P_N + P_N^2}. \tag{13.3}$$

Similarly, N_1 and N_2 are estimated:

$$\hat{N}_1 = w_3 X_1 + w_4 X_2$$
$$= w_3(S + N_1) + w_4(AS + N_2),$$
$$\hat{N}_2 = w_5 X_1 + w_6 X_2$$
$$= w_5(S + N_1) + w_6(AS + N_2). \tag{13.4}$$

The optimal weights are:

$$w_3 = \frac{A^2 P_S P_N + P_N^2}{(A^2 + 1)P_S P_N + P_N^2},$$

$$w_4 = \frac{-A P_S P_N}{(A^2 + 1)P_S P_N + P_N^2},$$

$$w_5 = \frac{-A P_S P_N}{(A^2 + 1)P_S P_N + P_N^2},$$

$$w_6 = \frac{P_S P_N + P_N^2}{(A^2 + 1)P_S P_N + P_N^2}. \tag{13.5}$$

To compute the weights, the values for A, P_S, and P_N need to be known. These are computed as follows. Given the subband representation of the stereo signal, the power (P_{X_1}, P_{X_2}) and the normalized cross-correlation are computed. The normalized cross-correlation between the L and R channels is:

$$\Phi(i, k) = \frac{\text{Re}\{E\{X_1(i, k)X_2(i, k)\}\}}{\sqrt{E\{X_1^2(i, k)\}\, E\{X_2^2(i, k)\}}}. \tag{13.6}$$

A, P_S, and P_N are computed as functions of the estimated P_{X_1}, P_{X_2}, and Φ. Three equations relating the known and unknown variables are:

$$P_{X_1} = P_S + P_N,$$
$$P_{X_2} = A^2 P_S + P_N,$$
$$\Phi = \frac{aS}{\sqrt{P_{X_1} P_{X_2}}}. \tag{13.7}$$

These equations, solved for A, P_S, and P_N, yield:

$$A = \frac{B}{2C},$$
$$P_S = \frac{2C^2}{B},$$
$$P_N = X_1 - \frac{2C^2}{B}, \tag{13.8}$$

with

$$B = P_{X_2} - P_{X_1} + \sqrt{\left(P_{X_1} - P_{X_2}\right)^2 + 4P_{X_1} P_{X_2} \Phi^2},$$
$$C = \Phi\sqrt{P_{X_1} P_{X_2}}. \tag{13.9}$$

Direct Sound Rendering

At each time k and frequency i, direct sound is rendered using this algorithm:

1. Given the amplitude ratio $A(i,k)$, compute the direction $-30° \leq \alpha(i,k) \leq 30°$ at which the direct sound appears in the stereo image. If matrix decoding is used, negative amplitude ratios map to surround direct sound directions (Faller, 2007b) ($\alpha(i,k) < -30°$ and $\alpha(i,k) > 30°$).
2. Mix the direct sound $S(i,k)$ with a specific gain into the output channels. The gains are obtained by using $\alpha(i,k)$ with a multichannel amplitude panning law.

Ambient Sound Rendering

At each time k and frequency i, two ambient sound channels are available. For 5.1 surround we use the following algorithm for rendering the ambient sound to the L, R, LS, and RS channels:

1. Add $N_1(i,k)$ and $N_2(i,k)$ to the L and R channels of the surround output signal, respectively (front stereo ambience is preserved directionally).
2. Delay $N_1(i,k)$ and $N_2(i,k)$ by about 20–30 ms, apply a 6 dB per octave low-pass filter with a cut-off frequency of about 2–6 kHz, and add the resulting signals to the LS and RS channels of the surround output signal, respectively.

For 7.1 surround or other surround formats with more channels, different delays are applied to the various surround channels. Alternatively, reverberators can be used to further decorrelate the surround channels.

Figure 13.3 Schoeps SuperCMIT 2U digital shotgun microphone. Source: Courtesy of Schoeps. Reproduced with permission of Schoeps.

13.3 Digitally Enhanced Shotgun Microphone

13.3.1 Product Description

The Schoeps SuperCMIT 2U is shown in Figure 13.3. It is a digital shotgun microphone, which by means of time–frequency signal processing enhances the directivity of an acoustic beamformer at low frequencies. It features a digital AES 42 interface and provides two beamforming presets. Its second digital output channel is a digital version of the unmodified shotgun microphone signal. Its applications are mostly film and location sound and sports broadcasting.

While for a broadcast upmix a few milliseconds delay are no issue, the desire was to make the delay of the SuperCMIT as low as possible. This is important for real-time monitoring and applications where sound is reproduced in real time. This is why the SuperCMIT does not use FFT-based processing. It employs a bank of filters based on full-rate infinite impulse response (IIR) filters, as described below.

13.3.2 Concept

Shotgun microphones (Tamm and Kurtze, 1954) are an acoustic implementation of a delay-and-sum beamformer (Veen and Buckley, 1988), based on a single microphone element. They feature a tube with a large number of acoustic openings on the sides. Sound enters these openings and travels within the tube to the microphone element located at one end of the tube.

Properly designed shotgun microphones do not suffer greatly from the inconsistencies and sound coloring artifacts found in digital beamformers. The artifacts in digital beamformers are caused by the use of only a limited number of microphone elements, when compared to the number of openings in acoustic beamformers. Furthermore, shotgun microphones can afford a higher-quality microphone element than multi-capsule delay-and-sum beamformers. Commercially available shotgun microphones typically feature an acoustic design that transitions towards a super-cardioid pattern at low frequencies, where the interference tube is not effective.

Shotgun microphones are still the state of the art when the goal is to achieve the highest possible directivity with high signal fidelity. The SuperCMIT is a conventional shotgun microphone with a second microphone element. An adaptive beamformer is used to improve the directivity at middle and low frequencies to match the interference tube's higher directivity at high frequencies. Additionally, the shotgun microphone's rear lobe is attenuated at low frequencies.

13.3.3 Signal Processing

The SuperCMIT is a conventional shotgun microphone with an additional backward-facing cardioid and an integrated digital signal processor (DSP) which processes the

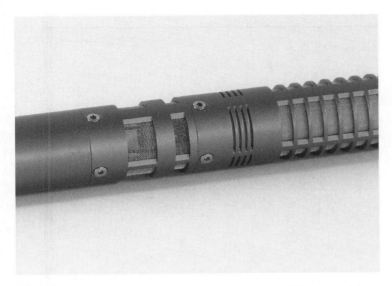

Figure 13.4 Two microphone elements used in the SuperCMIT. Source: Courtesy of Schoeps. Reproduced with permission of Schoeps.

signals of both microphone capsules. The two capsules are mounted nearly coincidently, that is, as closely spaced as possible, as illustrated in Figure 13.4. Two digital output channels are featured: Channel 1 is a beamformed output signal with two possible settings (Preset 1 and 2), and Channel 2 is the unmodified shotgun output signal.

The signal processing used to enhance the shotgun signal is described below. The shotgun signal is denoted $f(n)$ (forward-facing microphone signal), and the cardioid signal is denoted $b(n)$ (backward-facing microphone signal). Time–frequency processing is used to simulate, based on $f(n)$ and $b(n)$, a highly directive microphone with controllable directivity and diffuse response.

Filter Bank

A low-delay IIR filter bank was developed without using any downsampling, based on techniques described in Neuvo and Mitra (1984), Regalia *et al.* (1987), and Favrot and Faller (2010). The filter bank yields nine subbands. The top panel in Figure 13.5 shows the magnitude frequency response of the subbands and the all-pass response of the sum of all subbands (bold). The group delay response of the unmodified filter bank output signal (sum of all subbands) is shown in the bottom panel of the figure.

For the signal $f(n)$, the corresponding subband signals are denoted $F_i(n)$, where subband index $i = 0$ corresponds to the lowest-frequency subband and $i = 8$ corresponds to the highest-frequency subband. The subbands are similarly defined for the signal $b(n)$. The subbands of the filter bank are doubly complementary (Regalia *et al.*, 1987):

$$\left| \sum_0^8 S_i(z) \right| = 1 \quad \text{and} \quad \sum_0^8 |S_i(z)|^2 = 1, \tag{13.10}$$

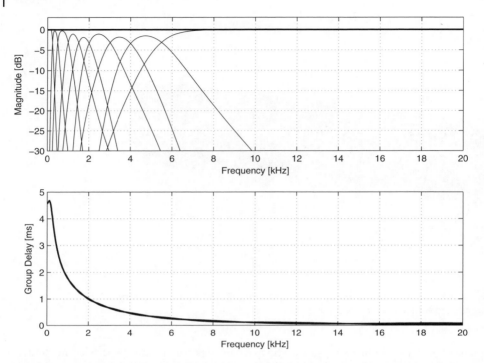

Figure 13.5 Magnitude response of the subbands (thin, top), filter bank output (bold, top), and group delay (bottom) are shown. Source: Wittek 2010. Reproduced with permission of the Audio Engineering Society.

where $S_i(z)$ is the z-transform of the impulse response of subband IIR filter i. The first property in Equation (13.10) ensures all-pass behavior of the synthesis (sum) output signal, and the second property implies frequency-separating subbands (the data in the top panel of Figure 13.5 implies both properties).

Directivity Enhancement and Diffuse Attenuation Processing

Processing is applied in the eight lower subbands, whereas the ninth high-frequency subband is not processed, since the previously mentioned weaknesses of shotgun microphones appear only at lower frequencies.

The beamforming technique described in Faller (2007a) uses two cardioid microphone signals, facing forward and backward, to generate a "virtual microphone" signal facing forward, which is highly directive and has controllable diffuse response. This technique was specifically adapted for the SuperCMIT. Figure 13.6 illustrates the processing that is applied to the signals $f(n)$ and $b(n)$. First, various IIR filters (optional low-cut and high-shelving filters, and so on) are applied to the $f(n)$ signal. Then, the previously described IIR filter banks (FB) are applied, resulting in the subband signals $F_i(n)$ and $B_i(n)$. The eight lower subbands, $0 \leq i < 8$, are processed as follows:

1. A predictor is used to predict $F_i(n)$ from $B_i(n)$. The predictor's magnitude is limited to achieve the desired directivity and rear-lobe attenuation (see Faller, 2007a, for more details). The limited predictor is denoted $p_i(n)$.

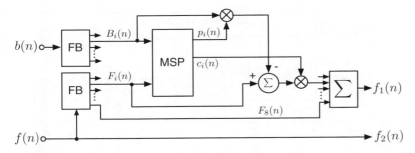

Figure 13.6 Schematic diagram of the processing that is applied to the forward- and backward-facing signals $f(n)$ and $b(n)$ to generate the output signals. The labels FB, IIR, and MSP denote filter bank, IIR filters, and microphone signal processing, respectively. Source: Wittek 2010. Reproduced with permission of the Audio Engineering Society.

2. The limited predictor $p_i(n)$ is applied to the signal $B_i(n)$ and the signal obtained is subtracted from the forward-facing signal $F_i(n)$:

$$F_i(n) - p_i(n)B_i(n).$$

3. A post-scaling factor $c_i(n)$ is computed such that a certain gain is applied to the diffuse sound. The post-scaling factor is applied to the previously computed signal:

$$F_{1,i}(n) = c_i(n)(F_i(n) - p_i(n)B_i(n)).$$

This allows the lowering of the shotgun microphone's diffuse response.

The processed subbands $F_{1,i}(n)$ are summed and added to the non-processed high-frequency subband $F_8(n)$ to generate the enhanced shotgun output signal $f_1(n)$. The second output channel $f_2(n)$ is the shotgun input signal $f(n)$.

Note that the parameters of the processing (such as directivity and diffuse-field response) are chosen in each subband individually to achieve similar processed results despite the variations in the two microphone elements' frequency characteristics.

13.3.4 Evaluations and Measurements

Figure 13.7 shows free-field magnitude responses of the unmodified shotgun signal (Channel 2) and the beamformed signals (Channel 1 with Presets 1 and 2). As desired, the forward free-field response of the beamformed signals is the same as the unmodified shotgun signal. The 90° and 180° free-field responses of the beamformed signals are as smooth as without beamforming, but more attenuated.

The polar pattern and diffuse response of the two beamformer presets and unmodified shotgun signal are shown in Figures 13.8 and 13.9, respectively. The polar pattern of the beamformed signals have significantly higher directivity, and Preset 2 almost completely removes the rear lobe. The diffuse response of the beamformed signals is lower by 6 dB and 9 dB than the unmodified shotgun signal. The SuperCMIT has a higher directivity index than standard shotgun microphones, as indicated in Figure 13.10.

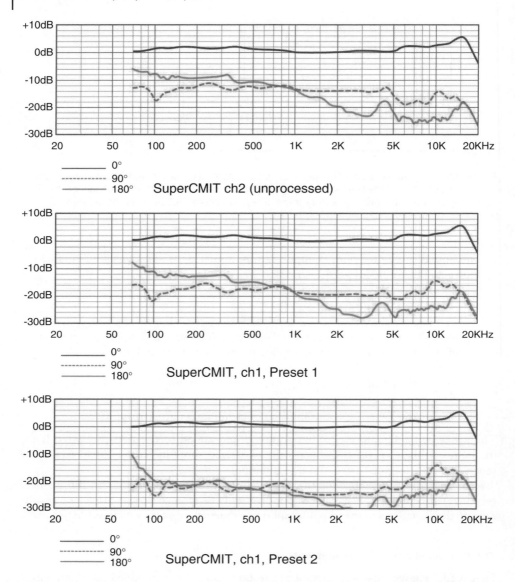

Figure 13.7 Free-field responses of the SuperCMIT at three sound incidence angles (black solid curve at 0°, dashed curve at 90°, solid curve at 180°) for the unprocessed shotgun signal (top) and the two beamformer presets (middle and bottom). Source: Wittek 2010. Reproduced with permission of the Audio Engineering Society.

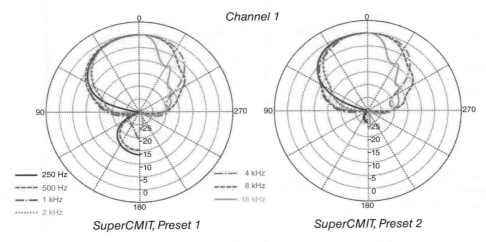

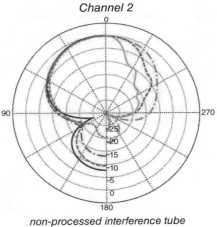

Figure 13.8 Polar patterns of the SuperCMIT beamformer (Channel 1, Preset 1 and 2) and unmodified shotgun signal (Channel 2). Source: Wittek 2010. Reproduced with permission of the Audio Engineering Society.

13.4   Surround Microphone System Based on Two Microphone Elements

13.4.1   Product Description

The Sennheiser Esfera is a microphone system for recording 5.1 surround signals. Figure 13.11 shows the Esfera setup using two cardioid microphones in coincident positioning with 90° angular spacing, and the processors' front and back panels. The processor features a two-channel phantom-powered analog input, stereo digital AES input, and six digital output channels (three AES outputs).

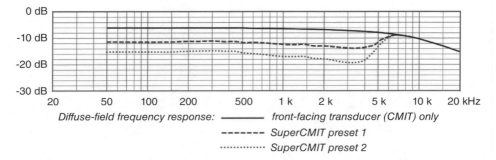

Figure 13.9 Diffuse-field responses of SuperCMIT Channel 2, SuperCMIT Channel 1 Preset 1, and SuperCMIT Channel 1 Preset 2. Source: Wittek 2010. Reproduced with permission of the Audio Engineering Society.

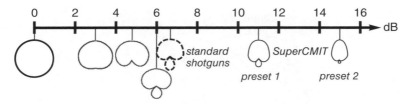

Figure 13.10 Directivity index for first-order directivities, conventional shotgun microphones, and the SuperCMIT beamformer Presets 1 and 2. Source: Wittek 2010. Reproduced with permission of the Audio Engineering Society.

Figure 13.11 Sennheiser Esfera microphone and processor. Source: Courtesy of Sennheiser. Reproduced with permission of Sennheiser.

The Esfera's unique feature is that it employs only two microphones for recording 5.1 surround. The two recorded microphone signals can be stored and transmitted prior to conversion to 5.1 with the Esfera processing unit. Advantages of this system are:

- Convenient and existing stereo transmission lines (also wireless) and recording/ storage can be used.
- The raw stereo microphone signals are high-quality stereo signals.
- Only two microphones are needed. This is a relevant cost factor and enables the use of better cardioid microphones at a given product price point.

Note that the Esfera does not simply do a "blind upmix" on the two microphone channels. The precise arrangement of the microphones and their directivity pattern are considered, enabling surround rendering within a sector of 300°. The Esfera's microphones are precisely matched and their measured directivity data are incorporated into the algorithm, which further improves spatial precision.

The processor connects to the network and a web application can be used to control conversion to 5.1 surround. Conveniently, a multichannel compressor and limiter are included in the processor. The graphical user interface of the web application is shown in Figure 13.12. Also, the delay of the processor can be selected, from low delay (6 ms total) to higher delays.

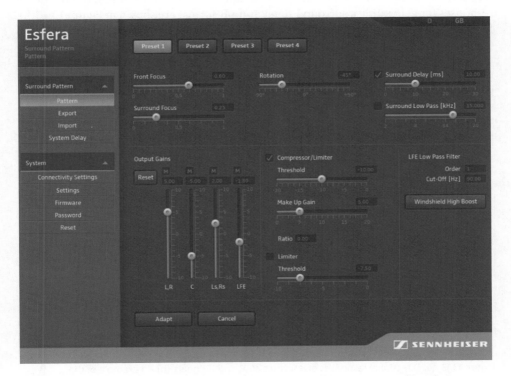

Figure 13.12 Sennheiser Esfera processor control user interface. Source: Courtesy of Sennheiser. Reproduced with permission of Sennheiser.

13.4.2 Concept

Many different techniques and philosophies are applied for multichannel surround recording and mixing (Rumsey, 2001; Eargle, 2004). While techniques from stereo recording and mixing are often adapted for more channels, in order to fully explore the possibilities of surround sound, new techniques have been and are being developed (Holman, 2000).

Here, we are concerned about a specific topic related to surround sound: recording of surround sound with small sound-capturing setups, such as used for recording surround sound for broadcasting or for home video. To fulfill the requirement for a small sound-capturing setup, coincident microphones are used.

The surround sound capturing described here (Faller, 2011) is based on two coincident cardioid signals (XY stereo microphone). While B-format enables total flexibility in terms of defining a first-order directional response for each surround sound signal channel, the limitation to first order and the corresponding low directivity compromise the surround sound quality due to low channel separation (perceived as compromised localization and spaciousness). The technique described does not provide total flexibility in terms of directional response steering, but generates signal channels with a higher corresponding directivity resulting in higher channel separation than linear B-format decoding.

Figure 13.13(a) shows the directional responses of B-format W, X, and Y signals, corresponding to one omnidirectional and two dipole responses. Examples of directional responses for generating a surround signal from B-format are shown in Figure 13.13(b). In this case, a cardioid response was chosen to generate each of the five main audio channels.

Figure 13.14(a) shows the directional responses of the input XY cardioid signals. Figure 13.14(b) shows an example of the achieved effective directional responses of the generated multichannel surround channels.

The previous examples imply that while rear center sound cannot be captured with unity gain by the proposed technique, the directional responses are more directive than B-format-based (linear) responses.

(a)   (b)

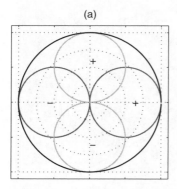 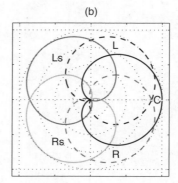

Figure 13.13 (a) Directional responses of B-format W, X, and Y signals. (b) An example of the directional responses of surround audio channels generated from a linear combination of signals with responses as in (a). Source: Faller 2011. Reproduced with permission of the Audio Engineering Society.

(a)

(b)

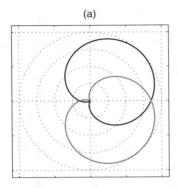
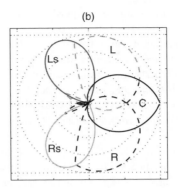

Figure 13.14 (a) Directional responses of a coincident cardioid pair. (b) An example of the directional responses of surround audio channels generated from the signals with responses as in (a). Source: Faller 2011. Reproduced with permission of the Audio Engineering Society.

The approach described can be viewed as a simulation of different microphone signals for the different multichannel surround audio channels with specific directional and diffuse responses. These different microphone signals are generated with an adaptive beamformer, similar in concept to the one used in the SuperCMIT (Section 13.3). The narrower the chosen beam, the more its angle goes towards the rear, because the two source microphones have an angle of 90°. When the angle between the source microphones is 180°, as in the SuperCMIT, then a change in the beam width does not change the beam direction. The front left and right beams are made more narrow towards the rear, to reduce overlap with the rear beams (Faller, 2011).

13.5 Summary

This chapter has presented three commercial products that utilize parametric time–frequency-domain processing of spatial audio in applications of stereo upmixing, adaptive beamforming, and spatial sound reproduction. Also discussed were the quality considerations in professional audio that are taken into account in the development of products, such as single-channel quality and delay.

References

Avendano, C. and Jot, J.M. (2002) Ambience extraction and synthesis from stereo signals for multi-channel audio up-mix. *Proceedings of the IEEE International Conference on Acoustics, Speech, and Signal Processing (ICASSP)*, Orlando, Florida, vol. 2, pp. 1957–1960.

Dressler, R. (2000) Dolby Surround Prologic II Decoder: Principles of Operation. Technical report, Dolby Laboratories, San Francisco, CA.

Eargle, J. (2004) *The Microphone Book*. Focal Press, Waltham, MA.

Faller, C. (2004) *Parametric coding of spatial audio*. PhD thesis, Ecole Polytechnique Fédérale de Lausanne (EPFL), Switzerland. http://library.epfl.ch/theses/?nr=3062 (accessed 1 June 2017).

Faller, C. (2006) Multi-loudspeaker playback of stereo signals. *Journal of the Audio Engineering Society*, **54**(11), 1051–1064.

Faller, C. (2007a) A highly directive 2-capsule based microphone system. *123rd Convention of the Audio Engineering Society*. Audio Engineering Society.

Faller, C. (2007b) Matrix surround revisited. *Proceedings of the 30th International Convention of the Audio Engineering Society*. Audio Engineering Society.

Faller, C. (2011) Surround recording based on a coincident pair of microphones. *Audio Engineering Society 42nd International Conference: Semantic Audio*. Audio Engineering Society.

Faller, C. and Baumgarte, F. (2003) Binaural cue coding – Part II: Schemes and applications. *IEEE Transactions on Speech and Audio Processing*, **11**(6), 520–531.

Favrot, C. and Faller, C. (2010) Complementary N-band IIR filterbank based on 2-band complementary filters. *Proceedings of the International Workshop on Acoustic Echo and Noise Control (IWAENC)*.

Holman, T. (2000) *5.1 Surround Sound Up and Running*. Focal Press, Waltham, MA.

Neuvo, Y. and Mitra, S.K. (1984) Complementary IIR digital filters. *Proceedings of the IEEE International Symposium on Circuits and Systems*, pp. 234–237. IEEE.

Pulkki, V. (1997) Virtual sound source positioning using vector base amplitude panning. *Journal of the Audio Engineering Society*, **45**, 456–466.

Pulkki, V. (2006) Directional audio coding in spatial sound reproduction and stereo upmixing. *Proceedings of the Audio Engineering Society 28th International Conference*. Audio Engineering Society.

Regalia, P.A., Mitra, S.K., Vaidyanathan, P.P., Renfors, M.K., and Neuvo, Y. (1987) Tree-structured complementary filter banks using all-pass sections. *IEEE Transactions on Circuits and Systems*, **34**, 1470–1484.

Rumsey, F. (2001) *Spatial Audio*. Focal Press, Waltham MA.

Tamm, K. and Kurtze, G. (1954) Ein neuartiges Mikrophon großer Richtungsselektivität. *Acustica*, **4**(5), 469–470.

Van Veen, B.D. and Buckley, K.M. (1988) Beamforming: A versatile approach to spatial filtering. *IEEE ASSP Magazine*, pp. 4–23.

Wittek, H., Faller, C., Langen, C., Favrot, A., and Tournery, C. (2010) Digitally enhanced shotgun microphone with increased directivity. *129th Convention of the Audio Engineering Society*. Audio Engineering Society.

Zwicker, E. and Fastl, H. (1999) *Psychoacoustics: Facts and Models*. Springer, New York.

14

Spatial Sound Scene Synthesis and Manipulation for Virtual Reality and Audio Effects

Ville Pulkki,[1] Archontis Politis,[1] Tapani Pihlajamäki,[2] and Mikko-Ville Laitinen[2]

[1]*Department of Signal Processing and Acoustics, Aalto University, Finland*
[2]*Nokia Technologies, Finland*

14.1 Introduction

When a spatial sound scene mimicking an existing or imaginary acoustical condition is synthesized for a subject, the perception of physical presence in locations elsewhere in the real world or in imaginary worlds is created, and the resulting perception is called a virtual reality (Sherman and Craig, 2003). The simulated virtual environment can be similar to the real world in order to create a lifelike experience – for example, for training purposes or for virtual tourism. Alternatively, it can differ significantly from reality, as in virtual reality games. The virtual reality often includes sound sources, and the sound emanated by them should be made audible to the listener in the same manner as a virtual representative of the listener, an avatar, would perceive it if the sound scene was real. In addition to the virtual sound sources, the virtual worlds may also consist of enclosed spaces producing reflections and reverberation, and acoustically reflective objects, whose effect on the sound should be taken into account in the sound synthesis.

When synthesizing a virtual sound scene plausibly for a listener, a number of tasks has to be performed, as discussed by Pulkki and Karjalainen (2015). The signal of a virtual source should be radiated in different directions following the directivity of the source. Furthermore, in a case where different parts of a volumetric source emanate non-coherent signals, that should also be taken into account. A volumetric virtual sound source should also be perceived as spatially broad in reproduction. Usually, the early response from a room is modeled using some kind of ray-based acoustic model where the direct and boundary-reflected paths from the source to the listener are modeled as individual rays. The distance attenuation, delay, and atmospheric effects on the propagation of sound are modeled by individually processing each modeled ray. Each sound ray arriving at the avatar is reproduced to the listener using a virtual source positioning technique summarized by Pulkki and Karjalainen (2015). The reverberant part of the room response is typically modeled separately with a dedicated digital signal processor

Parametric Time–Frequency Domain Spatial Audio, First Edition. Edited by Ville Pulkki,
Symeon Delikaris-Manias, and Archontis Politis.
© 2018 John Wiley & Sons Ltd. Published 2018 by John Wiley & Sons Ltd.
Companion Website: www.wiley.com/go/pulkki/parametrictime-frequency

structure that generates a response that is perceived similarly to reverberation (Savioja *et al.*, 1999).

This chapter will cover DirAC-based parametric time–frequency domain (TF domain) audio techniques where the virtual sound scene is synthesized based on a geometric description of it around the avatar. Methods are discussed that synthesize the sound arriving directly from point-like or volumetric sources and that synthesize the room effect. Furthermore, the parametric techniques allow the augmentation of the virtual reality by reproducing real spatial recordings either as background ambient sounds, or by projecting them onto the surfaces of virtual objects. Furthermore, the transmission of spatial audio over channels with limited capacity is discussed.

Time–frequency domain parametric spatial audio techniques also have applications in audio engineering, in addition to applications in virtual reality. They allow flexible manipulation of recordings of spatial sound. For example, the spatial properties of a recording can be changed either according to some heuristics, or according to directional properties analyzed from other spatial recordings. Furthermore, the level of the reverberant sound can be altered, or the reverberant sound can be recorded with a very low level of direct sound contamination. This chapter also presents a review of such techniques.

14.2 Parametric Sound Scene Synthesis for Virtual Reality

The synthesis technique should be able to render the most relevant sound components occurring in a virtual sound scene. An example of such a sound scene is shown in Figure 14.1. The subject should perceive virtual sources from the correct directions and distances, and with the correct spatial distribution. Furthermore, the virtual rooms should be perceived as having the responses that they would exhibit in reality.

14.2.1 Overall Structure

The present work was initiated by the question of (i) how DirAC processing (and parametric spatial TF processing in general) could be formulated as a virtual reality audio engine, and (ii) what benefits such a formulation might have when compared to traditional time domain methods. The basic principle of using DirAC-based reproduction of sound in virtual worlds is presented in Figure 14.2, as proposed by Laitinen *et al.* (2012). Each direct sound path and strong reflection path is realized as a virtual source created by using a separate spatial synthesis block for each, called DirAC-monosynth and explained in the next section. The block synthesizes both the direction and spatial extent in the resulting monophonic DirAC streams. Merging of multiple sound sources is performed by first transforming the mono-DirAC streams into B-format streams, and then summing the resulting B-format signals together to a B-format bus. In parallel, the late room response is generated using a B-format reverberator. A direct sum of the source signals is taken, and the resulting monophonic signal is fed into N reverberators, where N denotes the number of channels in the B-format stream. The signals are then also added to the same B-format bus. The bus is then reproduced over loudspeakers or headphones using DirAC. Note that the order of the B-format bus can be any

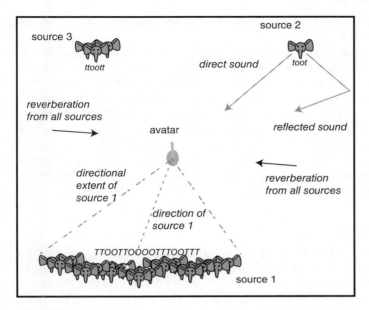

Figure 14.1 A virtual world with three virtual sound sources. The subject should localize the sources and perceive the room effect as the avatar would perceive it if the virtual world were real. The single elephant should be perceived as point-like, and the elephant triple and the elephant herd should be perceived as spatially spread.

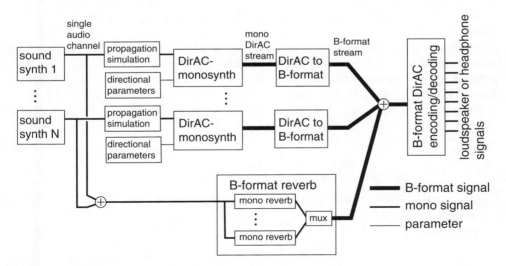

Figure 14.2 Virtual audio synthesis with DirAC-based processing.

positive integer; however, this chapter mostly discusses the realization of the system with a first-order B-format bus.

The formulation provides several possibilities for the transmission of spatial audio. The transmission of a first-order B-format bus with three (for 2D) or four (for 3D) channels already provides a compact and generic representation. If the transmission still reserves too many resources, only a reduced mono-DirAC stream can be transmitted, with shortcomings in the reproduction of reverberant spaces. A number of alternatives exist. For example, in a hypothetical implementation the direct sounds from virtual sources are summed to a B-format bus and compressed to a mono-DirAC stream, which is then delivered over the medium. Additionally, a signal channel representing the reverberated signal, presented in Figure 14.2 as the signal fed into the B-format reverberator, is also delivered, and the B-format reverberation takes place at the receiver side.

14.2.2 Synthesis of Virtual Sources

The basic processing block of DirAC-based spatialization in virtual worlds is the DirAC-monosynth, as presented in Figure 14.3. The block does not perform any signal processing, it only attaches directional metadata to the input audio signal containing the spatial information for the virtual source without changing the audio signal itself. In the figure, θ_{base} represents the direction of the sound source, and θ_{range} represents the angular half-width of the source, both relative to the coordinate system of the avatar. Furthermore, ψ is the diffuseness value that is attached to the output stream. The block can be seen as a simulated B-format microphone with subsequent DirAC encoding. The output of DirAC-monosynth is called a mono-DirAC stream.

The representation of point-like sound objects is trivial with this approach. The direction of a virtual sound source is computed from the relative orientations and positions of the avatar and the sound object. This frequency-independent direction value is attached together with a diffuseness value of zero to the input audio signal using the DirAC-monosynth block.

The spatial extent of virtual sound sources also can be synthesized using the DirAC-monosynth block. It has been shown that the listener perceives a source with a desired width when a monophonic signal with broad spectrum and smooth temporal structure is used, and a different direction value for each frequency band is assigned in the monosynth block (Laitinen *et al.*, 2012; Pihlajamäki *et al.*, 2014). The assignment of

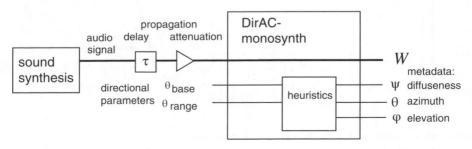

Figure 14.3 A sound signal emitted by a source in a virtual world is transformed to a mono-DirAC stream by attaching directional metadata to it.

Figure 14.4 Transformation from mono-DirAC stream into first-order B-format.

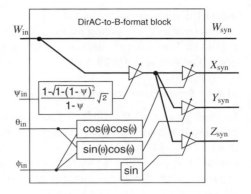

directions for each frequency is a free choice, though an automatic method is desirable for virtual reality applications. Different methods have been utilized, such as random assignment of directions (Pulkki *et al.*, 2009), minimal directional change between adjacent frequency channels (Laitinen *et al.*, 2012), and perhaps the most robust results were obtained with Halton sequences by Pihlajamäki *et al.* (2014). Physically, it is clear that the broader the angle the source covers from the viewpoint of the microphone, the higher the analyzed diffuseness. However, in perceptual tests it has been found that the diffuseness parameter ψ does not need adjustment to control the width of the virtual source, the source seems to be broader already if different frequencies of the signal are reproduced in different directions while the diffuseness maintains a value of zero. However, at least in some cases it has been found that when synthesizing very broad sources, increasing the value may help in mitigating spectral aliasing artifacts in the processing.

The next step in the processing is to convert the mono-DirAC streams into B-format. A method for conversion was presented by Pulkki *et al.* (2009) and Del Galdo *et al.* (2009), and is shown in Figure 14.4. The principle of the processing is that the monophonic signal is assumed to originate from a far-field point source or a plane wave, and that it is captured with gain factors that implement the directional characteristics of the B-format microphone, or spherical harmonics as discussed more generally in Section 6.2. In the figure, the conversion is shown for first-order B-format signals, but the processing is the same for any higher-order B-format stream.

The block should additionally synthesize the effect of sound field diffuseness to the output B-format signal, as in the case of spatially broad sources where a non-zero diffuseness has been set in the DirAC-monosynth block. This means that in a subsequent energetic analysis the output B-format signal should result in the same diffuseness parameter as ψ_{in}. A possible method to obtain this would be to partly decorrelate the B-format signals at a suitable level to reproduce ψ_{in} as desired. However, this would potentially cause an increased level of artifacts in the subsequent processing due to multiple cascaded decorrelators. A method has been developed without use of decorrelators that mixes the signal into the first- and zeroth-order components depending on ψ_{in}. The diffuseness value of zero is simply reproduced by the point-source simulation as described in the previous paragraph. On the other extreme, a diffuseness value of unity is in turn simply reproduced by feeding the monophonic signal only to the W channel. This can easily be verified with Equation (5.7) in Chapter 5, assuming that in this case the length of the active intensity vector takes the value of zero, which subsequently forces the diffuseness to the value of unity. Consequently, any value of diffuseness can be obtained by cross-fading between the point-source simulation and sound-only-to-W methods. Two different derivations of the gain factor for cross-fading have been developed (Del Galdo *et al.*, 2009), of which the simpler approach is shown in Figure 14.4.

14.2.3 Synthesis of Room Reverberation

Laitinen *et al.* (2012) proposed a method to synthesize room effects for virtual sources utilizing the B-format bus structure in DirAC-based audio synthesis, as shown in Figure 14.2. It is based on the idea of synthesizing the response of the virtual room as what would be recorded by a virtual B-format microphone in the position of the avatar. A separate monophonic reverberator is then used for each channel of the B-format signal, which implements the room response. The impulse response of the B-format reverberator in principle equals the reverberant portions of the impulse response of the virtual room captured in the position of the avatar, that is, the response without the components of the direct sound and most prominent reflections. This, in principle, allows the simulation of the spatial distribution of reverberant energy with such accuracy as the given B-format order permits.

A block diagram of the system is shown in Figure 14.2 for a relatively simple setup, where only a single channel of audio is fed into the B-format reverberator. Here, it is assumed that reverberation is diffuse, and thus arrives evenly from all directions to the avatar. The assumption of a diffuse reverberant field allows us to feed the same monophonic signal to each reverberator. The temporal structures and average output levels of each reverberator are equal in this case, and only the phase responses of the reverberators have to be independent from each other. In practice, the virtual source signals are summed and the resulting monophonic signal is sent to monophonic reverberators, whose outputs are sent to the dipole channels in the B-format audio bus. The reason why no signal is applied to the omnidirectional W channel is to force the subsequent DirAC analysis to see the stream as totally diffuse, since the active intensity vector vanishes when the pressure signal (W) is zero. As a consequence, this B-format reverberation finally reproduces loudspeaker signals as being totally diffuse, as desired.

An important feature of this technique is that the required number of reverberators does not depend on the number of loudspeakers used for reproduction. In the case of first-order B-format reproduction with the assumption of evenly surrounding reverberation, two reverberators are needed for any horizontal setup, and three for any 3D setup. In principle, any number of loudspeakers can be used in reproduction, since the decorrelation process of DirAC decoding that takes place when the sound is rendered to loudspeakers ensures that the coherence of the loudspeaker signals stays low enough to produce high-quality reverberation effects. This differs from traditional techniques, where the number of reverberated channels equals the number of loudspeakers.

14.2.4 Augmentation of Virtual Reality with Real Spatial Recordings

When producing audio for virtual reality, often some recorded speech, music, and/or environmental sounds are used as signals for virtual sources. The signals are typically recorded monophonically, and then spatialized with a virtual source positioning method. This can be interpreted as augmentation of virtual reality with point-like sources emanating a monophonic sound signal. The directional radiation patterns of a

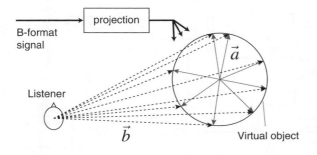

Figure 14.5 A B-format recording is projected onto the surface of a cylinder. Left: Schematic presentation of the process, where the vectors $\vec{a}$ define the surface positions as seen from the origin of the object, and the vectors $\vec{b}$ define the corresponding positions from the viewpoint of the avatar. Right: An illustration of the result, where a spatial audio recording and a 360° image captured from a square are projected onto a cylinder around the avatar. Source: Pihlajamäki 2012. Reproduced with permission of the Audio Engineering Society.

virtual source can then be implemented by simple direction-dependent filtering of the signal (Savioja *et al.*, 1999) for each simulated propagation path.

A background sound scene produced as a stereo or a surround audio track can also be added to virtual reality. Here, the virtual listening setup can be implemented as a set of plane waves arriving from the directions of the assumed listening setup. For example, in the case of two-channel stereophonic playback, two plane waves arriving from the azimuth directions of ±30° would implement this.

These methods cover only the cases of point-like and plane wave sources. Parametric methods also make it possible to augment finite-sized virtual objects with recorded spatial sound. This is realized by spatially projecting the recording onto the surface of an object as shown in Figure 14.5, similar to texture mapping on virtual objects in visual rendering. For example, the ambience of a large crowded hall recorded with a B-format microphone can be projected onto the walls of a virtual hall, or the B-format sound recorded from inside an empty liquid tank can be projected onto the surface of a virtual tank.

A method for such projection was developed by Pihlajamäki and Pulkki (2015), and is called "spatially modified synthesis." The method takes the spatial sound analysis and synthesis of DirAC and combines it with the algorithm used in the spatial extent synthesis (Pihlajamäki *et al.*, 2014). A block diagram showing all the processing is shown in Figure 14.6. This resembles normal DirAC B-format reproduction. However, there are several additions, namely the transformations, the gain control, and the diffuseness control that are utilized to project the TF domain atoms of sound onto the surface of the virtual object. The method shown in the figure separately produces non-diffuse and diffuse streams already synthesized for loudspeakers, as discussed in the next section. If B-format output is desired, the loudspeaker signals can be further downmixed. Alternatively, the processing in Figure 14.6 can also be redesigned for direct B-format output.

14.2.5 Higher-Order Processing

The techniques described above were for first-order DirAC decoding, which is an appealing solution due to its relatively low computational complexity. However, when

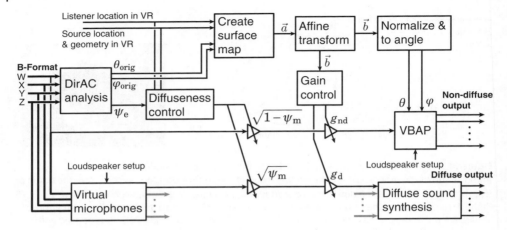

Figure 14.6 DirAC-based method to project spatial sound recording onto the surface of a virtual object as shown in Figure 14.5. Source: Pihlajamäki 2015. Reproduced with permission of the Audio Engineering Society.

the inherent assumption of a single plane wave mixed with a diffuse field in each frequency band does not hold, as may happen in spatially complex sound scenes, artifacts may occur similar to those described in Section 5.6. Nevertheless, the choice of the order of the B-format bus is free, as the DirAC-monosynth blocks and the B-format reverberator can be extended to higher orders. The use of higher-order DirAC decoding described in Chapter 6 can enhance the sound quality in spatially complex sound scenarios. In this case, transmitting the higher-order B-format stream (usually from 9 to 25 channels) may still be advantageous in terms of bandwidth compared to an object-based representation, which may contain many more source signals, while achieving almost perceptual transparency. However, such higher-order DirAC enhancement has not been evaluated so far.

14.2.6 Loudspeaker-Signal Bus

To avoid artifacts caused by spatially complex scenarios, besides using a higher-order B-format bus, the processing of different sound source types can also be separated. In Section 14.2.1 the utilization of separate B-format buses for direct signals and ambient/reverberant signals was discussed. In cases when all the processing is local, which means that no transmission of spatial audio is needed, the B-format bus can be replaced with separate loudspeaker-signal buses for direct and ambient signals. The principle, presented by Pihlajamäki *et al.* (2013), is based on the fact that most of the artifacts occur when the single B-format bus contains all virtual source signals and reverberation, which can in practice result in direct sound being routed to the diffuse stream, and reverberant sound to the non-diffuse stream. This causes timbral artifacts since the direct sound signals should not be decorrelated. Additionally, it causes directional bias effects on the perceived virtual source directions, due to interaction between multiple sources and reverberation during the direction-of-arrival analysis.

In virtual reality audio, the source signals and their spatial types are known a priori, such as direct sounds from sources, background B-format ambience signals, and

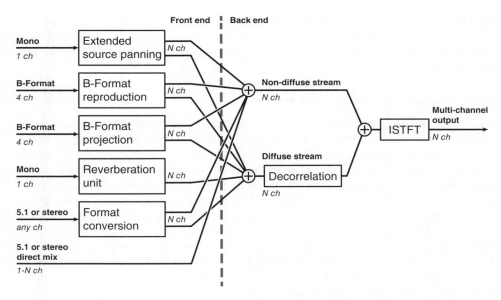

Figure 14.7 Inputs with different spatial characteristics are rendered into diffuse and non-diffuse loudspeaker signal streams. Source: Pihlajamäki 2013. Reproduced with permission of the Audio Engineering Society.

reverberated signals. In the method proposed by Pihlajamäki *et al.* (2013), each spatial type is individually rendered with DirAC to loudspeaker signals. The output of the renderer is modified in such a way that both diffuse and non-diffuse loudspeaker signals are produced separately, and the decorrelation of diffuse stream signals is conducted in processing outside the renderers.

The loudspeaker signals are then added in dedicated global non-diffuse and diffuse streams, as shown in Figure 14.7. The decorrelation process is applied to the diffuse stream, and it has been moved outside of the DirAC renderers to save computational resources by utilizing only a single decorrelation process instead of many. With this structure, the aforementioned synthesis artifacts do not appear at all, since the B-format bus is not used to sum all the spatial sound components together. However, on the downside, the system is not suitable for transmission since a compact intermediate format is not used, and it produces a somewhat higher computational load.

14.3   Spatial Manipulation of Sound Scenes

The DirAC processing principles discussed earlier in this book also have applications in audio engineering as a tool for the production of spatial audio content. This section presents the application of parametric analysis and synthesis to spatial audio effects, using the DirAC framework. These effects are divided into three broad categories by Politis *et al.* (2012): (i) effects that mainly modify the existing directional content of spatially encoded material or a recording; (ii) effects that modify the reverberation and ambience of a spatial sound recording; (iii) effects that synthesize spatial cues for monophonic sources, based on user-provided parameters or metadata. Examples of such parametric effects include angular compression/expansion of the soundfield, spatial modulation for recordings or impulse responses, diffuse-field level control, and ambience extraction.

14.3.1   Parametric Directional Transformations

Since each time–frequency bin in DirAC is associated with an analyzed direction, it is trivial to apply linear directional transformations, such as rotation of the sound scene around some arbitrary axis, by directly modifying the estimated direction of arrival. It should be noted that it is also possible to apply such rotations directly onto the B-format signals by multiplying the velocity components with appropriate standard rotation matrices. Nevertheless, the parametric domain offers more flexibility for non-linear transformations such as companding or expanding directions around some axis (Politis *et al.*, 2012). In general, the directional transformation can be defined as a mapping $T(\theta, \varphi)$ from the original DOA estimate (θ, φ) to the modified one $(\theta', \varphi') = T(\theta, \varphi)$. After the transformation has been defined, its operation is applied on the metadata of each frequency bin or subband. It is also possible to define frequency-dependent mappings $T(\theta, \varphi, k)$ for more complex directional control.

14.3.2   Sweet-Spot Translation and Zooming

Physically inspired directional transformations may be useful in specific application scenarios. One such is the illusion of movement inside a single-point-recorded sound scene. A general strategy to achieve this effect was already presented in Section 14.2.4 in the context of virtual reality; however, it is discussed here in the domain of audio engineering. Schultz-Amling *et al.* (2010) developed a technique for a teleconferencing application termed "acoustical zooming," where the audio effect automatically follows the zooming of an actual camera coincident with the microphone array. Translation of the listening point results in a new spatial relation for the listener with respect to the projection surface, which in this case is assumed to be at the distance of the loudspeakers during reproduction, and consequently to a new set of DOAs. In addition to the directional transformation, a more coherent sense of translation can be achieved by simultaneous modification of the diffuseness and of the amplitudes of each time–frequency bin, following an inverse distance law from the reference distance of the projection surface.

These approaches provide the opportunity for the audio engineer to manipulate the spatial attributes of audio content. Applying just directional transformation results in a perceptual sense similar to the optical analogy of narrowing or widening the field of view inside the recording. A sense of proximity to sound sources can be created, which means that the listening point can be virtually translated inside the recorded scene.

14.3.3   Spatial Filtering

Another possible effect in the parametric domain is spatial filtering, where the level of sound in reproduction is affected depending on the estimated direction of arrival. Kallinger *et al.* (2009) presented a DirAC-based method in the context of teleconferencing. A slightly modified implementation was presented by Politis *et al.* (2012), intended as a spatial effect with more control given to the user over the effect parameters. The effect can be achieved by defining a desired arbitrary directional gain function which results in time–frequency masks applied to each analysis frame. These target functions can be designed based on common analytical formulas for beam patterns, such as higher-order gradient patterns or spherical modal beamforming ones, or they can be described freely by the user by means of some suitable graphical interface.

The method starts by obtaining a first-order (or higher-order, if available) virtual microphone signal from the B-format towards the desired direction, which achieves some of the spatial filtering without time–frequency processing artifacts. The virtual microphone signal can be a simple cardioid or supercardioid, for example, or can be computed as the closest linear beamforming approximation to the desired directional gain. More specifically, in a two-dimensional formulation, if the pattern of the virtual microphone is given by $D(\theta)$ and that of the desired directional gain by $G(\theta)$, and the filtering is applied towards the angle θ_0, then the output S_{sf} of the effect is given as

$$S_{sf}(k, n) = S_{vm}(k, n)M(k, n), \tag{14.1}$$

where $S_{vm}(k, n)$ is the virtual microphone signal at each bin or subband, and $M(k, n)$ is a real-valued time–frequency mask derived as

$$M(k, n) = \frac{G(\theta_k - \theta_0)}{D(\theta - \theta_0)}, \tag{14.2}$$

where θ_k is the analysed DOA at subband k. The above formulation can be directly extended to the three-dimensional case as well.

Even narrow brick-wall spatial filters can be defined in the method; however, it is advantageous to use continuous directional patterns to avoid artifacts due to abrupt transitions in the spectral gains. Moreover, a portion of the diffuse sound should be added to the output to avoid loss of sound power. The diffuse contribution in this case is generated from a decorrelated version S_{diff} of the virtual microphone signal (Politis *et al.*, 2012). To achieve the correct power for the diffuse stream, it is scaled according to the power output of the spatial filter $G(\theta)$ and the virtual microphone $d(\theta)$ in a diffuse field, or, similarly, by the ratio of their directivity factors Q_{sf} and Q_{vm}:

$$S_{sf+diff}(k, n) = S_{vm}(k, n)M(k, n) + \sqrt{\frac{Q_{vm}}{Q_{sf}}} S_{diff}(k, n). \tag{14.3}$$

The addition of the diffuse stream helps in masking artifacts that may arise from fast fluctuations of directions inside and outside of the beam.

14.3.4 Spatial Modulation

A method to modify the directional (and/or diffuseness) parameters of one DirAC stream based on external information has been proposed by Politis *et al.* (2012), termed "spatial modulation." This operator imprints the spatial cues of one signal (control) on the spatial sound of another signal (carrier). Within the parametric domain of DirAC, this is a relatively easy operation as the metadata computed from the control signal can be directly combined with the carrier audio signal before synthesis. Furthermore, instead of direct replacement, arbitrary interpolation of the parameters of the two DirAC streams can be performed to produce a mix of the spatial cues. This procedure can achieve an effect similar to the spatial morphing utilized in computer graphics

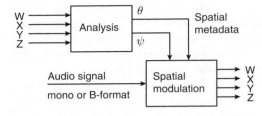

Figure 14.8 The spatial modulation effect. An audio signal is spatialized using spatial metadata analyzed from a separate B-format spatial sound recording. Source: Politis 2012. Reproduced with permission of Archontis Politis.

(Gomes and Velho, 1997). Spatial modulation can be applied in many creative ways. An example is the use of a signal with highly varying but strongly directional content as the control signal, and a strongly ambient surrounding sound as the carrier signal. The resulting ambient sound will have the directional properties of the control signal. The process is depicted in Figure 14.8, where the "Spatial modulation" block refers to applying or mixing the parameters of the control stream with the carrier, before the synthesis.

Different variants of the technique are presented in Politis *et al.* (2012), and related software has been used to produce demonstrations. For example, a compelling effect has been obtained when the spatial metadata of a B-format recording of a drum set was used to modulate a recording of a continuous choir singing. The rhythm of the drumming was audible as spatial movement in the sound of choir. Such an impression cannot be easily obtained with a non-parametric approach to spatial audio effect design.

14.3.5 Diffuse Field Level Control

The DirAC model of the decomposition of a recording into directional and diffuse parts permits manipulation of their relative power by the user. This can be done in a meaningful way as an adjustment of the direct-to-reverberant ratio (DRR). As shown in Equation (5.11), there is a direct relationship between the diffuseness and the DRR:

$$\psi = \frac{1}{1 + 10^{\Gamma_{dB}/10}}. \tag{14.4}$$

Here, Γ_{dB} is the DRR in dB. DirAC analysis gives an estimate ψ of the ideal diffuseness and, respectively, an estimate Γ_{dB} of the DRR in each time–frequency tile.

Based on this formulation it is straightforward to apply either a global DRR modification $\Delta\Gamma_{dB}$ to all subbands or a modification $\Delta\Gamma_{dB}(k)$ per subband k. The new DRR will be

$$\Gamma_{mod}(k, n) = \Gamma_{dB}(k, n) + \Delta\Gamma_{dB}(k), \tag{14.5}$$

and based on Equations (14.4) and (14.5), we can derive the modified diffuseness with respect to $\Delta\Gamma_{dB}$ as

$$\psi_{mod}(k, n) = \frac{\psi(k, n)}{\psi(k, n) + 10^{\Delta\Gamma_{dB}(k)/10}(1 - \psi(k, n))}, \tag{14.6}$$

which can be applied directly at the synthesis stage. One characteristic of the above formulation is that although it results in a diffuseness modification, the overall loudness is preserved.

The specific effect can be directly used in cases where the sound engineer wishes to suppress reverberation in certain frequency bands and to produce a stronger directional impression on the listener. The opposite can also be achieved, with directional sounds being suppressed in favor of the diffuse stream. For practical use, some upper limits should be imposed on the modification, since the diffuse stream naturally masks artifacts from rapid directional variations in the directional stream.

14.3.6   Ambience Extraction

The ambience extraction method attempts to make foreground sounds with a clear direction softer, while preserving background sounds or ambience (the sounds analyzed as mostly diffuse). It differs from the diffuse stream in that instead of scaling the input signals with the diffuseness value and decorrelating the result, it mainly tries to suppress the directional sounds in a direct way (Politis *et al.*, 2012). For each frequency band k, a virtual microphone is generated from the B-format components pointing in the opposite direction to the analyzed direction of arrival $\vec{u}_k$. The virtual microphone output can be expressed as

$$S(k, n) = a_k W(k, n) - (1 - a_k)\vec{u}_k^{\mathrm{T}} \cdot \vec{X}'(k, n),\qquad(14.7)$$

where a_k is the directivity coefficient ranging from 0 to 1, adjusting the ratio between the omnidirectional and dipole components, and $\vec{X}'(k, n)$ is the vector of the B-format velocity components. Since for diffuse sound no suppression should be performed, the directivity of the virtual microphone is adapted to range from cardioid for completely non-diffuse sound to omnidirectional for diffuse sound. More specifically, the coefficient is computed as

$$a_k = \frac{1}{2} + \frac{\psi_k}{2},\qquad(14.8)$$

where ψ_k is the analyzed diffuseness at band k.

The above procedure is presented schematically in Figure 14.9. The ambience extraction effect works effectively for recordings with mild to strong reverberation, where it produces an even reverberant output with strong foreground sounds suppressed. In cases of dry recordings with multiple concurrent sources, artifacts start to appear related

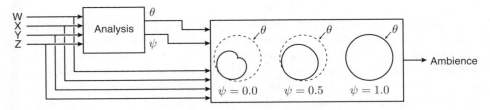

Figure 14.9 The ambient component of a sound field is captured by pointing a virtual microphone in the opposite direction to the direction of arrival of the sound, and modulating its directional pattern with the analyzed diffuseness parameter. Source: Politis 2012. Reproduced with permission of Archontis Politis.

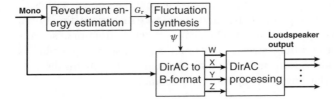

Figure 14.10 Upmixing a monophonic signal to a B-format stream by estimating the diffuseness parameter based on dereverberation methods. Source: Pihlajamäki 2013. Reproduced with permission of the Audio Engineering Society.

to fast fluctuations in the analyzed direction and its deviation from the actual source direction.

14.3.7 Spatialization of Monophonic Signals

Spatialization of a monophonic audio channel, either to a multichannel audio reproduction setup or as background sound in virtual reality, is a useful task in many audio engineering applications. Pihlajamäki and Laitinen (2013) proposed a method that takes advantage of the DirAC-monosynth block discussed in Section 14.2.2. The method divides a monophonic signal that contains both the source signal and room reverberation into diffuse and non-diffuse streams, and positions the non-diffuse stream in a user-defined direction, while the diffuse stream is spatialized evenly around the listener. The aim of the processing is to produce the perception of a virtual source in the user-defined direction, while the reverberation should be perceived evenly around the listener.

The method estimates the fluctuation of the diffuseness parameter from a monophonic signal based on the estimation of the reverberant energy in the signal, similarly to commonly known dereverberation methods. The estimated short-time ratio of reverberant energy G_r is then used to synthesize the diffuseness ψ, as shown in Figure 14.10. A formal listening test has been performed to compare the relative quality of the proposed method to constant diffuseness cases (Pihlajamäki and Laitinen, 2013). The results show that the proposed method increases the perceptual audio quality.

14.4 Summary

This chapter has discussed the utilization of time–frequency domain parametric processing of spatial audio in the context of virtual reality and the production of audio content. The work was motivated by the question of whether DirAC processing principles would be beneficial in such processing environments. It was found that DirAC-based processing brings certain advantages in virtual reality audio processing, such as control of the spatial width of virtual sources, efficient and listening-setup-agnostic generation of reverberation, efficient transmission of spatial sound, and computationally efficient implementation. In audio content production, the methods provide new and creative tools for composing spatial effects for artistic purposes. Although all of the effects were implemented in the domain of DirAC processing, they can also use other parametric time–frequency domain spatial reproduction methods.

References

Del Galdo, G., Pulkki, V., Kuech, F., Laitinen, M.V., Schultz-Amling, R., and Kallinger, M. (2009) Efficient methods for high quality merging of spatial audio streams in directional audio coding. *Audio Engineering Society 126th Convention*, Munich, Germany.

Gomes, J. and Velho, L. (1997) Warping and morphing, in *Image Processing for Computer Graphics*, pp. 271–296. Springer, New York.

Kallinger, M., Del Galdo, G., Kuech, F., Mahne, D., and Schultz-Amling, R. (2009) Spatial filtering using directional audio coding parameters. *IEEE International Conference on Acoustics, Speech, and Signal Processing (ICASSP)*, Taipei, Taiwan.

Laitinen, M.V., Pihlajamäki, T., Erkut, C., and Pulkki, V. (2012) Parametric time–frequency representation of spatial sound in virtual worlds. *ACM Transactions on Applied Perception*, **9**(2), 8.

Pihlajamäki, T. and Laitinen, M.V. (2013) Plausible mono-to-surround sound synthesis in virtual-world parametric spatial audio. *49th Audio Engineering Society International Conference: Audio for Games*. Audio Engineering Society.

Pihlajamäki, T., Laitinen, M.V., and Pulkki, V. (2013) Modular architecture for virtual-world parametric spatial audio synthesis. *49th International Conference of the Audio Engineering Society: Audio for Games*, London, UK.

Pihlajamäki, T. and Pulkki, V. (2012) Projecting simulated or recorded spatial sound onto 3D-surfaces. *Proceedings of 45th International Conference of the Audio Engineering Society: Applications of Time–Frequency Processing in Audio*, Helsinki, Finland.

Pihlajamäki, T. and Pulkki, V. (2015) Synthesis of complex sound scenes with transformation of recorded spatial sound in virtual reality. *Journal of the Audio Engineering Society*, **63**(7/8), 542–551.

Pihlajamäki, T., Santala, O., and Pulkki, V. (2014) Synthesis of spatially extended virtual source with time–frequency decomposition of mono signals. *Journal of the Audio Engineering Society*, **62**(7/8), 467–484.

Politis, A, Pihlajamäki, T., and Pulkki, V. (2012) Parametric spatial audio effects. *Proceedings of the 15th International Conference on Digital Audio Effects*, York, UK.

Pulkki, V and Karjalainen, M. (2015) *Communication Acoustics: An Introduction to Speech, Audio and Psychoacoustics*. John Wiley and Sons, Chichester.

Pulkki, V., Laitinen, M.V., and Erkut, C. (2009) Efficient spatial sound synthesis for virtual worlds. *Audio Engineering Society 35th International Conference*, London, UK.

Savioja, L., Huopaniemi, J., Lokki, T., and Väänänen, R. (1999) Creating interactive virtual acoustic environments. *Journal of the Audio Engineering Society*, **47**(9), 675–705.

Schultz-Amling, R., Kuech, F., Thiergart, O., and Kallinger, M. (2010) Acoustical zooming based on a parametric sound field representation. *128th Audio Engineering Society Convention*, London, UK.

Sherman, W. and Craig, A. (2003) *Understanding Virtual Reality: Interface, Application, and Design*. Morgan Kaufmann, Burlington, MA.

15

Parametric Spatial Audio Techniques in Teleconferencing and Remote Presence

Anastasios Alexandridis, Despoina Pavlidi, Nikolaos Stefanakis, and Athanasios Mouchtaris

Foundation for Research and Technology-Hellas, Institute of Computer Science (FORTH-ICS), Heraklion, Crete, Greece

15.1 Introduction and Motivation

In this chapter, applications of time–frequency parametric spatial audio techniques are presented in the areas of teleconferencing and remote presence. Specifically, the focus is on using circular microphone arrays in order to capture the sound field in these two different application areas. The methods presented here, consequently, have the scope to process multi-microphone signals so that the spatial properties of the original sound field can be reproduced using a multichannel loudspeaker setup or headphones as faithfully as possible, keeping as a priority computational efficiency, so that the proposed approaches can be implemented in real time using a typical PC processor.

The approaches presented are not simply extensions of established work into particular application areas, but are based on algorithms that are novel and innovative. It is relevant at this point to explain the main difference between the two application areas examined in this chapter. On the one hand, teleconference applications typically involve a small number of simultaneous sound sources at each remote participant site (the speakers at a given time instant), and the objective is to estimate their directions with respect to some reference point and reproduce the sources at these estimated directions for the receiving remote participants. On the other hand, the remote presence application area involves a more complex soundscape, where individual audio sources may not be considered as distinct, and thus estimating their direction and number is not feasible. In particular, in this chapter we focus on *crowded* soundscapes, such as those encountered at athletic events (for example, at a soccer game), where the objective is to reproduce the fans' reactions for the remote participant (for example, a TV viewer). The sports event application area is an important one in terms of commercial interest, and so far has not received sufficient attention from the spatial audio community. For

Parametric Time–Frequency Domain Spatial Audio, First Edition. Edited by Ville Pulkki, Symeon Delikaris-Manias, and Archontis Politis.
© 2018 John Wiley & Sons Ltd. Published 2018 by John Wiley & Sons Ltd.
Companion Website: www.wiley.com/go/pulkki/parametrictime-frequency

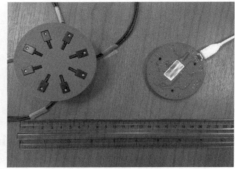

Figure 15.1 The analog microphone array with all the necessary cables and the external multichannel sound card along with its counterpart using digital MEMS microphones. All the bulky and expensive equipment can be replaced by our digital array, which requires only one USB cable. Source: Alexandridis 2016. Reproduced with permission of the IEEE.

both application areas, real-time demonstrations of the described approaches have been developed and extensively tested in real user scenarios.[1]

An important aspect of the approaches presented is the focus on microphone arrays, especially of circular shape (however, the algorithms presented can easily be extended to other microphone array configurations). For the particular applications under consideration, the circular shape provides a 360° coverage around the point of interest, which can be conveniently reproduced using headphones (for example by creating a binaural signal using the corresponding head-related transfer functions – HRTFs) or via multiple loudspeakers around the listener using some amplitude panning method. A significant advantage of uniform microphone arrays made of omnidirectional microphones is the availability of several algorithms for spatial audio processing based on direction of arrival (DOA) and beamforming approaches, which often promote the use of a compact size for the microphone array, a highly desirable feature in many practical cases. At the same time, the recent introduction of digital MEMS (microelectromechanical systems) microphones (typically omnidirectional) has made feasible the design of microphone arrays that are extremely low cost, compact, and portable (since an analog to digital converter is no longer required). A picture depicting a recently built MEMS microphone array and its analog counterpart is shown in Figure 15.1.[2] A digital circuit combining the MEMS microphone array with a local processor and wireless transmitter results in a microphone sensor that is an ideal device to be used in the applications under consideration in this chapter, and more generally comprises a compact and inexpensive sensor that can operate individually or as a node of a wireless acoustic sensor network (WASN).

In the following, we initially present the state of the art in parametric spatial audio capture and reproduction, followed by a description of our approach in the two application areas of interest.

---

[1] The demonstrations can be viewed at https://www.youtube.com/channel/UCi5pjihBIZqWg52biUg4MTQ. For more information, please refer to http://www.spl.edu.gr.

[2] The MEMS digital microphone array depicted in Figure 15.1 was built by the Signal Processing Laboratory team at FORTH-ICS. For more information, the interested reader may refer to http://www.spl.edu.gr and Alexandridis *et al.* (2016).

15.2   Background

In this section we briefly present the state of the art in parametric spatial audio techniques, emphasizing those methods that are based on uniform microphone arrays as the capturing device, since this is the device of interest for this chapter. The description provides the main research directions, and is by no means an exhaustive bibliographic review. See Chapter 4 for a broader introduction to the field.

Directional audio coding (DirAC; Pulkki, 2007), see also Chapter 5, represents an important paradigm in the family of parametric approaches, providing an efficient description of spatial sound in terms of one or more audio signals and parametric side information, namely the DOA and the diffuseness of the sound. While originally designed for differential microphone signals, an adaptation of DirAC to compact planar microphone arrays with omnidirectional sensors has been described by Kuech *et al.* (2008) and Kallinger *et al.* (2008), and an adaptation to spaced microphone arrangements by Politis *et al.* (2015). In the same direction, the approach in Thiergart *et al.* (2011) presents an example of how the principles of parametric spatial audio can be exploited for the case of a linear microphone array. It is also relevant to mention at this point the pioneering work of binaural cue coding (BCC; Baumgarte and Faller, 2003; Faller and Baumgarte, 2003), due to its innovative approach of encoding a spatial audio scene using one or more audio signals plus side information, which is a coding philosophy followed by several of the methods encountered in this chapter. In the context of binaural reproduction, Cobos *et al.* (2010) presented an approach based on a fast 3D DOA estimation technique. While not explicitly calculating any parameter related to diffuseness, the authors claimed that diffuseness information is inherently encoded by the variance in the DOA estimates. The capture and reproduction of an acoustic scene using a circular microphone array has been presented by Alexandridis *et al.* (2013b), and to a large extent the presented approach for the teleconference application area in this chapter is based on that work. The authors were able to show an advantage in terms of perceived spatial impression and sound quality, in a scenario with a limited number of discrete sound sources – thus suitable for applications such as teleconferencing – whose number and direction is provided by the DOA estimation and counting technique described by Pavlidi *et al.* (2013).

Demonstrating the applicability of parametric spatial audio techniques to large-scale sports events is certainly interesting, not only because of the great potential for commercial exploitation, but also because of the inherent technical challenges that such acoustic environments introduce. At each time instant there are hundreds of spectators cheering and applauding simultaneously from many different directions, and therefore the source sparseness and disjointness conditions that are assumed for DOA estimation are most of the time not met. Yet, techniques that attempt an accurate physical reconstruction of the sound field, such as Ambisonics (Gerzon, 1985) and wave field synthesis (Boone *et al.*, 1995), suffer from several practical limitations. The former technique suffers from a very narrow optimal listening area, while the latter requires a prohibitively large number of loudspeakers, which is impractical for commercial use. An interesting approach that tackles these problems has been presented by Hacihabiboğlu and Cvetković (2009), using a circular array of first-order differential microphones. Essentially, the authors propose to use simple array processing in order to emulate microphones with directivity responses that conform to stereophonic panning laws. As it does not rely on any parameter estimation, this approach presents an interesting example of a non-parametric approach to sound scene capture and reproduction.

In the following sections, we present the proposed approaches for the two application areas of interest: teleconferencing, and remote presence with an emphasis on crowded acoustic environments.

15.3 Immersive Audio Communication System (ImmACS)

The Immersive Audio Communication System (ImmACS) is a spatial audio system for teleconferencing applications. ImmACS utilizes a uniform circular microphone array in order to extract spatial information from an audio scene of interest, that is, the number and the DOAs of the active audio sources. The estimates are then used for efficient encoding and transmission of the audio scene to the multiple distant participant sites.

15.3.1 Encoder

The microphone array configuration utilized in ImmACS is depicted in Figure 15.2; Figure 15.3 depicts a block diagram of the ImmACS encoder. The time-domain signals $x_1(t), x_2(t), \ldots, x_M(t)$ are transformed into the short-time Fourier transform (STFT) domain signals $X_1(\tau, \omega), X_2(\tau, \omega), \ldots, X_M(\tau, \omega)$, where τ denotes the time frame index

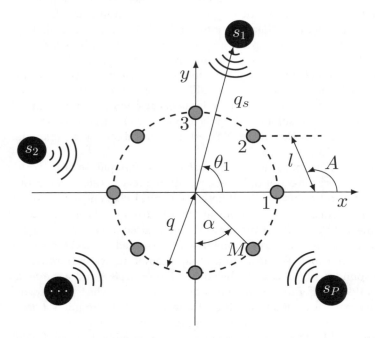

Figure 15.2 Uniform circular microphone array configuration. The microphones are numbered 1 to M and the sound sources are s_1 to s_P. The array radius is denoted as q, α and l are the angle and distance between adjacent microphones, and A is the obtuse angle formed by the chord defined by microphones 1 and 2 and the x-axis; θ_1 indicates the DOA of source s_1. Adapted from Pavlidi 2013.

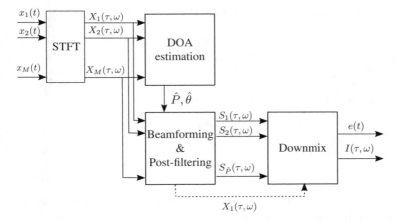

Figure 15.3 Block diagram of ImmACS encoder. Source: Alexandridis 2013a. https://www.hindawi.com/journals/jece/2013/718574/. Used under CC BY 3.0.

and ω denotes the radial frequency. The STFT domain signals are fed to the DOA estimation processing block, which returns the estimated number of active sources, $\hat{P}$, and the vector $\hat{\theta} \in \mathbb{R}^{\hat{P} \times 1}$ of DOA estimates. In parallel, the STFT signals are provided to the beamforming and post-filtering block which, given the number and DOAs of the sources, produces $\hat{P}$ separated signals $S_1(\tau, \omega), S_2(\tau, \omega), \ldots, S_{\hat{P}}(\tau, \omega)$. The separated signals are efficiently downmixed into a single audio channel, $e(t)$, and side information $I(\tau, \omega)$.

Counting and DOA Estimation of Multiple Audio Sources

For the counting and estimation of the DOAs, ImmACS adopts a relaxed sparsity assumption on the time–frequency (TF) domain of the source signals (Puigt and Deville, 2007; Pavlidi *et al.*, 2012). It is assumed that for each source at least one constant-time analysis zone can be found where the source is isolated, that is, it is dominant over the others. The constant-time analysis zone is defined as a series of Ω frequency-adjacent TF points at time frame index τ, denoted by (τ, Ω), and it is characterized as a single-source zone (SSZ) if it is dominated by one audio source. Such a sparsity assumption allows the signals to overlap in the time–frequency domain. It is therefore a much weaker constraint compared to the W-disjoint orthogonality assumption (Rickard and Yilmaz, 2002), where every TF point is considered to be dominated by one source. In ImmACS, DOA estimation of multiple sources relies on the detection of all SSZs and the derivation of local DOA estimates, deploying a single-source localization algorithm.

To detect an SSZ, the cross- and auto-correlation of the magnitude of the TF transform need to be estimated (in what follows, the time frame index, τ, is dropped for clarity):

$$R'_{i,j}(\Omega) = \sum_{\omega \in \Omega} |X_i(\omega) X_j(\omega)|. \tag{15.1}$$

The correlation coefficient between a pair of microphone signals (x_i, x_j) is then defined as

$$r'_{i,j}(\Omega) = \frac{R'_{i,j}(\Omega)}{\sqrt{R'_{i,i}(\Omega)R'_{j,j}(\Omega)}}. \tag{15.2}$$

A constant-time analysis zone is an SSZ if and only if (Puigt and Deville, 2007)

$$r'_{i,j}(\Omega) = 1 \qquad \forall i, j \in \{1, \ldots, M\}. \tag{15.3}$$

In practice, every zone that satisfies the inequality

$$\overline{r'}(\Omega) \geq 1 - \epsilon \tag{15.4}$$

is characterized as an SSZ, where $\overline{r'}(\Omega)$ is the average correlation coefficient between pairs of observations of adjacent microphones and ϵ is a small user-defined threshold.

The SSZ detection stage is followed by the DOA estimation at each SSZ. In general, any suitable single-source DOA estimation algorithm could be applied. In ImmACS, a modified version of the algorithm in Karbasi and Sugiyama (2007) is deployed, mainly because this algorithm is designed specifically for circular apertures, is computationally efficient, and has shown good performance under noisy conditions (Karbasi and Sugiyama, 2007; Pavlidi *et al.*, 2013).

The phase of the cross-power spectrum of a microphone pair is first evaluated over the frequency range of an SSZ as

$$G_{i,i+1}(\omega) = \angle R_{i,i+1}(\omega) = \frac{R_{i,i+1}(\omega)}{|R_{i,i+1}(\omega)|}, \qquad \omega \in \Omega, \tag{15.5}$$

where the cross-power spectrum is

$$R_{i,i+1}(\omega) = X_i(\omega)X_{i+1}(\omega)^* \tag{15.6}$$

and $*$ stands for complex conjugate.

The evaluated cross-power spectrum phase is used to estimate the phase rotation factors (Karbasi and Sugiyama, 2007),

$$G_{i \to 1}^{(\omega)}(\phi) \triangleq e^{-j\omega\tau_{i \to 1}(\phi)}, \tag{15.7}$$

where $\tau_{i \to 1}(\phi) \triangleq \tau_{1,2}(\phi) - \tau_{i,i+1}(\phi)$ is the difference in the relative delay between the signals received at pairs $\{1, 2\}$ and $\{i, i + 1\}$. The time difference of arrival for the microphone pair $\{i, i + 1\}$ is evaluated according to

$$\tau_{i,i+1}(\phi) = l \sin\left(A + \frac{\pi}{2} - \phi + (i - 1)\alpha\right)/c, \tag{15.8}$$

where $\phi \in [0, 2\pi), \omega \in \Omega, \alpha$ and l are the angle and distance between the adjacent microphone pair $\{i, i + 1\}$ respectively, A is the obtuse angle formed by the chord defined by the microphone pair $\{1, 2\}$ and the x-axis, and c is the speed of sound (see also Figure 15.2). Since the microphone array is uniform, α, A, and l are given by

$$\alpha = \frac{2\pi}{M}, \qquad A = \frac{\pi}{2} + \frac{\alpha}{2}, \qquad l = 2q \sin\frac{\alpha}{2}, \tag{15.9}$$

where q is the array radius.

Next comes the estimation of the circular integrated cross spectrum (CICS), defined by Karbasi and Sugiyama (2007) as

$$\text{CICS}^{(\omega)}(\phi) \triangleq \sum_{i=1}^{M} G_{i\rightarrow1}^{(\omega)}(\phi) G_{i,i+1}(\omega). \tag{15.10}$$

The DOA associated with the frequency component ω in the SSZ with frequency range Ω is estimated as

$$\hat{\theta}_{\omega} = \arg\max_{0\leq\phi<2\pi} |\text{CICS}^{(\omega)}(\phi)|. \tag{15.11}$$

In order to enhance the overall DOA estimation, the aforementioned algorithm is applied on "strong" frequency components, that is, those frequencies that correspond to the indices of the d highest peaks in the magnitude of the cross-power spectrum over all microphone pairs in an SSZ. Figure 15.4 illustrates the DOA estimation error versus signal to noise ratio (SNR) for various choices of d. It is clear that using more frequency bins leads in general to a lower estimation error. However, increasing d increases the computational complexity, which has to be taken into consideration for a real-time system. In ImmACS, a standard choice for d is $d = 2$.

The next step in counting and estimating the final DOAs of active audio sources constitutes the processing of estimated DOAs in SSZs in a block-based manner. To do so, a histogram is formed from the set of estimations in a block of B consecutive time frames, which slides one frame each time. The histogram is smoothed by applying an averaging filter with a window of length h_N, providing

$$y(\nu) = \sum_{i=1}^{N} w\left(\frac{\nu - \zeta_i}{h_N}\right), \qquad 1 \leq \nu \leq V, \tag{15.12}$$

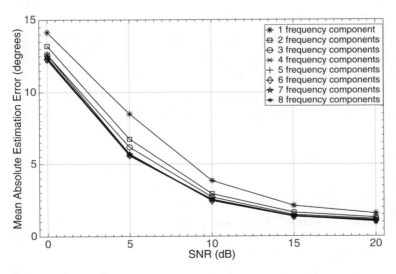

Figure 15.4 Direction of arrival estimation error vs. SNR in a simulated environment. Each curve corresponds to a different number of frequency components used in a single-source zone. Source: Pavlidi 2013. Reproduced with permission of the IEEE.

where $y(v)$ is the cardinality of the smoothed histogram at each histogram bin v, V is the number of bins in the histogram, ζ_i is the ith estimate out of N estimates in a block, and $w(\cdot)$ is the rectangular window of length h_N.

The smoothed histogram is processed following the principles of matching pursuit, that is, it is correlated with a source atom in order to detect the DOA of a possible source, the contribution of which is estimated and removed from the histogram. The process is repeated iteratively until the contribution of a potential source fails to satisfy a user-defined threshold.

A narrower-width pulse is used to accurately locate a peak in the histogram, the index of which denotes the DOA of a source. A wider-width pulse is used to account for the contribution of the source to the overall histogram and to provide better performance at lower SNRs. Each source atom is modeled as a smooth pulse utilizing a Blackman window. The correlation of the histogram with the source atom is performed in a circular manner as the histogram "wraps" from 359° to 0°. Thus, a matrix has to be formed whose rows contain wrapped and shifted versions of the source pulse. We denote by C_N and C_W the matrices for the peak detection (denoted by "N" for narrow) and the contribution estimation (denoted by "W" for wide), respectively. The rows of matrix C_N contain shifted versions of the narrow source atom with width Q_N, and, correspondingly, the rows of matrix C_W contain the shifted wide source atom with width equal to Q_W. This dual-width approach is illustrated in Figure 15.5.

In more detail, at each iteration denoted with loop index j, $\mathbf{u} = C_N \mathbf{y}_j$ is formed, where $\mathbf{y}_j$ is the smoothed histogram at the current iteration. $\mathbf{u}$ denotes the correlation of the histogram with the narrow pulse, and thus the index of its maximum value denotes the DOA estimated at the current iteration j, that is, $i^* = \arg\max_i u_i$, where u_i are the elements of $\mathbf{u}$. The contribution of the source located at iteration j is estimated as

$$\delta_j = \left(\mathbf{c}_W^{(i^*-1)} \right)^{\mathrm{T}} \frac{u_{i^*}}{E_{c_N}}. \tag{15.13}$$

Observe that the contribution is actually the row of the matrix C_W that contains a circularly shifted version of the source atom by $(i^* - 1)$ elements, that is, $\mathbf{c}_W^{(i^*-1)}$, weighted by $\frac{u_{i^*}}{E_{c_N}}$. In order to decide if the detected DOA and its contribution correspond to a true source and ultimately estimate the number of active sources, $\hat{P}$, we create γ, a length-P_{MAX} vector whose elements are some predetermined thresholds, representing

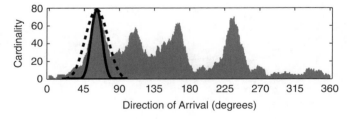

Figure 15.5 A wide source atom (dashed line) and a narrow source atom (solid line) applied to the smoothed histogram of four sources (speakers). Source: Pavlidi 2013. Reproduced with permission of the IEEE.

the relative energy of the jth source. Thus, if $\sum \delta_j \geq \gamma_j$ then a source is detected and its contribution is removed from the histogram:

$$\mathbf{y}_{j+1} = \mathbf{y}_j - \delta_j. \tag{15.14}$$

The algorithm then proceeds with the next iteration. In the opposite case, the algorithm ceases and the estimated DOAs and their corresponding number are returned.

Beamforming and Downmixing

Based on the estimated number of sources and their corresponding DOAs, ImmACS utilizes a spatial filter that consists of a superdirective beamformer and a post-filter in order to separate the sources' signals. The superdirective beamformer is designed to maximize the array gain, keeping the signal from the steering direction undistorted, while maintaining a minimum constraint on the white noise gain; the beamformer's filter coefficients are given by (Cox *et al.*, 1987)

$$\mathbf{w}(\omega, \theta) = \frac{[\epsilon \mathbf{I} + \mathbf{\Gamma}(\omega)]^{-1}\mathbf{d}(\omega, \theta)}{\mathbf{d}(\omega, \theta)^{\mathrm{H}}[\epsilon \mathbf{I} + \mathbf{\Gamma}(\omega)]^{-1}\mathbf{d}(\omega, \theta)}, \tag{15.15}$$

where $\mathbf{w}(\omega, \theta)$ is the $M \times 1$ vector of complex filter coefficients, θ is the beamformer's steering direction, $\mathbf{d}(\omega, \theta)$ is the steering vector of the array, $\mathbf{\Gamma}(\omega)$ is the $M \times M$ noise coherence matrix, $(\cdot)^{\mathrm{H}}$ is the Hermitian transpose operation, $\mathbf{I}$ is the identity matrix, and ϵ is used to control the white noise gain constraint. Assuming a spherically isotropic diffuse noise field, the elements for the noise coherence matrix are calculated as

$$\Gamma_{ij}(\omega) = \mathrm{sinc}\left(\frac{\omega d_{ij}}{c}\right), \tag{15.16}$$

where ω denotes the radial frequency, c is the speed of sound, and d_{ij} is the distance between the ith and jth microphones. The beamformer is signal independent, facilitating its use in the real-time implementation of ImmACS, as the filter coefficients are calculated offline and stored in the system's memory.

Hence, for each source s, the microphone signals at frequency ω are filtered with the beamformer filter coefficients $\mathbf{w}(\omega, \theta_s)$ that correspond to the source's estimated direction θ_s. This procedure results in the beamformer output signals:

$$B_s(\tau, \omega) = \mathbf{w}^{\mathrm{H}}(\omega, \theta_s)\mathbf{X}(\tau, \omega), \quad s = 1, \dots, \hat{P}. \tag{15.17}$$

A post-filtering operation is then applied to the beamformer output signals; each beamformer output signal is filtered with a binary mask. The binary masks are constructed by comparing the energies of the beamformer output signals for each source at a given frequency. The binary mask for the sth source is given by (Maganti *et al.*, 2007)

$$U_s(\tau, \omega) = \begin{cases} 1 & \text{if } s = \arg\max_p |B_p(\tau, \omega)|^2, \quad p = 1, \dots, \hat{P}, \\ 0 & \text{otherwise.} \end{cases} \tag{15.18}$$

The final separated source signals are given by:

$$\hat{S}_s(\tau, \omega) = U_s(\tau, \omega)B_s(\tau, \omega), \quad s = 1, \dots, \hat{P}. \tag{15.19}$$

From Equation (15.18), it is obvious that the binary masks are orthogonal with respect to each other. This means that if $U_s(\tau, \omega) = 1$ for some frequency ω and frame index τ,

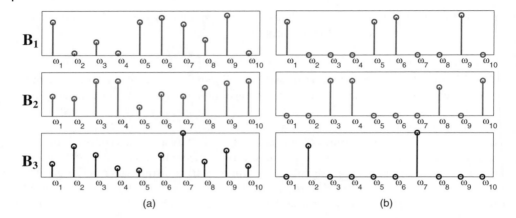

Figure 15.6 Examples of: (a) beamformer output signals, and (b) post-filtered signals, for a scenario of three active sound sources.

then $U_{s'}(\tau, \omega) = 0$ for $s' \neq s$, a property which also holds for the final separated source signals. This means that for each frequency only one source is maintained – the one with the highest energy – while the other sources are set to zero. The effect of the post-filtering operation is illustrated with an example of three active sound sources in Figure 15.6: for each frequency, only the source with the highest energy is kept, for example the first source for frequency ω_1, the third source for frequency ω_2, and so on. Looking at the frequency domain representation of the resulting post-filtered signals in Figure 15.6(b), the orthogonality property of the binary masks is evident.

The orthogonality property of the binary masks results in a very efficient downmixing scheme: the sources' signals can be downmixed by summing them in the frequency domain to create a full-spectrum monophonic audio signal:

$$E(\tau, \omega) = \sum_{s=1}^{\hat{P}} \hat{S}_s(\tau, \omega). \tag{15.20}$$

The downmix signal is then transformed back to the time-domain signal $e(t)$. To extract the estimated source signals from the downmix, additional information must also be attained, describing to which source each frequency of the downmix belongs. For this reason, ImmACS creates a side-information channel $I(\tau, \omega)$, which attains the assignment of frequencies to the DOAs of the sources, which also specifies the direction from which each frequency of the downmix signal will be reproduced in the decoder. Hence, the entire soundscape is encoded with the use of one monophonic audio signal and side information. These two streams are transmitted to the decoder in order to recreate and reproduce the soundscape.

ImmACS also offers the potential to include a diffuse part in the encoded soundscape. This is done by setting a cut-off frequency ω_c, which defines the frequencies up to which directional information is extracted: for the frequencies below ω_c, the aforementioned procedure of spatial filtering is applied in order to separate the active sources' signals. In contrast, the frequencies above ω_c are assumed to be dominated by diffuse sound. For these frequencies the spectrum from an arbitrary microphone of the array is included in the downmix (the dashed line in Figure 15.3), without further processing and without the need to keep any side information for this part, as it represents the part of the soundscape with no prominent direction. Thus, the side-information channel only attains the directional information for the frequencies up to ω_c. Processing only a certain range of frequencies may have several advantages, such as reduction in the computational complexity – especially when the sampling frequency is high – and reduction in the bitrate requirements for the side-information channel. But, more importantly, treating the higher frequencies as diffuse sound offers a sense of immersion to the listeners. Finally, applications like teleconferencing, where the signal content is mostly speech, can tolerate setting ω_c to approximately 4 kHz without degradation in the spatial impression of the recreated soundscape.

To reduce the bitrate requirements, coding schemes can be applied to both the audio and the side-information channels. Since the audio channel consists of a monophonic stream, any monophonic audio encoder can be utilized, such as MP3 (ISO, 1992), or the ultra-low-delay OPUS coder (Valin *et al.*, 2013; Vos *et al.*, 2013) that is used in ImmACS. Encoding the downmix signal with any of the aforementioned codecs has been found not to affect the spatial impression of the recreated soundscape (Alexandridis *et al.*, 2013b).

To encode the side-information channel, lossless coding schemes need to be used in order to attain the assignment of the frequencies to the sources. The side-information channel depends on Equation (15.18), and thus it is sufficient to encode the binary masks. The encoding scheme that ImmACS employs is based on the orthogonality property of the masks, as explained below.

The active sources at a given time frame are sorted in descending order according to the number of frequency bins assigned to them. The binary mask of the first (the most dominant) source is inserted into the bitstream. Given the orthogonality property of the binary masks, it follows that we do not need to encode the mask for the sth source at the frequency bins where at least one of the previous $s - 1$ masks is equal to one (since the rest of the masks will be zero). These locations can be identified by a simple OR operation between the $s - 1$ previous masks. Thus, for the second up to the $(\hat{P} - 1)$th mask, only the locations where the previous masks are all zero are inserted into the bitstream. The mask of the last source does not need to be encoded, as it contains ones in the frequency bins in which all the previous masks had zeros. A dictionary that associates the sources with their DOAs is also included in the bitstream. The encoding algorithm is presented in Algorithm 15.1 and is illustrated with an example in Figure 15.7. In Algorithm 15.1, find_positions_with_zeros($\cdot$) is a function that returns the indices of zero values of the vector taken as argument, and | denotes the concatenation operator. The resulted bitstream is further compressed with Golomb entropy coding (Golomb, 1966) applied on the run lengths of ones and zeros.

15.3.2 Decoder

The decoder recreates the soundscape for the remote participants according to the block diagram depicted in Figure 15.8. Based on the available equipment, ImmACS supports multichannel reproduction using multiple loudspeakers, Figure 15.8(a), or binaural reproduction using headphones, Figure 15.8(b).

Algorithm 15.1 Proposed encoding algorithm for the side information.

Input: U_s, $s = 1, \ldots, \hat{P}$: the estimated binary masks for the kth time frame, sorted in descending order according to the number of frequencies assigned to them.

Output: The bit sequence of the encoded masks, denoted by *Bitstream*.

1:      $Bitstream \leftarrow U_1$

2:      $next\_mask\_indices_{(1)} \leftarrow U_1$

3:      **for** $j = 2$ to $\hat{P} - 1$

4:         $pos \leftarrow \text{find\_positions\_with\_zeros}(next\_mask\_indices_{(j-1)})$

5:         $Bitstream \leftarrow \{Bitstream \mid U_j(pos)\}$

6:         $next\_mask\_indices_{(j)} \leftarrow (next\_mask\_indices_{(j-1)}) \text{ OR } (U_j)$

7:      **end for**

In the case where the downmix signal or the side-information channel have been encoded, the respective decoder is first applied. For decoding the side information, the mask of the first source is retrieved first. For the mask of the sth source, the next n bits are read from the bitstream, where n is the number of frequencies for which all the previous $s - 1$ masks are zero. This can be identified by a simple NOR operation. In terms

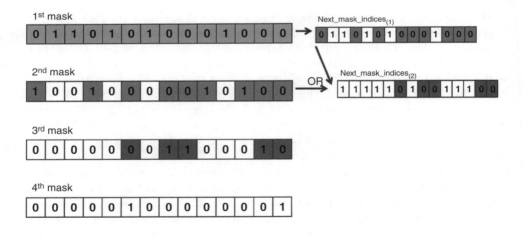

Figure 15.7 Example of the side-information coding scheme for a scenario of four active sound sources. All frequencies of the mask of the first source (the most dominant) are stored. For the second mask, only the frequencies where the first mask is zero need to be stored, as the other frequencies will definitely have the value of zero in the second and all other masks (orthogonality property). For each next mask, only the frequencies where all previous masks are zero need to be stored. Source: Alexandridis 2013. Reproduced with permission of the University of Crete.

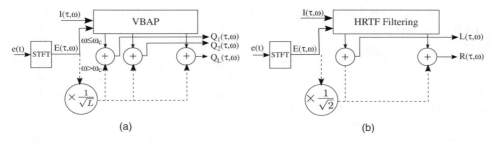

Figure 15.8 Block diagram of ImmACS decoder for (a) loudspeaker reproduction and (b) binaural reproduction. Source: Alexandridis 2013a. https://www.hindawi.com/journals/jece/2013/718574/. Used under CC BY 3.0.

of complexity, this scheme is computationally efficient, as its main operations are simple OR and NOR operations.

Then, for every incoming frame, the downmix signal $e(t)$ is transformed to the frequency domain and the spectrum is divided into the non-diffuse part and diffuse part, based on the beamformer cutoff frequency ω_c: the frequencies $\omega \leq \omega_c$ belong to the non-diffuse part, while the diffuse part—if it exists—includes the frequencies $\omega > \omega_c$.

For loudspeaker reproduction, the non-diffuse part is synthesized using vector base amplitude panning (VBAP; Pulkki, 1997) at each frequency, where the panning gains for each loudspeaker are determined according to the DOA at that frequency from the side-information channel $I(\tau, \omega)$. If a diffuse part is included, it is played back from all loudspeakers (dashed line in Figure 15.8(a)). Assuming a loudspeaker configuration with L loudspeakers, the signal for the lth loudspeaker is thus given by

$$Q_l(\tau, \omega) = \begin{cases} g_l(I(\tau, \omega))E(\tau, \omega) & \text{for } \omega \leq \omega_c, \\ \dfrac{1}{\sqrt{L}} E(\tau, \omega) & \text{for } \omega > \omega_c, \end{cases} \qquad (15.21)$$

where $g_l(I(\tau, \omega))$ refers to the VBAP gain for the lth loudspeaker in order to play back the time–frequency point (τ, ω) from the direction specified by $I(\tau, \omega)$.

For binaural reproduction using headphones, ImmACS utilizes an HRTF database in order to position each source at its corresponding direction. Hence, each frequency of the non-diffuse part is filtered with the HRTF of the direction specified by the side information. The optional diffuse part is included in both channels – the dashed line in Figure 15.8(b). Summarizing, the two-channel output is given by

$$Y_{ch}(\tau, \omega) = E(\tau, \omega)\, \text{HRTF}_{ch}(\omega, I(\tau, \omega)), \quad \omega \leq \omega_c,$$

$$Y_{ch}(\tau, \omega) = \frac{1}{\sqrt{2}} E(\tau, \omega), \qquad\qquad \omega > \omega_c, \qquad (15.22)$$

where $ch = \{L, R\}$ denotes the left and right channels.

15.4 Capture and Reproduction of Crowded Acoustic Environments

Typically, capturing large sports events is accomplished with several microphones placed around the pitch or inside the crowd, so that each microphone focuses on a particular segment of the event (Cengarle *et al.*, 2010). A great amount of equipment needs to be carefully distributed all around the playing field, requiring a lot of preparation time and attendance during the game. Then, it depends on the experience and subjective judgment of the sound engineer to mix all the signals into the final stereo or surround format that is transmitted by the broadcaster. The approach of using one or just a few compact sensor arrays to capture and reproduce sound from such large-scale events presents an interesting alternative; it may reduce the cost of equipment and implementation, allow flexibility in the processing and manipulation of the captured spatial information (Stefanakis and Mouchtaris, 2015, 2016), and allow for efficient encoding of the data to reduce bandwidth requirements during transmission (Alexandridis *et al.*, 2013a).

In this section we present two different approaches, a parametric and a non-parametric one, for the capture and reproduction of a sound scene in 2D using a circular sensor array of omnidirectional microphones. Such a study is interesting as it further extends the range of applications in which circular microphone arrays find a use, but also because it puts two inherently different approaches to the test. The techniques presented are applied to a real recording, produced inside a crowded football stadium with thousands of spectators.

15.4.1 Sound Source Positioning Based on VBAP

Sound source positioning using VBAP (Pulkki, 1997) is traditionally linked to virtual sound source placement and to parametric spatial audio applications, where a DOA θ with respect to some portion of the signal is always defined or estimated. VBAP requires that the loudspeakers are distributed around a circle of fixed radius around the listening area, but their number and direction might be arbitrary, a fact that provides important flexibility in designing the loudspeaker arrangement. Given the incident angle θ, this is mapped to the L loudspeaker gains $\mathbf{g}(\theta) = [g_1(\theta), \ldots, g_L(\theta)]^{\mathrm{T}}$ in accordance with the physical loudspeaker distribution around the listening area. In Figure 15.9 we show an example of how such directivity patterns look for the cases of $L = 4$ and 8 uniformly distributed loudspeakers in Figure 15.9(a) and Figure 15.9(c), and for $L = 5$ non-uniform loudspeakers in Figure 15.9(b), assuming that the problem is confined to 2D.

It is worth spending a few words here in order to explain how the aforementioned mapping between incident angle and loudspeaker gain is also useful in the case of the non-parametric approach which we intend to consider. Hacihabiboğlu and Cvetković (2009) explain how the physical loudspeaker angular distribution dictates the optimum directivity patterns for an equal (to the number of loudspeakers) number of directional microphones to capture the directional properties of the sound field. Given such microphones, one could then send the signal captured by each microphone to the corresponding loudspeaker, without further processing, to reproduce a convincing panoramic

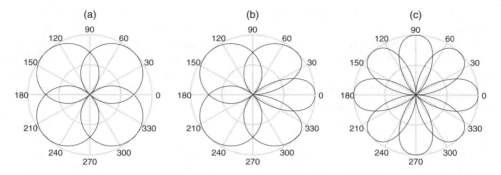

Figure 15.9 Desired directivity patterns for (a) a four-channel, (b) a five-channel and (c) an eight-channel system. Source: Stefanakis and Mouchtaris 2016.

auditory reality. Respecting the requirements stated in that paper, VBAP dictates optimal loudspeaker panning gains with the following characteristics:

- the panning gains are frequency-invariant;
- the problem of unwanted interchannel crosstalk is efficiently addressed by ensuring that given a single plane wave incident at a certain angle (in the azimuth plane), only two loudspeakers will be activated during reproduction (the two that are adjacent to the estimated angle);
- the sum of the squares of all loudspeaker gains along θ is equal, meaning that there is no information loss.

This implies that we can use the loudspeaker gains provided by VBAP as the desired directivity response for a set of beamformers, which we can then use in order to capture the acoustic environment in different directions. The signal at the output of each beamformer can then be sent directly to the corresponding loudspeaker, a process that is mathematically represented in Equation (15.23). As long as the desired beam patterns are accurately reproduced, the non-parametric approach represents the simplest way to capture and reproduce an acoustic scene, promising accurate spatialization of both the directional and diffuse components of the captured sound field. As will be seen, however, the accurate reconstruction of these patterns using beamforming is not an easy task, and deviations between the desired and actually reproduced directivity patterns are expected to degrade the sense of direction transmitted to the listener.

15.4.2   Non-Parametric Approach

In the time–frequency domain, the process of beamforming can be expressed as

$$Y_l(\tau, \omega) = \mathbf{w}_l(\omega)^H \mathbf{X}(\tau, \omega), \tag{15.23}$$

where τ is the time-frame index, ω is the radial frequency, $\mathbf{w}_l(\omega) = [w_{1l}(\omega), \ldots, w_{Ml}(\omega)]^T$ is the vector with the M complex beamformer weights responsible for the lth loudspeaker channel, $\mathbf{X}(\tau, \omega)$ is the $M \times 1$ vector with the signal at the microphones, and $Y_l(\tau, \omega)$ represents the lth loudspeaker signal in the time–frequency domain. This process is repeated for all the different loudspeaker channels $l = 1, \ldots, L$; the signals are transformed to the time domain and sent to the corresponding loudspeakers for playback.

Using the panning gains dictated by VBAP, we demonstrate in what follows an approach for calculating the beamformer weights associated with each loudspeaker channel. Consider a grid of N uniformly distributed directions θ_n in $[-180°, 180°)$ with the loudspeaker panning gains $g_l(\theta_n)$ prescribed by VBAP. The problem becomes finding the weights $\mathbf{w}_l(\omega)$ in order to satisfy

$$\mathbf{D}^H(\omega)\mathbf{w}_l(\omega) \approx \mathbf{G}_l, \; l = 1, \dots, L. \tag{15.24}$$

Here, $\mathbf{G}_l = [g_l(\theta_1), \dots, g_l(\theta_N)]^T$ is the desired response provided by VBAP and $\mathbf{D}(\omega) = [\mathbf{d}(\omega, \theta_1), \dots, \mathbf{d}(\omega, \theta_N)]$ is the matrix with the array steering vectors which model the array response to a plane wave incident at angle θ_n. Without loss of generality we may consider here that $\mathbf{G}_l \in \mathbb{R}^N$ $\forall l$ as the VBAP function does not dictate phase variations across the different channels. Also, for the case of a circular array of radius R the propagation model can be written as (Alexandridis *et al.*, 2013a)

$$d_m(\omega, \theta) = e^{jkR\cos(\phi_m - \theta)}, \tag{15.25}$$

where ϕ_m denotes the angle of the mth sensor with respect to the sensor array center and k is the wavenumber. Assuming that $L < N$, the linear problem of Equation (15.24) is overdetermined and the solution can be found by minimizing, in the least squares (LSQ) sense, the cost function

$$J = \|\mathbf{G}_l - \mathbf{D}(\omega)^H\mathbf{w}_l(\omega)\|_2^2. \tag{15.26}$$

However, unconstrained minimization involves inversion of the matrix $\mathbf{D}(\omega)\mathbf{D}(\omega)^H$, which is ill-conditioned at low frequencies as well as at other distinct frequencies. An example of this behavior is shown in Figure 15.10, where we have plotted the condition number of $\mathbf{D}(\omega)\mathbf{D}(\omega)^H$ as a function of frequency, considering a circular array of $M = 8$ uniformly distributed microphones and the reproduction system of Figure 15.9(a). Direct inversion of $\mathbf{D}(\omega)\mathbf{D}(\omega)^H$ might thus lead to severe amplification of noise at certain frequencies, which is perceived as unwanted spectral coloration. In order to avoid

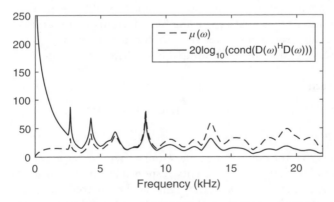

Figure 15.10 Condition number of matrix $\mathbf{D}(\omega)^H\mathbf{D}(\omega)$ in dB (black) and variation of the regularization parameter μ (gray) as a function of the frequency for an eight-sensor circular array of radius 5 cm. Source: Stefanakis and Mouchtaris 2016.

such a problem, we propose to use Tikhonov regularization by adding a penalty term proportional to the noise response in the previous cost function as

$$J = \|\mathbf{G}_l - \mathbf{D}(\omega)^{\mathrm{H}}\mathbf{w}_l(\omega)\|_2^2 + \mu(\omega)\mathbf{w}_l(\omega)^{\mathrm{H}}\mathbf{w}_l(\omega), \tag{15.27}$$

with $\mu(\omega)$ implying that the value of the regularization parameter varies with frequency. We have observed that this approach achieves a better trade-off between the noise gain and the array gain, as opposed to a constant value of the regularization parameter. In this chapter, we propose a varying value of the regularization parameter of the form

$$\mu(\omega) = \lambda\omega 20\log_{10}(\mathrm{cond}(\mathbf{D}(\omega)\mathbf{D}(\omega)^{\mathrm{H}})), \tag{15.28}$$

where λ is a fixed scalar and cond($\cdot$) represents the condition number of a matrix, that is, the ratio of its largest eigenvalue to its smallest one. The beamformer weights can then be found through LSQ minimization as

$$\mathbf{w}_l^o(\omega) = (\mathbf{D}(\omega)\mathbf{D}(\omega)^{\mathrm{H}} + \mu(\omega)\mathbf{I})^{-1}\mathbf{D}(\omega)\mathbf{G}_l, \tag{15.29}$$

where $\mathbf{I}$ is the $M \times M$ identity matrix.

Finally, we consider an additional normalization step, which aims to ensure unit gain and zero phase shift at the direction of maximum response for each beamformer. Letting θ_l^0 denote this direction for the lth beam at frequency ω, the final weights are calculated as

$$\hat{\mathbf{w}}_l^o(\omega) = \frac{\mathbf{w}_l^o(\omega)}{\mathbf{w}_l^o(\omega)^{\mathrm{H}}\mathbf{d}\left(\omega, \theta_l^0\right)}, \tag{15.30}$$

and the signal for the lth loudspeaker is obtained as in Equation (15.23). The beamformer weights need only be calculated once for each frequency and then stored to be used in the application phase.

In Figure 15.11 we present plots of the actual directivity versus the desired directivity pattern, for the same reproduction system as considered in Figure 15.9. Observe the increment in the amplitude of the side-lobes at 2660 Hz, which is close to a problematic frequency according to Figure 15.10. Also, the subplot corresponding to 7 kHz (in Figure 15.11) is indicative of the spatial aliasing problems that occur at higher frequencies.

15.4.3 Parametric Approach

As shown in Figure 15.11, it is difficult to obtain exactly the desired directivity patterns relying on simple beamforming. Looking, for example, at the sub-figure corresponding to 2660 Hz, we see that a sound source at $-135°$ would also be played back by the loudspeaker at $45°$, something that may blur the sense of direction transmitted to the listener. On the other hand, a parametric approach avoids this problem by defining the loudspeaker response as a function of the estimated DOA.

However, it is questionable what type of processing is applicable to the particular type of array that we typically focus on in our work. While the method works sufficiently well for an application with a limited number of speakers, such as the one described in Section 15.3, it is inappropriate for the considered acoustic conditions due to the enormous number of potential sound sources that participate in the sound scene. On the other hand, DirAC does not impose a limitation on the number of sound sources comprising the sound scene, but it is typically implemented with smaller arrays – in

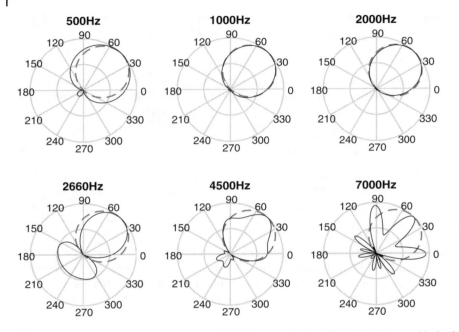

Figure 15.11 Actual directivity patterns (solid black) versus desired directivity patterns (dashed gray) at different frequencies for an eight-element circular sensor array. The directivity patterns shown correspond to the first loudspeaker at 45°, considering a symmetric arrangement of four loudspeakers at angles of 45°, 135°, −135°, and −45° degrees. Source: Stefanakis and Mouchtaris 2016.

terms of both radius and number of sensors – than the one considered in this application (Kuech *et al.*, 2008).

The parametric approach that we present in this section is based on the method introduced by Ito *et al.* (2010). We exploit this previous study here in order to perform direct-to-diffuse sound field decomposition, at the same time extending it for the purpose of DOA estimation. Assume that a sensor array comprised of M sensors is embedded inside a diffuse noise field and at the same time receives the signal from a single directional source. The observation model for sensor m can be written

$$X_m(\tau, \omega) = S(\tau, \omega)d_m(\omega) + U_m(\tau, \omega),\tag{15.31}$$

where $s(\tau, \omega)$ is the directional source signal, $d_m(\omega) = e^{-j\omega\delta_m}$ is the transfer function from the source to the sensor at frequency ω, and $U_m(\tau, \omega)$ is the diffuse noise component, which is assumed to be uncorrelated with the source signal. The observation interchannel cross-spectrum between sensors m and n can thus be written

$$\phi_{X_m X_n}(\tau, \omega) = E\left\{X_m(\tau, \omega)X_n^H(\tau, \omega)\right\}\tag{15.32}$$
$$= \phi_{ss}(\tau, \omega)d_m(\omega)d_n^*(\omega) + \phi_{U_m U_n}(\tau, \omega),\tag{15.33}$$

where ϕ_{ss} is the source power spectrum and $\phi_{U_m U_n}$ represents the component of the cross-spectrum due to noise. Assuming that the noise field is isotropic, then $\phi_{U_m U_n} \in \mathbb{R}$, as explained in Ito *et al.* (2010). This means that the imaginary part of the cross-spectra observation is immune to noise, and therefore depends only on the directional source location. According to this, one may write

$$\Im\{\phi_{X_m X_n}(\tau, \omega)\} = \phi_{ss}(\tau, \omega) \sin(2\pi f(\delta_m - \delta_n)), \tag{15.34}$$

where $\Im\{\cdot\}$ denotes the imaginary part of a complex number. We may now use this observation in order to create a model of the imaginary cross-spectra terms through the vector

$$\mathbf{a}(f, \theta) = [\sin(2\pi f(\delta_m(\theta) - \delta_n(\theta)))]_{m>n}, \tag{15.35}$$

which contains all the $M(M-1)/2$ pairwise sensor combinations, with delays $\delta_m(\theta)$ and $\delta_n(\theta)$ defined according to the incident plane wave angle θ. This vector can be seen as an alternative steering vector with the property $\mathbf{a}(\omega, \theta \pm \pi) = -\mathbf{a}(\omega, \theta)$, for θ in radians. Consider now the vector constructed by the vertical concatenation of $M(M-1)/2$ imaginary observation cross-spectra terms as

$$\mathbf{z}(\tau, \omega) = \Im\{[(\Phi_{mn}(\tau, \omega))]_{m>n}\}, \tag{15.36}$$

where $\Phi_{mn}(\tau, \omega)$ corresponds to the element at the mth row and nth column of the signal covariance matrix

$$\Phi(\tau, \omega) = E\{\mathbf{X}(\tau, \omega)\mathbf{X}(\tau, \omega)^H\}. \tag{15.37}$$

Now, $\mathbf{z}(\tau, \omega)$ in Equation (15.36) represents an immune-to-noise observation of the sound field, which should be consistent with respect to the model in Equation (15.35), given that the sound field is a plane wave with an incident angle θ. The most likely DOA at frequency ω and time frame τ can be found by searching for the plane wave signature $\mathbf{a}(\omega, \theta)$ that is most similar to $\mathbf{z}(\tau, \omega)$:

$$\theta_{\tau, \omega} = \arg\max_{\theta}(\hat{\mathbf{a}}(\omega, \theta)^H \mathbf{z}(\tau, \omega)), \tag{15.38}$$

where $\hat{\mathbf{a}}(\omega, \theta) = \mathbf{a}(\omega, \theta)/\|\mathbf{a}(\omega, \theta)\|_2$ implies normalization with the Euclidean norm so that all plane wave signatures have the same energy. The source power spectrum at the particular TF point is then found by projecting $\mathbf{a}(\omega, \theta_{\tau, \omega})$ onto $\mathbf{z}(\tau, \omega)$ as

$$\phi_{ss}(\tau, \omega) = \frac{\mathbf{a}(\omega, \theta_{\tau, \omega})^H \mathbf{z}(\tau, \omega)}{\mathbf{a}(\omega, \theta_{\tau, \omega})^H \mathbf{a}(\omega, \theta_{\tau, \omega})}. \tag{15.39}$$

On the other hand, an estimation of the total acoustic power can be found by averaging the power across all microphones as $\phi_{yy}(\tau, \omega) = \text{tr}(\Phi)/M$, where $\text{tr}(\cdot)$ is the sum of all the diagonal terms of a matrix. The ratio $q(\tau, \omega) = \phi_{ss}(\tau, \omega)/\phi_{yy}(\tau, \omega)$ then represents a useful metric that can be associated with the diffuseness of the sound field; intuitively, $0 \le q(\tau, \omega) \le 1$ should hold, with a value close to 1 dictating a purely directional sound field whereas a value close to 0 implies a purely noisy sound field. Furthermore, we may use this metric in order to establish a relation with the so-called diffuseness of the sound field:

$$\Psi(\tau, \omega) = \min\{0, 1 - q(\tau, \omega)\}, \tag{15.40}$$

where the function $\min\{\cdot\}$ returns the minimum of a set of numbers and is useful here in order to ensure that the diffuseness does not take negative values, if $q(\tau, \omega) > 1$.

The input signal at the lth loudspeaker is then synthesized as $Y_l(\tau, \omega) = Y_l^{\text{dir}}(\tau, \omega) + Y_l^{\tau, \text{dif}}(\omega)$, where

$$Y_l^{\text{dir}}(\tau, \omega) = X_{\text{ref}}(\tau, \omega)\sqrt{1 - \Psi(\tau, \omega)}g_l(\theta_{\tau, f}), \tag{15.41}$$

$$Y_l^{\text{dif}}(\tau, \omega) = \sqrt{\Psi(\tau, \omega)}\hat{\mathbf{w}}_l(\omega)^H \mathbf{X}(\tau, \omega). \tag{15.42}$$

Here, $X_{\text{ref}}(\tau, \omega)$ is the captured signal at a reference microphone (the first microphone), g_l is the VBAP gain responsible for the lth loudspeaker and $\hat{\mathbf{w}}_l(\omega)$ are the beamformer weights derived in Section 15.4.3. As can be seen, at each TF point, the diffuse channels of this method are nothing but a weighted replica of the loudspeakers signals produced with the parametric technique of Section 15.4.2.

15.4.4 Example Application

Both the parametric and the non-parametric approaches presented above were applied to a real recording produced with the eight-microphone circular array of radius $R = 0.05$ m shown in Figure 15.12(a). The recording took place in a crowded open stadium during a football match of the Greek Super League. Figure 15.12(b) presents a sketch of the football stadium with the location of the array represented by a white dot. The array was placed at a height of 0.8 m in front of Gates 13 and 14, which were populated with the organized fans of the team which was hosting the game. These fans were cheering and singing constantly throughout the entire duration of the recording, thus providing most of the acoustic information captured by the array.

The reproduction system that we considered for the application phase consisted of four loudspeakers uniformly distributed around the azimuth plane, specifically at 45°, 135°, −135°, and −45° (see Figure 15.12), at a radius of 2.10 m. With respect to the listener's orientation, these loudspeakers were located at rear right (RR), front right (FR),

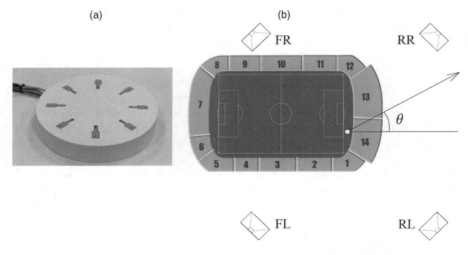

Figure 15.12 (a) The sensor array. (b) Sketch of the football stadium with the loudspeaker setup used for evaluation. The white dot in the lower-right corner of the football field denotes the array location. Source: Stefanakis and Mouchtaris 2016.

front left (FL), and rear left (RL). This configuration is of particular interest as it can easily be extended with a 5.1 surround system, adding one more channel to be used for the commentator. The panning gains derived from VBAP for this setup are identical to those depicted in Figure 15.9(a).

The signals were recorded at 44 100 Hz and processed separately with the parametric and the non-parametric approach. For the STFT we used a Hanning window of length 2048 samples and hop size 1024 samples (50% overlap). For the non-parametric method, we used the frequency-varying regularization parameter of Equation (15.28), with $\lambda = 0.003$. The variation of μ as a function of the frequency is illustrated with the gray line in Figure 15.10, while the polar plots of Figure 15.11 are illustrative of the deviation between the desired and the actual beamformer directivities.

Avoiding the problem with the imperfect beam shapes of the non-parametric method, one expects the parametric approach to produce better localization of the different acoustic sources. The opinion of the authors, who were also present inside the stadium during the recording, is that the parametric method indeed produced a more convincing sense of direction in comparison to beamforming, and was also more consistent with respect to changes in the orientation of the listener's head inside the listening area, as well as with reasonable displacements from the sweet spot. On the other hand, the non-parametric method provided a more blurred sense of direction but a much better sense of the reverberation in the stadium. As additional evidence for this contradiction between the two methods, we have plotted in Figure 15.13 the rectified loudspeakers' signal amplitudes in time, as derived by each technique, for a short-duration segment where the crowd at Gates 13 and 14 was by far the most dominant acoustic source inside the stadium. As the acoustic energy is concentrated at a particular part of the scene, one should expect an uneven distribution of the signal energy across the different channels, which is what we actually observe for the parametric method in Figure 15.13(b). On the other hand, the non-parametric method produces more coherent loudspeaker signals, resulting in reduced separation of different sound sources across the available

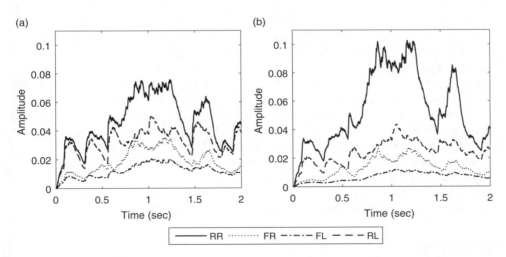

Figure 15.13 Rectified loudspeaker signal amplitudes for (a) the non-parametric method, and (b) the parametric method.

loudspeaker channels. In Figure 15.13(a) this is observed as an increased contribution from loudspeakers at irrelevant directions with respect to where Gates 13 and 14 actually are.

It is worth at this point spending a few words on the aspect of sound quality. As previously explained, the non-parametric method is subject to spectral coloration caused by the matrix inversion issues explained in Section 15.4.2. We believe, however, that this problem is sufficiently addressed by the regularization method introduced in this chapter, and interested readers are invited to judge for themselves by listening to the sound examples provided online at http://users.ics.forth.gr/nstefana/Stadium2016. On the other hand, the parametric method is prone to musical noise, caused by the rapid variations of the estimated angle across time at each frequency bin. More specifically, at each time frame, at most two loudspeakers reproduce a certain frequency component. As the assignment of frequency components to loudspeakers may change abruptly from one time instant to another, this is perceived as musical noise. However, while this type of degradation is indeed perceived when listening to each channel alone, informal listening tests revealed that in multichannel listening conditions these degradations are not strongly perceived. This is caused by the fact that as the sounds from all channels sum at the center of the loudspeaker array, the components responsible for the phenomenon of musical noise are masked by the superposition of sound from all the available channels. The interested reader may again verify this claim by listening to the sound files provided, but we recommend the use of a surround multichannel system for this purpose.

15.5 Conclusions

In this chapter, innovative algorithms and applications of time–frequency parametric spatial audio were presented, namely in teleconferencing and in remote presence, with an emphasis on crowded acoustic environments. In both cases, the objective is to provide the listener with an immersive audio experience, capturing and reproducing the spatial attributes of the soundscape accurately and efficiently. For the first application area, the number of discrete sound sources of interest is assumed to be limited and countable. The sparseness of the sources in the TF domain is exploited in order to estimate their number and corresponding DOAs, and to efficiently encode the spatial information in a single audio stream. In the case of a crowded acoustic environment, it is not possible to introduce source counting as the number of sound sources is very large. Here, the parametric approach is implemented by treating each time–frequency point as a different sound source. For both application areas, we used a uniform circular microphone array as the capturing device, given the fact that such a device can be made compact and at low cost due to the recent introduction of MEMS microphone technology. Using a digital circuit to implement the microphone array has the additional advantage of having a single device to perform the capture, processing, and transmission of the directional audio content, a desirable property for many applications.

References

Alexandridis, A. (2013) *Directional coding of audio using a circular microphone array.* Master's thesis. University of Crete.

Alexandridis, A., Griffin, A., and Mouchtaris, A. (2013a) Capturing and reproducing spatial audio based on a circular microphone array. *Journal of Electrical and Computer Engineering*, **2013**, 1–16.

Alexandridis, A., Griffin, A., and Mouchtaris, A. (2013b) Directional coding of audio using a circular microphone array. *IEEE International Conference on Acoustics, Speech, and Signal Processing (ICASSP)*, pp. 296–300. IEEE.

Alexandridis, A., Papadakis, S., Pavlidi, D., and Mouchtaris, A. (2016) Development and evaluation of a digital MEMS microphone array for spatial audio. *Proceedings of the European Signal Processing Conference (EUSIPCO)*, pp. 612–616.

Baumgarte, F. and Faller, C. (2003) Binaural cue coding – Part I: Psychoacoustic fundamentals and design principles. *IEEE Transactions on Speech and Audio Processing*, **11**(6), 509–519.

Boone, M., Verheijen E., and van Tol, P. (1995) Spatial sound-field reproduction by wave-field synthesis. *Journal of the Audio Engineering Society*, **43**(12), 1003–1012.

Cengarle, G., Mateos, T., Olaiz, N., and Arumi, P. (2010) A new technology for the assisted mixing of sport events: Application to live football broadcasting. *Proceedings of the 128th Convention of Audio Engineering Society*. Audio Engineering Society.

Cobos, M., Lopez, J., and Spors, S. (2010) A sparsity-based approach to 3D binaural sound synthesis using time–frequency array processing. *EURASIP Journal on Advances in Signal Processing*, **2010**, 415840.

Cox, H., Zeskind, R., and Owen, M. (1987) Robust adaptive beamforming. *IEEE Transactions on Acoustics, Speech, and Signal Processing*, **35**(10), 1365–1376.

Faller, C. and Baumgarte, F. (2003) Binaural cue coding – Part II: Schemes and applications. *IEEE Transactions on Speech and Audio Processing*, **11**(6), 520–531.

Gerzon, M. (1985) Ambisonics in multichannel broadcasting and video. *Journal of the Audio Engineering Society*, **33**(11), 859–871.

Golomb, S.W. (1966) Run-length encodings. *IEEE Transactions on Information Theory*, **12**(3), 399–401.

Hacihabiboğlu, H. and Cvetković, Z. (2009) Panoramic recording and reproduction of mutlichannel audio using a circular microphonoe array. *IEEE Workshop on Applications of Signal Processing to Audio and Acoustics (WASPAA)*, pp. 117–120. IEEE.

ISO (1992) *Coding of Moving Pictures and Associated Audio for Digital Storage Media at up to about 1.5 Mbit/s*. ISO/IEC JTC1/SC29/WG11 (MPEG) International Standard ISO/IEC 11172-3. International Organization for Standardization, Geneva, Switzerland.

Ito, N., Ono, N., Vincent, E., and Sagayama, S. (2010) Designing the Wiener post-filter for diffuse noise suppression using imaginary parts of inter-channel cross-spectra. *IEEE International Conference on Acoustics, Speech, and Signal Processing (ICASSP)*, pp. 2818–2821. IEEE

Kallinger, M., Kuech, F., Schultz-Amling, R., Del Galdo, G., Ahonen, J., and Pulkki, V. (2008) Analysis and adjustment of planar microphone arrays for application in directional audio coding. *Proceedings of the 124th Convention of the Audio Engineering Society*. Audio Engineering Society.

Karbasi, A. and Sugiyama, A. (2007) A new DOA estimation method using a circular microphone array. *Proceedings of the European Signal Processing Conference (EUSIPCO)*, pp. 778–782.

Kuech, F., Kallinger, M., Schultz-Amling, R., Del Galdo, G., and Pulkki, V. (2008) Directional audio coding using planar microphone arrays. *Proceedings of Hands-Free Speech Communication and Microphone Arrays*, pp. 37–40. IEEE.

Maganti, H.K., Gatica-Perez, D., and McCowan, I.A. (2007) Speech enhancement and recognition in meetings with an audio-visual sensor array. *IEEE Transactions on Audio, Speech, and Language Processing*, **15**(8), 2257–2269.

Pavlidi, D., Griffin, A., Puigt, M., and Mouchtaris, A. (2013) Real-time multiple sound source localization and counting using a circular microphone array. *IEEE Transactions on Audio, Speech, and Language Processing*, **21**, 2193–2206.

Pavlidi, D., Puigt, M., Griffin, A., and Mouchtaris, A. (2012) Real-time multiple sound source localization using a circular microphone array based on single-source confidence measures. *Proceedings of the IEEE International Conference on Acoustics, Speech, and Signal Processing (ICASSP)*, pp. 2625–2628. IEEE.

Politis, A., Laitinen, M., Ahonen, J., and Pulkki, V. (2015) Parametric spatial audio processing of spaced microphone array recordings for multichannel reproduction. *Journal of the Audio Engineering Society*, **63**(4), 216–227.

Puigt, M. and Deville, Y. (2007) A new time–frequency correlation-based source separation method for attenuated and time shifted mixtures. *Proceedings of the 8th International Workshop (ECMS and Doctoral School) on Electronics, Modelling, Measurement and Signals*, pp. 34–39.

Pulkki, V. (1997) Virtual sound source positioning using vector base amplitude panning. *Journal of the Audio Engineering Society*, **45**(6), 456–466.

Pulkki, V. (2007) Spatial sound reproduction with directional audio coding. *Journal of the Audio Engineering Society*, **55**(6), 503–516.

Rickard, S. and Yilmaz, O. (2002) On the approximate W-disjoint orthogonality of speech. *Proceedings of the IEEE International Conference on Acoustics, Speech, and Signal Processing (ICASSP)*, vol. 1, pp. 529–532. IEEE.

Stefanakis, N. and Mouchtaris, A. (2015) Foreground suppression for capturing and reproduction of crowded acoustic environments. *Proceedings of the IEEE International Conference on Acoustics, Speech, and Signal Processing (ICASSP)*, pp. 51–55. IEEE.

Stefanakis, N. and Mouchtaris, A. (2016) Capturing and reproduction of a crowded sound scene using a circular microphone array. *Proceedings of the European Signal Processing Conference (EUSIPCO)*, pp. 1673–1677.

Thiergart, O., Kallinger, M., Del Galdo, G., and Kuech, F. (2011) Parametric spatial sound processing using linear microphone arrays, in *Microelectronic Systems* (ed. Heuberger, A.), pp. 321–329. Springer, Berlin.

Valin, J.M., Maxwell, G., Terriberry, T.B., and Vos, K. (2013) High-quality, low-delay music coding in the Opus codec. *Proceedings of the 135th Convention of the Audio Engineering Society*. Audio Engineering Society.

Vos, K., Sørensen, K.V., Jensen, S.S., and Valin, J.M. (2013) Voice coding with Opus. *Proceedings of the 135th Convention of the Audio Engineering Society*. Audio Engineering Society.

Index

Parametric Time–Frequency Domain Spatial Audio, First Edition. Edited by Ville Pulkki,
Symeon Delikaris-Manias, and Archontis Politis.
© 2018 John Wiley & Sons Ltd. Published 2018 by John Wiley & Sons Ltd.
Companion Website: www.wiley.com/go/pulkki/parametrictime-frequency